SOCIAL ISSUES IN TECHNOLOGY

A Format for Investigation

Fourth Edition

Paul A. Alcorn
DeVry Institute of Technology
Decatur, Georgia

Upper Saddle River, New Jersey
Columbus, Ohio

Library of Congress Cataloging-in-Publication Data
Alcorn, Paul A.
Social issues in technology : a format for investigation/Paul A. Alcorn.—4th ed.
p. cm.
Includes bibliographical references.
ISBN 0-13-060257-4
1. Technology—Social aspects. I. Title.
T14.5.A43 2003
303.48'3—dc21 2001038748

Editor in Chief: Stephen Helba
Assistant Vice President and Publisher: Charles E. Stewart, Jr.
Assistant Editor: Delia K. Uherec
Production Editor: Tricia L. Rawnsley
Design Coordinator: Diane Ernsberger
Cover Designer: Robin Chukes
Cover art: Photoresearchers
Production Manager: Matthew Ottenweller
Electronic Text Management: Karen L. Bretz
Product Manager: Scott Sambucci

This book was set in Times Roman by D&G Limited, LLC. It was printed and bound by R.R. Donnelley & Sons Company. The cover was printed by Phoenix Color Corp.

Pearson Education Ltd., *London*
Pearson Education Australia Pty. Limited, *Sydney*
Pearson Education Singapore Pte. Ltd.
Pearson Education North Asia Ltd., *Hong Kong*
Pearson Education Canada, Ltd., *Toronto*
Pearson Educación de Mexico, S.A. de C.V.
Pearson Education—Japan, *Tokyo*
Pearson Education Malaysia Pte. Ltd.
Pearson Education, *Upper Saddle River, New Jersey*

10 9 8 7 6 5 4 3 2 1
ISBN: 0-13-060257-4

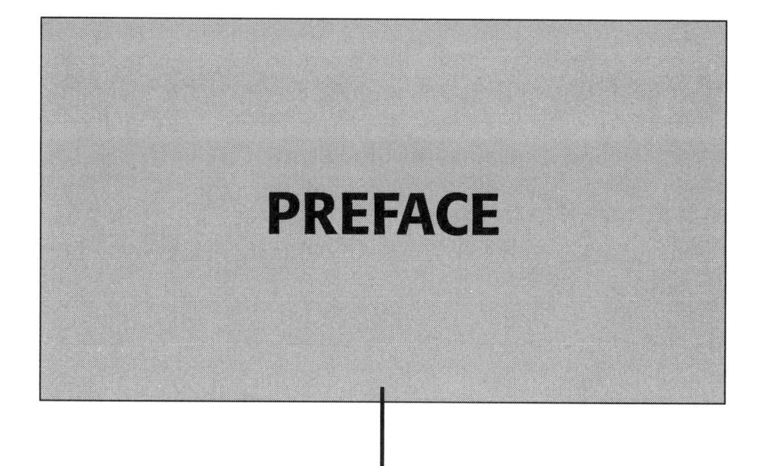

It has been only two short years since the third edition of *Social Issues in Technology* was published, yet once again I find myself writing another preface. For a book that covers specific techno-social issues, that might not seem strange, particularly considering the rate at which the world is changing. For a text that deals with fundamental theories on the subject, it may appear unusual. In reality, it is not so.

In the last two years alone we have seen surprising changes in the relationships in our world and in our paradigmatic understanding of that world as well. We have survived Y2K. We are on the verge of communications media revolutions—we may soon have a single device that will do everything from making phone calls to purchasing items from vending machines. We have experienced real-world applications of Bell's Theorem of Action at a distance. We have mapped the human genome in its entirety.

Because of this, the fourth edition clarifies specific points that support a better understanding of the theoretical thesis, and deals with exemplary changes that have occurred since the last version was completed. Thanks to encouraging and thoughtful input from my reviewers and the power of hindsight, I expanded Chapters 4 and 10 and added a new report on the effects of modernization on a traditional Maasai society. These, along with minor corrections, make up the bulk of the additions.

No book of this sort is ever complete. The future promises to hold even greater and more rapid change for us than the past, and as the future unfolds, our perceptions of how the system works will change. The principles, though, should remain the same.

It is my hope that the reader finds this edition to be more complete and more cohesive in its explanations, and that we all continue to have a most exciting and rewarding journey as we create our future.

I would like to express my thanks to those who helped in this revision, particularly Winnie Mukame, for her thoughtful review of the Maasai document; my reviewers, Steven Gerardi, New York City Technical College, Brooklyn; and Lynnette Mullins, University of Minnesota, Crookston; and my editor, Delia Uherec, for her patience, support, and expertise. Many thanks also to those colleagues who were willing to provide me with information that allowed me to make this a better text than before. They have all been a great help in this endeavor.

To Martel and Cindy Day
with my thanks and my love
because dreamers are the saviors of the world.

CONTENTS

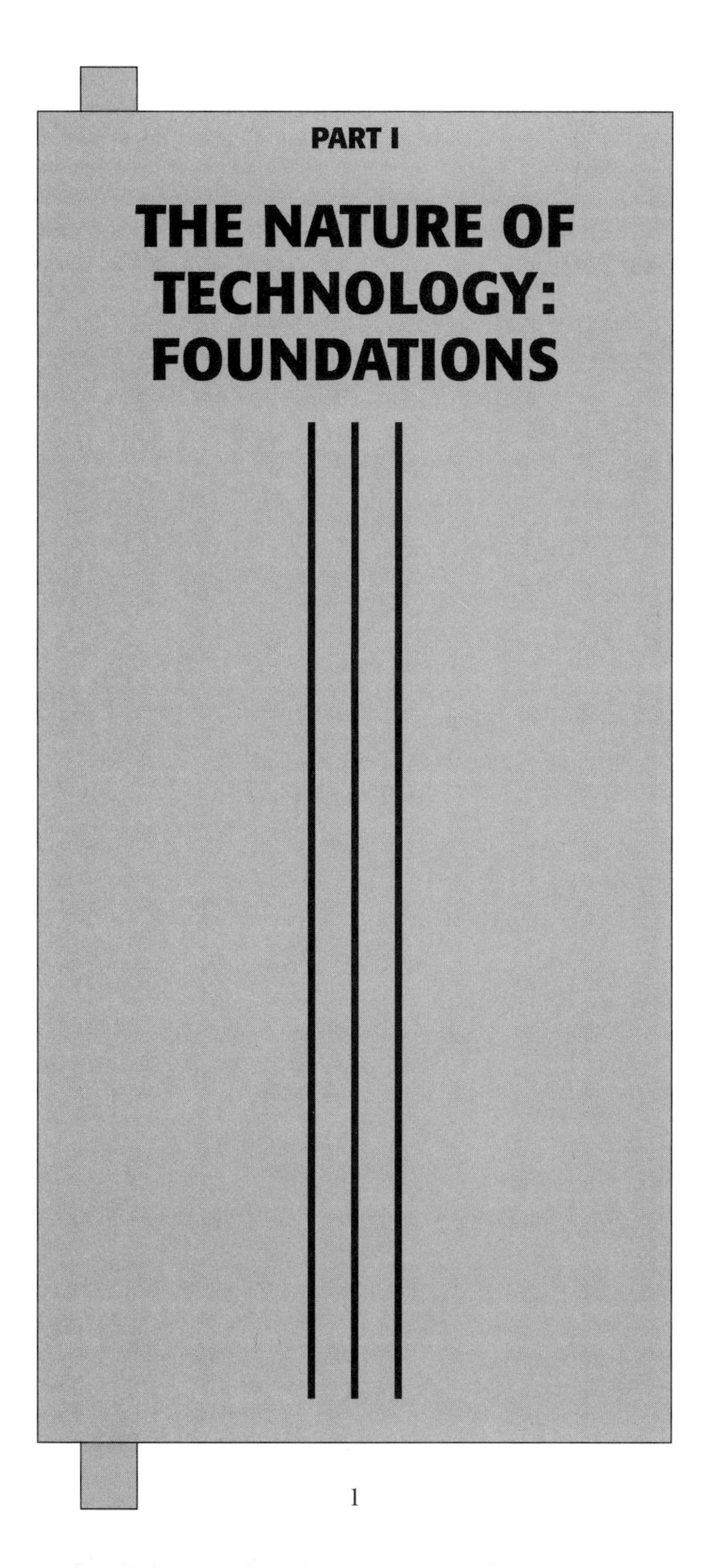

PART I

THE NATURE OF TECHNOLOGY: FOUNDATIONS

CHAPTER 1

TECHNOLOGY: A NATURAL PROCESS

He was a small creature, no more than forty inches in height, and he was, as usual, hungry. His brain was about half the size of a modern *Homo sapiens*, about 650 cubic centimeters, and he lived on a wide, flat plain in Africa, where the tall grass and clutches of low-branched trees made a hunter's paradise for him and his fellow creatures. His kind would one day be known as *Homo habilis*, but that was nearly 2 million years in the future. For our purposes, we will call him George.

As I have said, he was hungry. It was a characteristic of his species. George and his companions spent much of their time ranging out across the plain near their most recent campsite in search of food. These warm-blooded hunters were omnivorous, as likely to devour the tough nutty fruit of a nearby berry bush as they were the raw flesh of some small reptile or insect that failed to escape their notice in time. From day to day, George and his fellows satisfied their internal furnace with the fuel of whatever they could find, always searching for the great kill that would allow them to gorge themselves and replenish the dwindling supply of protein gathered from the last great kill some days or even weeks earlier.

Today they were near the high rock carapace to the east, though they had no concept of direction in those terms. It was merely the "high place over there, where the sun rises." George was scouting ahead of the pack, a chore he seemed to relish. A loner, he would often run ahead, somehow enjoying the prospect of being the first to sight a potential prey, hoping to be the first to wrestle it to the ground, to pound it to death with his clenched hands, or to tear its throat with his teeth.

The band was in the narrow passage that led into the center of the mound of rocks, close to where a fellow hunter had perished only days before at the hands of another predator, a huge cat creature with claws to tear at the throat and jaws to sink deeply into the flesh and break the victim's neck. George had seen it happen. He remembered it all too well. He was cautious, listening and sniffing the air, remembering what had happened to . . . who was it? His simple consciousness forgot those things easily, but the memory of the danger remained solidly in his mind.

The others were far behind him and out of sight as he turned into the natural bowl formed by the circle of high, flat rocks near the center of the carapace. He could feel the eyes on him, almost smell death in the air. Instinctively, he knew he was not alone. Turning quickly, he scanned the rocks above, seeking any telltale clue of whatever was lurking there. He spun so quickly and jerked his head about so violently in his panic that he nearly missed the low, flat, black furred head, the huge yellowish eyes that stared back at him.

Above George and a little to his left was the same sleek creature that had made a meal of his fellow hunter only days before. George panicked. He turned and leaped to the side of a sheer rock, clinging with his toes and fingertips as the cat made a lunging pass at him. The panther missed the small ape-man by inches. George scrambled toward the summit of the rock, churning his legs wildly in search of some foothold, reaching out blindly with his hands for any purchase further up the rock face. Lacerations appeared on his knees and thighs as he slid against the sharp black obsidian. His fingers numbed as they bit again and again into the narrow, knifelike crevices above. But he was making progress. Below him, the cat yowled and paced, panting heavily and leaping toward the fleeing figure.

Springing with all its might, the huge panther nearly reached George, who pulled forward with a last great effort and reached a wide ledge nearly halfway up the rock face. As he slid himself onto the strip of rock and flattened himself against the wall, a single stone slipped over the edge and fell, striking the huge cat squarely on the nose. With a howl, the panther retreated. George kicked another rock toward the beast, then another, missing both times. In panic, he grabbed several more and hurled them toward the beast, striking him again, this time dead center at the skull. The panther slumped to the ground, stunned by the blow. George grabbed for another loose rock and another, improving his aim with each throw, grasping larger and larger rocks until at last he found himself holding heavy slabs of obsidian over his head with both hands and hurling them down on the lifeless victim. He struck the creature again and again and again.

In the night, belly taut, legs splayed out before him, George lay with the other hunters in the natural bowl of the rocks, stuffed with the meat of the dead panther. He was smiling, staring up into the night sky at the bright starlit veil, a swath of white that spread across the sky like a river in the firmament. Absently, he licked his hand and passed the thick saliva over the crusted scratches on his belly and legs. Around him, the sound of night creatures echoed off the surrounding walls as predator and prey continued the struggle for survival.

In his right hand, George clutched at a single round stone, about three inches long across its short axis and weighing nearly half a pound. He felt safe now. He knew that he could fend off any attack. Tomorrow he would try his luck again with his newfound weapon. Tomorrow he would try it against one of the doglike scavengers of the plain or use it to bring down a bird near the river. Perhaps he would never need to be hungry again. Had he not slain the mighty panther single-handedly? Had he not proven himself the greatest hunter of them all? Who knew what he might be able to do the next time? Who could really know?

Was that the beginning of technology? It probably was not. Can we infer that George is the precursor of all humanity's inventiveness? This is definitely an erroneous assumption. Is this then only a story? No, it is a great deal more than that. It is a concept. It is an admittedly fanciful presentation of a process, a process that has inexorably led to the technology we enjoy today.

In the imaginary actions of this one *Homo habilis*, a creature who populated the broad fertile plains of Africa more than 2 million years ago, is the seed of all that we are as a civilization today. What happened in this brief incident is what happened again and again in our own real past. It was a first step in a chain of events leading from complication to complication, from one level of sophistication to new levels of sophistication. Now, in the twilight of the twentieth century, we find ourselves on the edge of ultimate power over nature—the power to destroy or enhance not only our own lives but also the very existence of every living creature on the planet. This is due in large part to what we call *technology*.

The question we need to ask first, the one that must be explored before we can ever hope to understand this process that has so totally enmeshed itself in every aspect of our lives, is simply whether technology and all that comes with it is a natural consequence of our humanity or is an artificial construct, separate from the natural way of doing things, and, perhaps, in direct opposition to nature itself. Are we dealing with an attempt to thwart nature, or are we expressing a part of what we naturally are? What of steel mills and electrical power, computers and steam engines, printing presses and gears, valves and mortar, and chisels and plows? Are all of these artifacts of our technological development unnatural? Are we dealing with man fighting against nature? Or are we merely seeing a *natural progression* of the species along an evolutionary path?

It is, as one television advertisement would have us believe, "not nice to fool Mother Nature." Fighting the entire universe, it would seem, is a hopeless task, at best an irrational idea. Is that what we are doing? We will initially address ourselves to this issue.

THE EVOLUTIONARY PROCESS

In the description of our fortunate *Homo habilis*, George was able to grasp the meaning of a combination of events and use them to his advantage in preserving his life. The fact that he was capable of doing that stems from the fact that nature had supplied him with a brain capable of making those vital connections. Without this ability, he would most likely have perished. The facility to reason, which is what we are really dealing with here, is a *survival trait,* that is, a characteristic of the organism that enhances the creature's chances of survival in a hostile environment. This is part of what nature does, and it is part of what has come to be known as evolutionary theory.

Evolutionary theory has become a hotly debated subject in recent times. Whether its tenets explain the development of humankind accurately is not always clear, particularly in the eyes of those who champion alternative theories. But it is a useful theory, one that tends to support the bulk of scientific evidence and one that is increasing in sophistication and refinement. It is not the purpose of this book to attack or defend the evolutionary theory, nor to argue the vagaries of that part of its content beyond the periphery of our understanding. The theory has proven itself to be an excellent method of explaining real-world phenomena and as such is here accepted in a simplified form as correct. Of particular interest to us in our study of technology and society are the concepts of "survival of the fittest," "natural selection," "specialization," and "adaptability."

Natural selection and survival of the fittest are passive phenomena, representing observations of the way in which nature appears to operate. Basically, these ideas say that there are a very large (if not infinite) number of ways in which organisms can exist but that only those best suited to surviving in an environment will tend to perpetuate themselves. Consider the many differences between a one-celled creature such as the amoeba and the complicated network of cells that go into the construction of a human being. Natural selection says that any other combination or any other path along a developmental scale is possible but that this one exists because of its superiority as a "format" that works. Humans could have had gills. We could have had webbed feet and five-foot-long necks and compound eyes and exoskeletons and a host of other features. But we do not. And the reason we do not is that they don't work as well as what we *do* have. The other features, the ones we do not have, simply do not exist because they perform no useful function for us. We do not *need* gills or webbed feet or compound eyes, and therefore we do not have them. For fish, gills are fine. For insects, compound eyes work beautifully. For a giraffe, a long neck is essential for survival. But for *Homo sapiens*, they are superfluous.

Natural selection says that as traits occur, if they are useful, the organism with the trait has an "edge" over similar organisms without the trait and, through time, more of the organisms *with* the trait survive than those without (survival of the fittest). The result is a predominance of organisms with the trait and the trait becomes "generalized" over the species.

For purposes of illustration, let's use an analogous example from our own culture. A given population contains a range of physical and psychological types. Some people are physically strong and others physically weak. Some are extremely intelligent and others not so intelligent. Some people are brilliant when it comes to music, whereas others, who cannot carry a tune in a bucket, have the capacity to understand complicated theoretical physical laws. (That is not to indicate that the two traits are mutually exclusive.) Now let's take a random sample of these people and put them into a specific environment, for instance, an accounting firm. For the sake of discussion, a representative sample of people all go to work for the same accounting firm on the same day. Ignore for the moment that the firm itself is selective.

Certain facts become rapidly apparent. People who are physically strong but not particularly capable when it comes to mathematics will find themselves at a distinct disadvantage. Those who are physically small or who have only normal physical strength find the lack of superiority in this trait of no consequence when it comes to doing their work. Those who are blind find that they probably cannot function at all unless artificially aided in some way. Musical geniuses, unless they can find some way to use their musical skills constructively for accounting, are going to find that the additional trait, while admirable, is of little or no use to them in their job. At the end of a year, assuming that all of these people are competing for a limited number of jobs within the firm, the only people left at the accounting firm from the original sample will be those whose traits lend themselves to the job at hand, that is, accounting.

We could have as easily made a case for fifty people going out for a football team or fifty people entering college or fifty people aspiring to be mystery writers. Those who survive in these "environments" are those who are best suited to carry out the activities connected with the environment.

This is conceptually no different from what takes place in nature. The world is filled with creatures that compete for survival in an environment that is filled with the possibility of failure and destruction. Only those whose traits are such that they have an edge over their fellow contestants can survive.

The concept of specialization carries this process a step further. Given that those best able to survive in an environment are the ones who will survive there, we find that nature will select more and more succinctly for traits that fit in with the environment in question. This is called *specialization,* and it is the tendency of an organism, through time, to retain only those traits that lend themselves to survival and to give up or forego those traits that are not useful to this end. Economy is a basic law in nature, or so it appears. Waste is not a thing that nature tends to support. Everything is done in such a way that efficiency is maximized. It is one of the reasons that technology mimics nature so closely in principle. Nature seems to know how to get the most out of a system. In terms of specialization, this means that traits that are useful remain through time and those that are not fall by the wayside, thus maintaining the efficiency of the organism.

But specialization is a double-edged sword. On the one hand, it increases the ability of an organism to survive in a given environment by selecting those traits that ensure success. This is its positive side. On the other hand, the process of specialization removes those traits that do not lend themselves to survival in a specific environment. *As an organism becomes more specialized and therefore more efficient at survival within a given environment, it also becomes more dependent on the specific environment for its survival.* That means that outside the specific environment, it has difficulty surviving.

Let's return to our analogy of the accounting firm. We have seen that within the firm, the employees remaining after a year are the ones best suited to the environment, that is, those who have traits that match the needs of the firm. A good accountant will still be employed, whereas one who is only marginally successful will have fallen by the wayside. Now assume that the environment has changed. Suppose the accounting firm is absorbed by a larger firm that replaces 90 percent of the jobs with computer technology and transfers most of the staff to another sector of the parent company specializing in forecasting and predicting future markets. What happens to the employees of the original accounting firm? Obviously, some will survive. Those whose traits are equally useful in fields of accounting and in forecasting will feel right at home. Some of them may even advance more rapidly in their jobs because of some unused traits that now come into play, such as intuition and the ability to think both inductively and deductively. Others, however, will find that some traits that they do not have are essential to survival in the new job, and they will eventually leave the company, either voluntarily or through being fired. Those who are whizzes with numbers and details, for instance, but have very poor social skills, may not survive in the firm due to their inability to express their opinions in meetings, interact with other departments, or deal with the public. As long as these individuals were insulated from such activities, they thrived. Faced with new responsibilities and requirements, however, they are forced out of the organization.

The same phenomenon occurs when a natural environment changes. Specificity in an organism created through the process of natural selection ties an animal tightly to its environment. If the environment changes, the results may be disastrous. The key to long-term survival appears to be the ability to change with the times, to adapt over time to a changing environment. If the environment changes too rapidly, or if the organism is too tightly specialized, it may become extinct before it can adjust. This explains the importance of the last concept of the evolutionary theory that we need concern ourselves with—"adaptability."

At this point, it is useful to note a concept that will be dealt with throughout the text. It is the concept of "balance." From what we have seen so far, two things are necessary for an organism to survive in nature: the ability to fit into the structure of an environment through natural selection and specialization and the ability to adapt to changes in the environment. These two characteristics may appear to be somewhat in opposition, specialization defining a very tight system and adaptability describing the necessity of the ability to despecialize. In truth, what is necessary for survival is the proper combination of specialization and ability to change. Change implies the passage of time. Inasmuch as the environmental changes that occur are not too catastrophic or too revolutionary in character, an organism, though tightly specialized, may still adapt to the change through natural selection. It is a balanced combination of the two that allows a species to enjoy long-term survival.

Nature is always recreating a state of balance, or *equilibrium,* as it functions. Evolution is itself a function of equilibrium, constantly redefining life in such a way that it can harmoniously fit into the fabric of the biosphere. When change occurs, nature adjusts, and change is, after all, the one thing we can be certain will take place. Some organisms change. Others cease to exist. New life forms come from old forms to thrive side by side with successful organisms already extant.

The world is full of examples of animals and human cultures that have adapted to their environment. This, at least, is fundamentally undeniable. Insects exist in forms that fit with the advantages and shortcomings of their environments. Animals live where they are capable of surviving, having those traits that lend themselves to the animals' survival and void of those traits that would endanger their existence. One does not find Gila monsters in Alaska, nor does one encounter polar bears in Arkansas. The flora and fauna of a given environment are "selective" of what works.

There is a beetle in Africa that enjoys living on termites. That in itself would not be particularly amazing were it not for the manner in which the beetle has become specialized to the task. This particular beetle uses two methods of hunting that allow it to be extremely effective. One is its ability to disguise itself as a rock, and the other is its ability to use natural tools to enhance its prospects of success.

The beetle coats its shell with a sticky substance that it produces internally and then covers itself with small grains of sand and other debris until it resembles a pebble. It then positions itself near the entrance of the nearest termite nest and waits. Soon, worker termites near the entrance who, as part of their natural programming (what we call *instinct*), have a desire to keep the entrance to the nest clear, venture out to remove the pebble from the mouth of the opening. They unsuspectingly grasp the "pebble" and carry it to one side. At a distance sufficient to avoid notice, the beetle turns on the worker and kills it. And so we see the advantages of looking like a rock.

But our beetle doesn't stop there. Rather than devouring the freshly killed prey, it returns to the nest with the corpse and waves it over the entrance, attracting the attention of soldier termites, whose job it is to protect the integrity of the nest, enticing them out into the open. In this way, the beetle is able to supply itself with a wealth of victims, each coming willingly to the slaughter. Is this a natural process, or is it technology? Is it both?

Nature is filled with examples of creatures like the beetle that are able to adapt to their surroundings. Prairie dogs utilize natural law to create ventilation in their highly sophisticated towns. Beavers use engineering techniques as complicated as those discovered by humans to control the flow of water in rivers and streams. Birds build nests that depend on static tension and precise structural design for stability. Fish use lures ranging from phosphorescent lights to tongues that mimic worms in their search for food. Insects use camouflage for protection, as do lizards and mammals. Ants and bees and baboons and *Homo sapiens* use complicated social structures for mutual benefit and defense. Why?

In every case, the answer is the same. They do it because it works! That is the bottom line. That is the value of nature's pragmatic approach to the evolution of the various forms of life that inhabit this planet, and it is as true of humankind as it is of any other creature on the face of the earth. Nature simply does what works.

GENETICS AND EVOLUTION

And how does nature accomplish all this restructuring of organisms and ecosystems? For the most part, it does it quite effectively through a genetic device, DNA. DNA is the encoder, the messenger of an organism that transmits information from generation to generation to assure that each generation knows what it is supposed to look like, how it is supposed to behave, and what is expected of its constituent parts. It is in effect a blueprint of the organism in question. All that an organism is, including the changes that take place within that particular organism, and all that it is capable of being are encoded in the genes of the DNA present in every cell of the organism. As differences occur in a single organism's makeup, that which makes it different from others of its own kind, the differences are recorded in the DNA codes, ready for transmission to the next generation issuing from that organism. If a given trait adds to an organism's survival ability, there is a greater probability that the particular organism will survive long enough to reproduce itself, including the new characteristic. This is nature's method of transferring instinct from one generation to the next. Behavior patterns that are instinctual are transmitted to each new generation of an organism through the functioning of this encoding system. It is, as we have seen, extremely effective. Every living thing is capable of being and doing what is successful (what works) because of the instinctual programming transmitted through this process. Thus, a duck flies south for the winter due to an instinctual understanding that that is what it is supposed to do. No one needs to teach a moth how to weave a cocoon, or an anglerfish how to dangle its "bait" in front of its mouth to attract a potential victim. No one needs to explain to a spider monkey that remaining in the trees reduces the number of predators it need concern itself with, or to tell a spider how to weave a web in the most efficient fashion. These things are understood instinctually. This "biological programming" gives creatures an edge within their own realms.

And yet, instinct is restricting. If instinctual programming is thwarted, if it is somehow blocked, the creature cannot, in many instances, alter its behavior. The results may be disastrous.

Consider the sand wasp. Nature has supplied this insect with several instinctual behavior patterns that allow it to be highly successful as a species. The sand wasp digs its nest in the ground in, as its name indicates, loose, sandy soil. The nest itself is little more than a small hole in the surrounding landscape. It is advantageous for the sand wasp not to advertise the location of its nest. Predators can enter and devour its young. Others may take over the nest, awaiting the wasp's return to make a meal of it. It is not easily seen (as anyone who has inadvertently stepped on one when the resident was home can tell you).

Yet sand wasps always know where their nests are when returning from foraging expeditions in the surrounding territory. How can they be so sure? Research indicates that the sand wasp fixes the image of the nest in its mind by recording its relationship to surrounding landmarks. It has a record of the nest's location firmly fixed in its mind and can always return to the same spot after an expedition. This is confirmed by covering the hole. Even when a secondary false hole is added, the sand wasp will go directly to its real nest and uncover it, ignoring the false nest entrance. But if the landmarks around the nest are moved, if pebbles and plants are rearranged, the sand wasp can become confused, be unable to find its nest again, and become totally undone by the fact that the landmarks are not where they are supposed to be. It may never find its nest again. Because the method of locating its nest is instinctual, *it has no choice as to how it goes about the business of locating it.* And therein lies the key to the difference between humankind and other organisms.

Evolution and Artifact

So far, we have taken a brief look at some of the basic concepts of evolution. We have explored in a general way how evolutionary theory explains the diversity of life (humankind included) that exists on the planet. We have looked at the importance of specialization and natural selection to survival and have seen how instinct can aid a creature's survival capabilities. Yet we have said nothing of technology except by way of analogy. Or have we? What about the use of a dead termite to lure soldier termites from a nest? Is the beetle not using a tool in its search for food? Is this an artifact, that is, an "unnatural" device? Is this a primitive example of manipulation, or is it an adjustment to an environment? A beaver fells trees and builds elaborate dams to divert the course of a stream and create a home for itself and its mate. Is this the utilization of engineering principles to achieve an end? Is technology at odds with instinct?

Obviously, the beetle using a dead termite as a lure and the beaver building a dam are behaving instinctually. Neither of them have been to college to learn how to do these things, nor have they sat at the knee of an elder around a warm fire, learning what works (at least as far as we know). So what has all this to do with technology and the way in which humans operate in nature?

The difference between humans and animals lies in *the way humankind goes about the evolutionary process.* The transmission of information concerning what is successful and unsuccessful in a given environment is done in most cases through the genetic process, through DNA coding. We call this *instinct.* When adaptation is necessary, the vast majority of creatures adapt slowly, over many generations, allowing for a natural selection of positive survival traits. Human beings, on the other hand, have been given a survival trait that allows them to bypass this procedure and adapt with a speed that is incomparable when compared with other animals. Turtles have hard shells because shells help them survive. The shell protects them from predators. Birds have feathers and fly through the air because this gives them an advantage over earthbound creatures. In the case of the turtle, we see a trait so successful that it has remained for hundreds of millions of years. In the case of the bird, we see an adaptation that took millions of years to create.

Birds originated with a small coelurosaur known as an archaeopteryx, a theropod living during the late Triassic and early Jurassic periods. The small reptile, finding itself in competition with the emerging mammals, many of whom were nocturnal predators, developed feathery scales that eventually developed into feathers, allowing it to survive in the face of increased competition by escaping into the skies. It took nearly 35 million years to accomplish this! Humanity accomplished the same feat without growing the first feather.

The difference is that human beings evolved externally to their bodies as well as internally through the process of natural selection. Marshall McLuhan expressed it succinctly when he referred to wheels as extensions of the feet, cameras as extensions of the eyes, and microphones as extensions of the ears. It is this external evolution to which I am referring. Humanity is specialized in one respect—the ability to think abstractly, which allows the species to be at once specific and general to an environment. We do not have long, sharp teeth as do members of the cat family, yet we are much more efficient hunters. We cannot run at more than sixty miles per hour like the cheetah, yet we can move much more rapidly and for much longer periods of time. We are not the biggest, nor the strongest, nor the swiftest. We are not equipped with natural camouflage or repellent chemical sprays or hard-shelled armor. We have none of these specializations, yet we are one of the most successful creatures in nature. And that is because we are adaptive.

Rather than grow evolutionary changes, humanity manufactures them. In order to adapt, we *extend* ourselves out into the surrounding environment, not being restricted by internal change. With the spear as a weapon, we need not have sharp, extensive canines. With the bow and arrow and later the flintlock and rifle, we can extend our "fangs" far beyond the range of any other creature. With the wheel and steam and the internal combustion engine, we expand our mobility at a rate that surpasses any other beast on the planet. In each case, we are extending ourselves outward from our bodies into our environment, at the same time changing the environment within which we are operating in order to suit our own purposes. If it is too hot, we create refrigeration. If it is too cold, we learn to control fire, to build the hearth and chimney. Each extension increases our ability to adapt to change and to react to alterations in the environment that we may encounter. In effect, we have specialized in the ability to generalize.

Humankind does evolve. We do adapt to changing environments and changing conditions. But it is accomplished through the creation of artifacts, by a manipulation of the environment and an understanding of natural law that enables us to improve and re-create the environment in the form we find most advantageous.

The archaeopteryx developed from dinosaur to bird in 35 million years. Human beings progressed from Kitty Hawk to the moon in less than seventy years. As an evolving species, we seem to always be in a hurry. We seem to be too interested in progressing to wait for nature to supply us with the internal devices that we need to progress. We do not need specialization in the physical sense because we are specialized in one very important way—the ability to think. Conceptually, there is no difference between a turtle's shell and a human's "house," although the latter is more flexible in form and function. It is still a protective dwelling. The concept is the same although the application is greatly expanded in the human body. Conceptually, there is no difference between a bird's wings and an airplane. Both are means of flight. Yet the variety of designs and uses to which humankind has put the concept of flight is far beyond that of birds. We evolve rapidly, and we evolve externally. We use *artifacts,* that is, artificial constructs that mimic natural principles, to do the job, rather than physical changes. And because of that one fundamental difference, we have been able to conquer every environment and every circumstance the planet can offer, from the depths of the ocean to the cold of the poles. We generalize by being specific only in our ability to understand, to construct technology, and to adapt. And the evidence of how successful this approach can be is made clear by the huge numbers of human beings and their influence over the world's ecosystem.

CONCLUSION

In this chapter, we showed that humanity is as much a part of nature as any other species and that this includes its use of technology and artifacts and its ability to manipulate the environment to suit its own needs. Humankind's actions are often involved with the production of "unnatural" devices, called *artifacts.* They follow natural laws and perform useful work in accordance with the way the physical world is structured. They are artificial by definition in that they do not appear naturally in nature. This does not mean that technology is not a natural part of the evolutionary process. Nor does it mean that technology represents an unnatural refusal to live within the ecosystem or that what we do is in opposition to the equilibrium or balance of nature. We do what every other organism does: We evolve, we adjust, we change, we react. It is only in the *external* manner of our evolution and in the far greater efficiency of adaptation that we are separated from other animal life.

Technology is part of what it is to be human and, consequently, part of what it is to be part of nature. We are an experiment. We are an attempt on the part of nature to look for a better, more efficient way of doing things. We are a step on the road to order. So far, we have been a very successful experiment. Whether we will continue to be so remains to be seen. This is a large part of the subject matter that makes up the rest of this book. In the next chapter, we begin an investigation of this question that deals with the "flip side" of the picture, nature's system of checks and balances and how the human race controls its evolutionary process.

Thus far, we have explored the evolutionary nature of technology and how this ability to technologize greatly enhances the capacity of human beings to survive and find their niche in the ecosystem. From the discussion as well as from your personal observation of the world in which you live, it should be obvious that the act of technologizing has created a position of dominance for humankind, in terms of species success, numbers, and adaptability. The problem, of course, is that it also brings with it hazards. There is a natural *yin-yang* in the way nature carries out its plans, and with every benefit, there is a price to pay. Later in this text, we explore this concept more fully. For the moment, we must admit that the capacity to produce rapid evolutionary change through the use of externally produced, artificial constructs could very well lead to disaster if not intelligently and judiciously employed. Thus, nature has created another survival mechanism, offering a second comparative advantage for humans that keeps us from destroying ourselves with our own success. At the least, that is the case so far in our history. Chapter 2 concerns itself with the nature and appropriateness of that second survival mechanism.

KEY TERMS

Adaptability
Artifact
Change
DNA
Equilibrium
Evolutionary Process
Evolutionary Theory
Instinct
Natural Selection
Specialization
Survival of the Fittest
Technologize

REVIEW QUESTIONS

1. What is evolutionary theory and how does it relate to technology?
2. What is the difference between an artifact and technology?
3. What trait innate to humans does the text present as nature's way of creating our specific niche in the ecosystem?
4. What is the difference between internal and external evolution?
5. How is our species specialized and how does it adapt? How is this different from other life forms?

6. Give four examples of human adaptation through technology.
7. Give four examples of human adaptation through biology.

ESSAY QUESTIONS

1. Discuss how technology is natural and also how it is artificial. How do these two views conflict? How do they complement each other?
2. It would appear that primates, particularly *Homo sapiens*, embraced the use of artificial devices to enhance their chances for survival. In what way would our world be different if other mammals (a raccoon or an opossum) had embraced this approach?
3. It is argued that creationists who use the scientific method to further their own view of the world fail to remember that the world in which we live is not a closed system. This is seen as a fundamental flaw to their argument. Comment on this view, mentioning both pros and cons.
4. Given the scenario in the first chapter, do you find technology to be merely an extension of our innate abilities or to be a development of additional characteristics not seen in the species before now?

THOUGHT AND PROCESS

1. Primitive human beings experience their environment and learn from it, gaining the tools that will allow them to survive. What are some of the obvious connections between the probable observations of the primitive human being and the invention of (a) the sling, (b) the spear, (c) cooking, (d) the wheel, and (e) artificial shelter (houses)?

2. If invention and technologizing are environment specific, speculate on the difference in primitive tools created by people living (a) in a tropical rain forest, (b) on a heavily wooded tropical island, (c) on an arid, high-altitude plain, and (d) on flat tundra near the Arctic Circle (but not on a coast). How do your speculations compare with your knowledge of such cultures? Did you base your speculations on your knowledge of these cultures?

3. Suppose that you are marooned in an unfamiliar environment, for instance, having crash-landed on an apparently uninhabited earthlike planet. You are surrounded by trees, wooded valleys, and a water source in the form of a small stream. You have no knowledge of the seasonal variations of the planet, what season you are in, how long you will be marooned, what the fauna are like, or whether there are intelligent inhabitants on the planet. There are no hard minerals, but an abundance of sand, coal near the surface, and sandstone outcroppings. Describe your efforts to survive.

4. In light of the statement in this chapter that adapting to a single environment apparently dooms an animal to dependence on that environment, classify each of the following organisms in relation to their migratory capacities, arranging them in descending order of adaptability to a change in environment: (a) human beings, (b) elephants, (c) koala bears, (d) common honeybees, (e) mosquitoes, (f) grizzly bears, (g) bison, (h) monarch butterflies, (i) tree frogs.

Now make another list based on sources of information about each of these creatures and, in light of your extended knowledge of their habits, compare the second list with your original.

5. If humans mimic nature in creating technology, we can assume that the microphone is a mechanical mouth to the speaker and ear to the listener, the wheel is an improved foot, the telescope and camera represent extensions of the eye, and so forth. How do we extend our concept of the human thinking process through artificial means? How do we artificially mimic the procreation–gestation–birth process? Are there analogies in human technologies that closely parallel these aspects of human biological life?

CHAPTER 2

RESISTANCE TO CHANGE

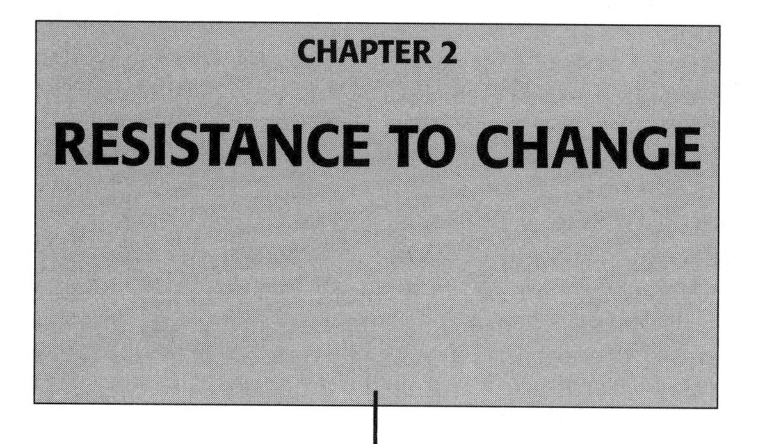

William Fletcher was not an evil man. Quite the contrary, he was one of the most highly respected men in the district. He had grown up among the people in Harrogate and was a part of everything that was typical of the traditional English countryside. Prior to this moment, he had never so much as considered an act that could be construed as unlawful, at least nothing beyond the usual exuberance of his youth. Yet here he was, skulking about the commons like some miscreant up to mischief. He was beginning to regret his decision already. Harrogate was, after all, his home.

He loved this quiet village along the upper reaches of the Thames. The river coursed gently through this section of the country, moving so slowly in places that the current could scarcely be detected at all. But along its near shore, eddies and subtle changes in the bottom would churn it into a fearful rush, and it was this fact that brought William Fletcher to be abroad this night. It was also this fact that had brought the mill to Harrogate.

William had lived his entire life among the simple folk of Harrogate. Indeed, his family had lived and worked in this village for as long as anyone could remember. His very name bespoke his ancestors' honorable profession, first as arrow makers to a noble lord and later as armorers to the king. But that was all before. That was before the machines, before the new ways, before the coming of the landlord's new enterprise. It was before the factory had come to the banks of their river. Everything was different now, and William was not alone in his fear of the changes.

William had spent considerable time making his way around the commons, avoiding the more direct path across the green that would have meant detection by the squire's livery. He slipped from wall to wall, skirting stone cottages and deserted lanes until he found himself directly opposite the Hanged Horse Tavern. A final skittering dash to the rear of the building and he had arrived. Three sharp raps and he was through the door to Jonathon Roberts's private rooms.

William looked about. He was in a long, narrow room with a low, timbered ceiling and heavy planked floor. A single small window—all diamonds and circles where scraps of glass had been rescued from the bottoms of discarded bottles to give precious little light on a winter's day—punctuated the otherwise blank wall to his right. Candles stood in wooden holders by the entrance and by the door to the tavern itself. There was not a sound from the common room. No man would be about this late, at least no honest man. The candles were unlit. The only light came from the burning log in the great hearth to his left. In its ruddy glow he could make out the silhouettes of nearly a dozen others, and he knew most of them well, though he could not clearly make out their features.

Jonathon Roberts, the silent host, stood by the fire, his crooked leg as good a signature as any to distinguish him from the others. William still remembered the day the hay wain had fallen on Jonathon, crushing that leg. To Roberts's left stood the Cooper brothers, Michael and Daniel, broad of shoulder and straight of limb, not yet bent by the work that would stoop their bodies in another ten years. To William's left was Allen Smithson, an ironmonger like himself, and he had brought the three from Hardmoore as he promised. The remaining two were William's own brother, Harold, and his uncle, Geoffrey. William smiled and nodded.

"Are we all here, then?" asked Jonathon Roberts.

"All that be coming, I suspect," said Allen Smithson. "There'll not be many willing to do what must be done."

"Is it come to that, then?" asked William. "Is there nothing else for it?"

Jonathon Roberts heaved a great sigh, and, as if in answer, the great log beside him on the hearth hissed and sizzled, sending a spray of red sparks into the air.

"You have seen as well as we, William Fletcher, what the squire's factory has meant to Harrogate. Nothing is the same. Nothing can ever be the same again if we do not act and soon."

William nodded assent. "Aye, Jonathon. We've said it all before. It is not a happy task we have before us now. I merely hoped—"

"That we could find some other way? Yer daft, man. Look about you. Our river's fouled with the waste of that evil enterprise. Aye, even down to Hardmoore, near seven miles away, it's fouled. We've suffered our peace disturbed, our lives disrupted, and our very livelihoods taken from us by those devil machines. There's no life to them, William. They're all wood and iron and stink and noise. There's no heart to them. The ways of our fathers have been destroyed. They're taking our young men to work and causing our women to yearn after rich-made goods. First they woo us with their promises of good wages and a better life, then, when new machines do our work faster, they send us on our way or lower our wages until we cannot support our families. There's not been an honest day's work done since their coming! We've not had a moment's peace since that day, and we'll have none until we've destroyed them."

"But it's simply not right, Jonathon," protested William. "This is not our right! The factory and all within it belongs to the squire, aye, and our homes and livestock as well. This is not the way."

"Then why have ye come if ye feel that way?" retorted Jonathon.

William stood for a long moment in silence, looking into the dark face of Jonathon Roberts. He could not hear a sound in the room but his own breathing and the crackle of the huge oak log in the fireplace. His shoulders finally slumped in resignation. He nodded and said, "Because I know of no other way to fight it, Jon. That is why I am here, as you well know."

"As I well know," Jonathon repeated. He crossed to the rough table in the center of the room and grasped the heavy iron bar laying there. He raised it into the air and brought it

down on the table with great force. A plank split, sending a cloud of splinters and dust into the air. "This is the only way to stop them, William. This and fire. And once we've started, others will join us, of that you can be certain! We'll stamp out these factories here and now and have done with them! Let the progress be gone! Progress indeed! It's a lord's way of tramplin' his people underfoot! We'll send the squire's devil machines to hell where they belong!"

"And if the squire opposes us?" William challenged.

Roberts looked up into William's eyes with a deathly grin. "Then we'll send him to hell, too!"

Allen Smithson grabbed an iron bar from the table. The three men he had brought from Hardmoore lifted the axes that had been resting by their sides. William shuddered inside, closed the last door on his doubts, and licked his dry lips. He raised the heavy iron-headed hammer he had brought from his shop.

One by one, the men filed from the small room and began the short journey along the river chase to the factory. William wondered what the squire's men would do if they caught them. They were armed these days. They had been armed since the riots in Birmingham. He wondered if he would ever see his Emma again—if he would live to see tomorrow's first light. But then he remembered the gentle river and the quiet, unchanging village and the way his neighbors had strayed from their peaceful, ordered lives, and nothing else was of consequence. Only the factory and the machines mattered—and their destruction.

The scene depicted here has a strange flavor of both abhorrence and sympathy. We are presented with a picture of violence about to be unleashed on society, yet we may be strangely moved to sympathize with the plight of these unfortunate people, caught in the tangle of their dilemma.

This is a story of resistance, of frustration, of violence. It is a story of desperate individuals about to destroy technology. How can such a reaction come about? What forces people to take such risks? What compulsion could be so strong that they would go against all that they had been taught by the society in which they live and commit crimes directly opposed to the strongest urgings of their ethical base?

These hypothetical workers are typical of a group that actually existed in the early years of the nineteenth century in England. They were called *Luddites,* a name derived from the actions of one Ned Lud of Leicestershire, who, around 1799, destroyed two textile frames belonging to his employer. Ned Lud might be excused of his excessive frustration over the introduction of laborsaving devices. But what of the Luddites to follow? Between 1811 and 1816, the Luddites routinely smashed laborsaving devices in the textile industry of England in protest of lower wages and unemployment stemming from the introduction of new technology.

The Luddites are an extreme example of a natural phenomenon that coexists with the development of technology, a phenomenon known as *homeostasis.* The word *homeostasis* comes from two roots, *homeo,* meaning "the same" or "like," and *stasis,* referring to the stoppage of the flow of change in some system. Hence, *homeostasis* refers to a resistance to change, a seeking of the status quo.

There are numerous examples of the homeostatic reactions of people to changes in society in general and in the technological base in particular. They appear throughout history and are probably far more common in our own lives than we suspect. Let's consider some other cases of homeostasis in action.

THE GREGORIAN CALENDAR

During the papal reign of Gregory XIII, a study revealed that the calendar system in use at the time, the Julian calendar, was inaccurate in its determination of dates. The inaccuracy resulted from the difference between solar time and the official calendar, a matter of one-fourth day per year. The difference had become so pronounced since the adoption of the Julian calendar in 46 B.C. that by the time Pope Gregory XIII looked into the matter in 1582, a total of ten days separated the calendar date from the true date. That is to say, it was ten days later than everyone thought it was. Much to Pope Gregory's credit, he instituted our present calendar system, complete with leap years and other refinements, to take care of the disparity and to ensure that the lapse could not occur again. But what of the extra ten days? Gregory was a practical and well-organized man. He solved the problem logically and simply by starting the new calendar ten days forward of the old Julian date, thus eliminating the difference forever. So much for the problem of calendars. Or was it?

Consider how you would feel if you got out of bed tomorrow and were told by your favorite early morning news commentator that it was Monday, not Friday, because some famous scientists had just discovered—after viewing the far reaches of the cosmos through their telescopes—that the entire universe was being misinterpreted by a matter of four days. Accordingly, you had to go to work for five more days before the weekend rather than one day. What would your reaction be? I don't blame you. I'd probably react the same way. And that's just how the common people of Europe reacted!

There were riots! There were burnings! There were *burgermeisters,* or mayors, put to the sword! And everywhere there was the cry of "Give us back our ten days!" And it was all because Pope Gregory wished to straighten out a long-standing problem with the calendar.

COUNTERCULTURES

In the 1920s they were called *Bohemians.* In the 1950s we called them *beatniks* or the *beat generation.* In the 1960s they were called *hippies.* And in the 1990s we have called them *Generation X.* All of these labels have been used, either by those involved in the movements or by those opposed to them, to describe a phenomenon known as the *counterculture.* In each case, a group of people, usually predominantly young, chose to separate themselves from the mainstream of their society and try another approach, another way of doing things. They created their own social structures, their own literature, and their own dialects, seeking to disassociate themselves from the rest of the culture in which they lived. And in each case, the roots of their "discontent" or desire to return to a simpler life can be traced to the technological upheavals of the times in which they live or lived.

Bohemians first appeared after World War I. They lived in a new world, totally different from the one that had existed prior to 1914, and, rather than move with the changing times, they chose to wander through Europe, living in the back streets of a city's old town and generally divorcing themselves from the progress going on around them. The beat generation sprang from the changes that took place in the United States after World War II, a period of redefining U.S. culture and lifestyle. The beat generation rejected the move to suburbia, the rise of the corporate world, and the move to higher and higher levels of affluence among the population epitomized by high-fidelity technology, television, and the suburb. They chose to spend their time philosophizing about the plight of the country from their basement coffeehouses, producing poetry and music that was marked, as much as anything else, by themes of alienation and forms that were very different from the established patterns of the times. Then, in the 1960s, the hippies, with their long hair and their insistence on a less complex life, free of imposed encumbrances, sought to produce in our culture a mixture of optimism and a desire to move toward simpler times, similar to that advocated by the transcendentalist poets and philosophers of the nineteenth century.

It is not the author's purpose to debate or pass judgment on the ideals and the trappings of these subcultures, each of which flourished for a time as a poignant statement of the troubles of our culture, but rather to view them as symptomatic of a phenomenon. It is the individual who refuses to accept the directions of society at large that is important to us here. The individuals in each counterculture had their reasons for their actions and for their attitudes. There are few people who can deny the effects of the hippie movement of the 1960s on our culture, both politically and practically. They were an instrumental part of the American reaction to the Southeast Asian war in Vietnam and neighboring countries. They undoubtedly contributed significantly to the ecology movement in the United States and elsewhere in the world. They enriched our art, our music, and our language, but above all, they demonstrated a fundamental refusal to follow the direction in which society was headed. In this sense, they demonstrated not only the truth of homeostasis but also its importance to the process of technological change and our survival as a species.

Other examples of reaction to change could be cited, from the refusal of Russian peasants to accept agricultural improvements introduced by Peter the Great in the seventeenth century to the fears of technocracy in the 1920s. In each of these cases, we see the same demonstration of a resistance to a forward movement in the society punctuated by technological innovation. It is the causes of this homeostasis that need to be explained.

SOURCES OF TECHNOLOGICAL HOMEOSTASIS

The fear of technology and the accompanying homeostatic reaction are rooted in no single cause. A number of factors seem to contribute to the rising feeling of helplessness and fear that often accompanies advances in technology. The following discussion describes some major causes.

Fear of the Unknown

There is a basic experience that everyone who has ever lived has shared. It is the birth experience. And like all personal experiences, the cognitive processes of the mind allow us to assimilate information about the experience, using the knowledge to make determinations about our world. That is, when we experience something, we become aware of it at some level and place the new data gained into the overall fabric of our understanding of our universe. In short, we learn.

At birth, of course, we have no conscious understanding of what is taking place. We have no way to place the incoming data into perspective or compare it logically and classify it. This does not mean that we are not aware of what is going on, only that we have no way of communicating about it with ourselves or others, beyond bawling our eyes out, yelling our heads off, or cooing and laughing gently, depending on one's personal experience of birth.

There is an entire approach to giving birth that centers on the importance of the event as a psychological experience for the newborn and the parents. It focuses on the comfort of the child and on reducing the possibility of fear in the child to an absolute minimum. It is called the Leboyer method, and it is being used more and more in the birthing process.

Why bother? Why be concerned with the feelings of the newborn infant? After all, it has just been born and will have no memory of the experience, no recollection of who did what to whom. Why not just take care of the process as efficiently and effectively as possible, ensuring the well-being of mother and child? All babies cry when they're born, don't they?

As it turns out, this is not exactly true. The birth experience is far too often a traumatic one, and the importance of that fact is being increasingly recognized as a source of later behavior.

Consider this. You are an unborn child, safe in your mother's womb. Your entire experience is wrapped up in the gestation process during which you have nothing to do but grow and develop. In the next instant you are forced to emerge into an entirely new world, filled with unexpected and unfamiliar experiences. You are possibly cold, possibly in some pain, and constantly jostled about by unseen hands, the reality of which has no meaning for you. There are new sounds and new feelings far beyond anything you have experienced to that point in time. This is one of your very first learning experiences, and often it is a matter of learning that the unknown is frightening and therefore to be feared.

There you have it. Right out of the womb you are taught to be afraid of the unknown. From this birth experience, human beings often have their first taste of anxiety, and whenever new experiences come along, they may subconsciously return to that first experience of life that says that the unknown is to be feared. In this sense, homeostasis can be viewed as an attempt to avoid the unknown by avoiding change. No new experiences means no new bouts with the unknown. We can maintain our reference points. We know how to react and what to expect. We can avoid those pangs of anxiety that have been with us since very soon after birth. It is nothing more than the id in action, seeking pleasure and avoiding pain. What could be more natural?

If the birth experience were the only experience of change common to humans, however, they would simply never change. Fortunately, other changes occur in our lives before we are able to take any conscious control of our own circumstances, and through the learning process the young child develops a sense of understanding that all change is not bad. Indeed, there are many pleasurable new experiences as time goes on, stemming from our interactions with our family and later with friends, and from our interactions with our environment through play. Play is, after all, a method of practicing for the real world to come. It's a training ground for how to handle the future, with all its surprises and decisions. Thus we shuttle the birth experience into some neat corner of our mind and continue to impress new experiences and reach new conclusions about the meaning of their content, both good and bad, as we grow up. And this brings us to a second source of homeostasis, the process of *adjustment*.

Anxiety and Adjustment

Life is not always easy. And that is probably a very positive thing. Without resistance in our lives, we would have no reason to strive to overcome problems and there would be no growth, either in society or in the individual. In a hostile environment, the tendency to seek to overcome obstacles is a survival mechanism that works. It is the cause of technology in the first place.

But what about obstacles that cannot be overcome? What about when we are unable to overcome the difficulties inherent in some undertaking? Without some way to handle the possibility of our efforts being thwarted, we would be like windup toys, blocked in our progress by a wall or table leg, yet still pushing with all the strength of our spring-powered motors, trying to force our way through the difficulty until we run down or break.

Fortunately, most people neither run down nor break. Nature has supplied our wandering, creative minds with a series of defense mechanisms that protect us from overload. An entire branch of psychology, the psychology of adjustment, is devoted to the study of this one subject.

When we feel frustration, we react by first exerting more effort toward the solution of the problem, causing more frustration. If this fails, we adjust in some way to the nonattainment of our goal. All of the various types of adjustment available to us are not equally acceptable in the eyes of society. This in no way prevents us from using them.

Resistance to change can be viewed as an adjustment process, by which the frustration and anxiety caused by the technological change are denied. People may feel helpless in the face of technology. It is new and it is undefined, an anxiety-causing condition that at first spurs them on to seek out knowledge that will alleviate their initial fears of the unknown. If and when this effort fails, then they must cope by adjusting to the change in some way to ease their feelings of helplessness. One of the forms that the adjustment can take is homeostasis. The Luddite-like villagers in the vignette at the beginning of the chapter were reacting in this manner. They were threatened and angered (another adjustment to anxiety) by the road progress was taking, and to combat it, they chose violent resistance. The countercultures of the twentieth century used withdrawal, a form of schizophrenic reaction, to deal with the changing environment within which they were living. They chose to withdraw not only from the technology but also from the changing culture that the technology was creating. They "dropped out" and "turned on" as a means of escape from the realities of the changes that were occurring. This is by no means to say that all hippies were schizophrenic. It merely indicates that withdrawal in the face of a technosociological change can be a defense mechanism designed to reduce anxiety and prevent the serious short-circuiting of a person's psyche.

Psychophysiological Restructuring

We know now that the mind stores data in a more or less random pattern, placing memories and information at numerous locations simultaneously. In a process that depends on a random search pattern, it is then able to retrieve that information from one or more of these locations. The more times we receive data about a given event or concept, the more places we store it and the more information we retrieve when remembering. There is a strengthening that takes place with consecutive instances of *imprinting*, or storing of information. The more we experience a phenomenon in a given way, the more often it is stored in exactly that way. When we experience conflicting concepts concerning a phenomenon, they too are stored, though not as often as the more abundant standard experiences. We therefore build up a mental image, ever increasing in complexity and detail, about the nature of our world. Through this process, we develop an overview that becomes our paradigm.

When changes occur that affect the way our world operates, it creates conflicting information. New technologies, new ideas, and new ways of looking at the nature of reality offer opportunities to grow and develop, yet our brains have many more stored experiences that support the old, established view of things. Changing wholesale the manner in which one thinks is not an easy process. We at first fight vigorously, then slowly the change takes place as we receive more and more supportive data and think of experiences more and more in new ways, replacing our old concepts with new ones.

When we are young, this is a relatively simple process, first because of the more limited past information to be reorganized and overcome, and also because of an innate plasticity in the brain as it assimilates and records new data, fitting into a mosaic of understanding that grows through time. As we grow older, we lose some of the plasticity of the mind, that is, some of its ability to rewire itself in the light of new information. As more and more experiences enter our life, we slip them into our consciousness (and possibly subconscious) through well-worn paths long used with success in running our lives. We compare and classify based on our experiences. We fit new data into neatly ordered pigeonholes in our minds.

What then of an alien concept that happens to come along? What if that concept is in opposition with our usual mode of thinking? What if it does not fit neatly into one of our little preconceived boxes? Are we able to understand it? Can we assimilate and place it into proper perspective? Psychologists tell us that we can but that as we age, it becomes more difficult.

We are blessed with plasticity. We are capable throughout our lives of adjusting to new circumstances as they arrive. But the speed and ease involved in the adjustment, that is, the degree of plasticity that we as individuals have, tend to drop as we grow older. Reinforced concepts are more easily used than new ones. As we age, our learning curve takes a turn for the worse, and we become increasingly intractable when it comes to change and to new ideas.

Consider how traditional a typical seventy-year-old person can seem to be, how tenaciously he or she sometimes sticks to old ways of doing things, retaining old habits and old belief systems in the face of even the most incontrovertible evidence to the contrary. On the other hand, consider young children. They are ready to accept anything. They are curious, questioning, and willing to seek new knowledge wherever they can find it. To them the world is an adventure, a playground full of new ideas and new experiences. They seek it eagerly. They ask why and how and who about everything, and they can assimilate and use the new knowledge with incredible speed. It is because of their plasticity that this occurs. They have not had their experiences reinforced yet, as have their adult counterparts.

How does this apply to homeostasis? It means that new concepts, new ways of life, new technologies, and new social structures that accompany change are hard to accept when one is older. The mind is still plastic, to be sure, but it has trouble converting its belief systems. It is experiencing difficulty changing gears. And for that reason, much of the resistance to change that appears in the culture occurs in those who are older rather than younger. For a change to truly last in our world, it must fight and win over the objections of the adult population. They are the ones who will have the greatest trouble accepting it.

The computer and all it brings with it is the obvious example of this type of homeostatic reaction. A young child can learn a programming language such as JAVA in a very short period of time. In two or three sessions, the child can be merrily programming in the language, not at all put off by the strange jargonish nature of the commands or the strict step-by-step logic of the program's construction. Adults can, and many do, learn to use a computer as well, but the task is far harder. They have had no preparation in the sense that the child has. They have strongly reinforced patterns of neuronic interaction in their brain that determine their predominant mode of thinking. In many cases this is not a mode that easily lends itself to learning computer programming. It is hard work for the adult, not play. It requires effort, time, and concentration in the adult just to load and execute a canned program, constantly adjusting the adult's thinking to what the computer is doing. For the child, it is, what else, "child's play."

People resist that which is difficult for them. In our seeking of an easier way to do things, we tend to reject those ideas that are difficult in favor of those that are easy, and the familiar is always easier than the unfamiliar, whether it is the computer, a new form of music, or the workings of a mechanical loom. Frustration can result from the effort, followed by anxiety, and finally by a refusal to accept it.

This plasticity of brain networking also goes a long way to explain the predominance of youth in social movements. Society is a living organism, a symbiotic creature whose mechanism consists of all the people within it. It is only logical to conclude that the changes that the organism undergoes will be created by those best able to make the change. In the case of society, those are the young.

Chauvinistic Conditioning

Still another source of resistance to change is *conditioning*. It is the result of the plasticity problem discussed above, though it deals not with the inability of the individual to accept the change but rather with the preconceptions that a person may have as a result of the way in which that person views the change. We are dealing with a matter of context rather than content.

The *content* of a new technology is its physical reality. The content of steam power is the steam engine and the machinery it can run. The content of the computer consists of the software and hardware and the techniques for implementing them. The *context* of the technology is another matter altogether. This is the way one sees the technology based on preconceived ideas. Much resistance to technological innovation can stem from this contextual cause. Consider the person who suddenly realizes that his or her world is about to be peopled with robots. Robots will build the automobiles and operate the steel mills. Large robots will dig ore and run cranes. Small ones will serve drinks to guests in private homes or keep track of the physical environment within an office building. They may park your car, clean your park, and deliver your cleaning. There is no end to the laborsaving uses to which they may be put. And yet this person may be horrified, even if he or she fully understands the *technical* aspects of the new technology. Why? Because, as everyone who grew up in the 1950s and 1960s knows, *robots are monsters!* Conditioning, reinforcement of ideas, tales from childhood and motion pictures designed to make your hair stand on end, science fiction novels created to fill your imagination with high adventure and great drama—all of it is ready to prove to you beyond a shadow of a doubt that a robot is a thing to be feared.

It does sound rather silly, doesn't it? It is. And it is also true. No matter how much you tell yourself that all that "stuff is bunk," you may still get a feeling of uneasiness when one of those mechanical monsters enters the room. If you have ever encountered a true robot under uncontrolled circumstances, you are probably already aware of this.

It would be wise, however, to note that industrial robots are not nearly as frightening as they otherwise might be, since they in no way resemble human beings. The traditional movie monster image of the robot is almost invariably based on the concept of a "mechanical man" or, as in the silent film *Metropolis* (1926), a "mechanical woman." Since the actual human form is inherently inefficient for most robotic applications today, the tendency to anthropomorphize these electronic and mechanical wonders is minimized. Fortunately, more recent work in the field of science fiction films and books has tended toward a more realistic presentation of the robotic tools of today and tomorrow, though many are just as diabolical in behavior.

And the robot is only a single example. What about genetic engineering? Are we creating Frankenstein's monsters? What about the peaceful use of nuclear forces? Are we dooming ourselves to extinction? What about the image of technology as the destroyer of jobs? Are we putting ourselves out of business?

Much of the concern we have about the burgeoning technologies of today are concerns born in ignorance and nurtured by the fears of others. Historically, technology has always resulted in progress. Historically, laborsaving devices both create and destroy jobs, though they create more jobs than they destroy. Historically, we have not created monsters through technological development, unless, as with the tank and the nuclear warhead, a monster is what we were looking for. Yet conditioning may predispose us to a chauvinistic denial of reality in favor of our belief systems and thus cause an unreasonable resistance to change. It is the *unreasonable* nature of the fear born of ignorance that causes the resistance from this element of conditioning.

Specialization

Specialization is a fundamental underpinning of the Western way of life. It is a principle of economics and of life that has actually created much of the advancement that the human race has undergone in the past ten thousand years or so. It is a simple concept. *Specialization* means breaking the work up among a large number of people (what we call the division of labor) so that each person can *specialize* in the performance of one job at which he or she is particularly good. The results are synergistic in that we gain more of everything for everyone, a higher quality in all goods, and a lower cost than we would have if everyone tried to do everything for themselves. But specialization is accompanied by certain shortcomings. There is a price to pay for the preponderance of goods and services and progress that specialization allows us to enjoy. That price can be in the form of boredom and often creates an alienation of people from their environment.

A separation comes about with specialization in which people are no longer intimate with the things that make their lives what they are. In the field of technology, it creates a separation of those who understand science and engineering and the accompanying disciplines that create our technological base from those who do not, that is, the general public. We can, through becoming a highly technological society, experience a separation of the society into castes as definitive and absolute as any created in history. We become intellectual haves and have nots.

The computer revolution has dramatically brought home this point. We are already divided into those who are computer literate and those who are computer illiterate. A very large sector of the population does not comprehend how a computer works or how to use one. And the separation is becoming even more pronounced between the younger and older generations, where the very basis of logic is shifting from one of simple linearity to one of systemics. This is the same type of shift that occurred with the advent of the printing press, when the printed word and the ability to read caused great social upheavals as the common individual's world view shifted from a holistic one to a linear one.

Those who do not understand or cannot use a new technology often feel alienated in their own homes. To the average twenty-year-old a television set is as natural as a tree or a light switch. To a person who grew up in the 1940s, it will never be quite the same device, and the two will never totally understand each other. The same is true of the average individual's understanding of space technology and the scientist's view, with all his or her specialized knowledge of what makes that technology tick. No matter how often a rocket ascends, I cannot help but stand in awe of humanity's accomplishments. Does a scientist or engineer who monitors the spaceship's readouts have the same thrill?

Attendant to this separation is the increase in speed with which we experience change that specialization causes. It is difficult to keep up with changes as they occur. Nor is there time to adjust to the changes. Around six thousand years ago someone invented the plow. The period from the time when that first crude stick was put into the dirt to when the first steel plow was used spans nearly 90 percent of that six thousand years. Such a revolution in agriculture is certainly slow enough for any culture to assimilate the advance in technology. (Though it should be noted that when Peter the Great, czar of Russia, tried to introduce modern plows to the serfs of his kingdom, they politely accepted them, thanked his representatives, and then burned them, refusing to change their traditional ways.) The McCormick reaper appeared in the latter half of the eighteenth century. The steam tractor soon followed, then the gasoline tractor, and finally the huge combines used on farms today. In each case, the time between major changes in the technology was drastically reduced.

Other technologies show the same pattern. Again and again we see a constant reduction in the time between one major development and the next. Our world is changing, and the rate at which it changes is changing. We progress and achieve new levels of sophistication at a higher and higher rate of speed. The result is less and less time to assimilate the changes as they take place and a rising resistance in the individual approaching overload. Do we specialize because of the rapidly increasing mountain of information available to the society? Or does this deluge of information come from the process of specialization? The answer in both cases is yes. And in the face of ever-mounting pressure for the individual in the society to keep up with rapid change, resistance mounts. Keeping up means higher and higher levels of specialization or more and more time spent relearning the nature of our world as that nature changes. Is it any wonder that there is resistance when radically new technologies are introduced?

THE ROOTS OF RESISTANCE

We have cited five causes of homeostatic reactions to technological change: fear of the unknown, adjustment to anxiety, psychophysiological restructuring, chauvinistic conditioning, and specialization and the telescoping of time. With this impressive list of causes, it may seem surprising that any technological progress takes place at all. Yet at the beginning of this discussion, it was suggested that homeostasis has a useful place in the scheme of things, that it is a natural and necessary part of the process. And this is, in fact, the case.

We operate in an ecosystem that includes every living creature on the face of the earth. We are a part of this symbiotic creature called earth just as we are part of the symbiotic creature that makes up our own culture. And all systems are either in a state of adjustment or they are in a state of balance. Adjustment to changing circumstances is in reality nothing more than a system that has lost its state of balance and is seeking to regain its natural equilibrium. If you stand at the apex of a seesaw, you can balance yourself, albeit precariously. If someone adds a bit of weight to one side or the other, you must *adjust* your distribution of weight to compensate for the change in the dynamics of the system. That is what we do as we experience the changes in our society. Progress and nonprogress are both involved in balancing this growing system called the human race. Without the slower, more cautious considerations of mature members of the society, what would be the result of the youthful exuberance to charge headlong into the future? And without the experimentation and creativity of our technological minds, where would the growth in our society come from? Both progress and caution are necessary, and nature has given us healthy doses of each to ensure not only our growth but also our survival. The time frame is too short to let survival of the fittest take care of mistakes in a natural way. A counterproductive genetic change can die out before it becomes too generalized. A technological mistake, unfettered by the caution of the society, might drag the entire human race into a debacle before anyone realizes what has occurred.

It is no accident that innovations must overcome the objections of the established society in order to survive. It is a test, a method by which humanity determines the usefulness of the new idea. The Bohemian experiment was short lived. It disappeared in less than a generation. The same is true of the beat generation. Yet the hippie movement remains with us in its influence on our music and our literature and is reflected in the increased concerns of our society for the well-being of our fellow humans and the environment.

Many technologies have faced the same test. Teflon® was invented long before it was put to practical use. Boolean algebra had to wait for the invention of the computer to find its place in the sun. And the automatic chewing gum–operated eyeglass windshield wiper never made it at all. Resistance to the changes presented in these innovations saw to it that only what was useful and beneficial was allowed to survive.

Homeostasis is a natural phenomenon. It is part of what it is to be human just as creativity and technology are part of what it is to be human. Both are necessary for our survival. The importance of homeostasis in understanding the social consequences of a technology is in determining what can be expected, what directions public and private opinion are likely to take, and in anticipating the causes of resistance that are bound to occur. Only in this way can we hope to understand the process of technosociological change as it unfolds.

CONCLUSION

Because of the differential manner in which *Homo sapiens* evolves, that is, by evolving externally to the physical body, there is increased opportunity for evolutionary change to be detrimental and yet unextinguishable before an unsuccessful trait becomes generalized over the entire species. Nature uses long time frames in experimenting with species while seeking to create and maintain perfect balance, the less favorable adaptations of a given organism simply dying out while successful adaptations survive and become dominant. With humankind, the telescoping of time becomes a problem in this respect. To help ensure that we do not allow our adaptations to outstrip their ability to control and select for positive adaptive value, humanity also has the survival trait of homeostasis, that is, a natural resistance to change. In this way, the homeostatic tendency can be used as a governor to ensure that adaptations are not generalized too rapidly and possibly allow for the destruction of the species. It functions as a system of checks and balances, necessary for our survival due to the rapidity with which we are able, as a species, to evolve through the development of technology.

We now have viewed two of the primary elements of the puzzle that leads to our dominance on the planet through technology, those being our ability to technologize and our homeostatic tendency to resist the changes that new technology presents. There is an inherent balance here, a push-me-pull-you that allows us to make progress but, it is hoped, not so quickly that it gets out of hand. What has not been addressed so far is the nature of this technologizing process—the functioning of the process itself. Where do these ideas come from? What causes one form of technology to be produced while others go unnoticed, even though all the elements that would lead to their creation are readily at hand? How do we actually come to produce new ideas and turn them into physical constructs or artifacts?

These are conceptual questions of the workings of the human mind in its natural drive to explore and utilize its world. The unraveling of this part of the puzzle is the subject of Chapter 3, as we look at how the creative process operates and how it serves to bring about the innovations that represent technological progress.

KEY TERMS

Adjustment
Anxiety
Chauvinistic Conditioning
Counterculture
Homeostasis
Luddites
Plasticity
Psychology of Adjustment
Psychophysiological Restructuring
Systemics

REVIEW QUESTIONS

1. What is homeostasis?
2. What are the five sources of technological homeostasis?
3. How is homeostasis a survival mechanism and why is it necessary?
4. What is a Luddite? Can you think of modern counterparts?
5. Is specialization a key determinant of homeostasis?

ESSAY QUESTIONS

1. What are some examples of homeostasis regarding leading edge technologies and what are their root causes? (i.e., What are the primary fears that are causing the resistance?)
2. Discuss your own homeostatic reactions to technological change, citing at least three examples of new technologies that have caused you brief or long-term anxiety.
3. How does technological homeostasis help to explain the current trend toward "politically correct" behavior in the United States of America? Do you foresee a homeostatic backlash to this movement?
4. Discuss historical examples from the past century (the 20th century) of homeostatic reactions to technological change in third-world countries and industrially burgeoning nations.

THOUGHT AND PROCESS

1. Have you ever resisted a change in the culture? If you said no, it is not surprising, particularly if you are of a relatively young age. But think again. Answer the questions below. Every yes answer is an example of your own homeostasis.

- **a.** Have you ever resisted a new clothing style? Did you change your mind as it became more accepted by the general public?
- **b.** Did you ever experience anxiety over the entry of a new boss or a new worker into your place of business, not being quite sure how to "take him or her," until you learned more about that person?
- **c.** Have you ever rejected a radical new automobile design as outlandish or ugly, only to change your mind later?
- **d.** What was your initial feeling about computers?
- **e.** Would you tend to be cautious if tomorrow you heard that an alien had landed on earth and was conferring with officials of the major powers?
- **f.** Have you ever felt uneasy about going to a party at the home of someone you did not know well?
- **g.** Have you ever worried about a blind date—even after you met the person and did not find anything immediately displeasing about him or her?
- **h.** Have you ever reacted with dismay at a friend's new hairdo?

i. Have you ever resisted playing a new game with someone who knows it well?

j. When you go out to dinner, do you resist trying new or exotic foods, preferring instead the old standbys?

k. Would you react with caution to a person of the opposite sex entering a job traditionally held by a male or female—if it directly affected your own life? For instance, would you feel uncomfortable working for a female boss or having a male housekeeper or secretary?

l. Would you react with fear or anxiety if your job were replaced by a machine and you had to be retrained?

2. For each of the questions above, classify the type of homeostatic reaction cited in terms of the five causes mentioned in the chapter (fear of the unknown, adjustment to anxiety, psychophysiological restructuring, chauvinistic conditioning, and specialization and telescoping of time).

3. All of the reasons for homeostasis given in this chapter center on fear. How does just good common sense as a reason to resist change fit into this process?

4. Suppose a method of supplying free energy to everyone on an unlimited basis were to be discovered tomorrow. What would be your concerns about the availability of such a technological breakthrough? How would you go about the process of making the technology available to the world, assuming that you had the power to determine the method by which it was to be done (or not done, if that is your choice)?

5. Can you think of any modern examples that parallel the Luddites other than the counterculture examples offered in the text? Do you know anyone with exhibited tendencies toward this form of reaction to change?

The printing press represents one of the technologically pivotal events in the history of humanity. It represents the first true breakthrough in mass communication, leading to the Renaissance, the Protestant Reformation, the Age of Reason and return to rational logic, and, eventually, the Industrial Revolution and the modern age. That is a great deal for one single invention. As you read the following report, note the wholesale and manifest changes across the spectrum of human activities that resulted from this single innovation in writing, and note as well the systemic nature of the approach used by those who created the study. Would you have organized such a report in a similar manner? How does this reflect the information we have discussed thus far in the book? Do you see any parallels between the cause and effect of the printing press and that of the computer?

The Printing Press and Its Consequences

Robert Burchfield and Don Steele*

INTRODUCTION

There were books long before the printing press was invented, but they were not in the form that we are familiar with. The ancient books or scrolls of the Chinese were handwritten on scrolls or codex and stored by the many scribes during the eighth century A.D. As early as the eleventh century, mold printing existed, and as late as the fourteenth century, cast metal type was known to exist in Korea. By the middle of the fifteenth century, in a period of religious awakening, there was a growing demand for books. To meet this demand, scholars and scribes joined together in the copying of books. Eventually, book factories employing fifty scribes could be found in Italy, Germany, France, and Holland, in which the scribes worked diligently to produce books. It was unfortunate, however, that these books were carelessly composed during the transition. It is often said that need necessitates invention, and it is a period of time in history when there was a growing need to produce vast numbers of books. Keep in mind that it all didn't happen overnight, but the spread of its success overtook the printing technique of that period.

Scientists and scholars disagree on who deserves credit for the invention of movable type, but many seem to give credit to Johann Gutenberg as the man who perfected the concept of movable type, earning him the title, "father of printing." Ironically, the ending of one life marked the beginning of a new idea, that of printing. After Gutenberg's death in 1469 and for the next two decades, the new art of printing spread throughout Europe like a raging fire out of control. By the end of the century, more than 40,000 different books had been printed ranging from Greek classics to arithmetic.

Again, as in the days of Gutenberg, demand created a climate for invention. Gutenberg's simple wooden press had survived until the eighteenth century without any major changes. This would change in 1804 when an English earl, Charles Stanhope, built a press of great size made of iron.

The Stanhope press still did not meet the demand because it was hand operated. Then, in 1812, German-born Friedrich Koenig built a power-driven cylinder press with an automatic inking system. However, the sheets of paper were still hand fed. In a joint venture in 1814, Koenig built a two-cylinder press that was automated and could print one

*Robert Burchfield and Don Steele, "The Printing Press and Its Consequences" (unpublished paper presented at DeVry Institute of Technology, Atlanta, Georgia, August 10, 1984).

thousand sheets per hour. Then, fourteen years later, the English team of Applegate and Cowper built the first four-cylinder press. In 1847, the American inventor Richard Hoe invented the rotary press, which mounted the type directly on the cylinder. This would later be modified by the use of duplicate plates. Still there remained a need for a small automatic press which would take up less space but print faster. In 1858, George Gordon made the press to order: his would be called the *platen press* and would become the standard of the print shop. The letter press as we have come to know it is grouped under three basic types: platen, cylinder, and rotary.

The platen press has two flat-surfaced jaws that open and close as it prints. When open, a set of inking rollers apply a film of ink to the form and at the same time paper is fed to the platen. The platen press is not suitable for printing books or magazines, but is ideal for flyers, brochures, and other circulars.

The cylinder press places the form on a flat bed, which moves back and forth under a rotating cylinder. As the cylinder rotates, the paper is fed to it and held by a set of steel clamps. The paper is then rolled over the form while the bed passes under the cylinder. As the bed moves back, the type form is reinked and the freshly printed sheet is fed onto a pile. The cylinder printer is preferred for the printing of books due to its versatility to print vertically or horizontally as a unit.

The final mechanical printing press is the rotary press. Both the impression surface and printing surface are cylindrical. The rotary press prints from duplicate plates, which are mounted onto the printing cylinder. The importance of the rotary press is that it can allow up to five plate cylinders, thus allowing five colors to be printed simultaneously. This is great for newspapers and magazines that favor this printing apparatus.

RELIGION

Before the idea of movable type on a printing press was established and before the Renaissance, the Church that ruled Europe was in its lowest plunge to decadence. The fifteenth century brought to Europe a revived interest in orthodox religion. Minds were stimulated and were reaching out for new knowledge. The renaissance of piety in the Church influenced worthwhile activities among the priests and monks as well.[1]

Because the mind was once again eager to receive new knowledge, the Renaissance also stimulated education. Books were becoming more easily available and a general desire to learn to read blanketed Europe. Only those people in the very lowest classes were the exception. The upper classes were no longer exclusive in the ownership of books. As people began to read, a popular demand for Martin Luther's translation of the *Holy Bible* increased.[2]

Religion as it existed then was not a personal religion such as is experienced presently. One's religion was inherited, traditional, and ritualistic in nature. The Middle Ages had allowed the Church to become a hierarchy of power, and it demanded, in some respects, many things of the people it governed. No one ever thought in terms of

[1]Douglas C. McMurtrie, *The Book: The Story of Printing and Bookmaking,* 3rd ed. (New York: Dorset, 1989), pp. 125–26.

[2]McMurtrie, *The Book,* p. 327.

questioning the Church. A mere verbal declaration from the Church was as good as written law. Because the Church went unchecked by its followers, corruption was widespread, and, prior to the Renaissance, literacy was held strictly by the clergy. The scribes hand wrote the only books the world knew, and great sums of money were spent on personal libraries. Common people had no means by which they could purchase books, and if they did fall into a sum of money and decided to purchase books, they would not be able to read them.

Gutenberg's idea of movable type turned the situation completely around and shed some light on a very shadowy organization. As mentioned earlier, the renaissance of piety and orthodoxy created a desire for knowledge. As Martin Luther was seeking to satisfy this newfound desire for knowledge, he thought he was living a life pleasing to God, because he followed the Church. As he studied the Bible manuscripts, he realized that not only had he not been living as his Bible told him, the Church was not doing so either. Martin Luther continued to uncover alarming new truths and quickly became one of the world's foremost theologians and translators of the Bible. He developed into one of the greatest leaders of the Protestant Reformation in Germany. His translation of the Bible went to the presses and into the hungry hands of the common people, who had a great demand for the Word.

Religious reformation was a first fruit of the printing press. Mass production of the *Holy Bible* allowed the average individual to gain access to something that was held very dear. The almost countless number of denominations of Christianity existing today are further extensions of Gutenberg's invention. His idea of movable type gave human beings a chance to think for themselves for the first time since the fall of Greek and Roman cultural heritages as the mainstay of reason.

GOVERNMENT

With the introduction of printing, the governments found that they had a potential disaster to deal with. The governmental structure at the time of the invention of movable type was almost completely controlled by the clergy. This period of medieval uprising and cultural impression formed a people's defense for the hierarchical government under which they lived. The opportunity for secular government was beginning to take form.

With the introduction of the press came books, and with books came the freedom of knowledge. Unwilling to face the undeniable cruelties and public unrest that such an invention would bestow on the heads of state, governments instituted plans to curb the people's appetite for books. The governments held public book burnings, destroying books that they found to be heretical in content, but in essence containing damaging proof of their tyrannical tactics. But when the public outcry turned to violence, the restrictions were lifted and governments turned to a different point of view, determining to join the people in the face of an inability to beat them. It was around the turn of the sixteenth century, after the death of Gutenberg, that government began to play an important role in the printing industry. Secretly funding town printers and openly accepting what was being printed, the government found an inside tactical method of baby sitting the presses while controlling what was being printed.

In the late 1600s, the British government placed intricate laws on the Colonial printers of the time. The government allowed itself to pick who was to be a printer, how often they printed, and what they printed in Colonial America. This proved to be disastrous, as printing had given rise to a cause for freedom and the people were sure to find that out. However, all was not lost as the government placed in use its own printing press and techniques. With the new press, the government could print public notices, money, proper documentation, and keep important records in nicely printed books.

In the American colonies, the printing press found uses never considered before. The government learned to use the printing press tactfully to achieve cohesion among the people whom they served. Today the government has its own printing facilities in Washington, DC, known as the U.S. Printing Office. It is encouraging to see an invention overwhelm even the governmental bodies in its practical applications as well as its ability to enhance the means of mass communication and public unity.

INDUSTRY

Industrially, Europe was fortunate. The idea of mass producing print did not catch the contemporary European industry by surprise. As with any invention, an inventor depends on and makes use of knowledge already established in related areas. This is the only way we are able to see growth in technology. Examples of this idea are evident in many inventions. Stagecoaches were reconstructed to serve as the first railroad cars. The first bodies for automobiles were carriages and buggies adapted to support the mechanical power placed in them. Combustion engines were installed in wooden box kites to serve as the world's first airplanes.[3] Obviously, there is an endless list.

Europe's related industrial arts had substantially progressed to a point where they were able to support the idea of mass-produced print. There were four requirements awaiting this new invention:

1. Paper was plentiful and easy to process.
2. The proper ink was available.
3. Presses for metal and paper were available.
4. Alloys were available to cast, construct, and mold type.

The absence of any of these would have made the development of printing in Europe impossible.[4]

As Gutenberg's application of the printing press progressed, it became a necessary factor in everyday life. One of the first benefits to industry was the obvious creation of jobs. A new type of business was introduced into the world, and it incorporated and stimulated related businesses as well. Printing has evolved and expanded within itself to create other more specific businesses and has boosted employment incredibly. It is not

[3]McMurtrie, *The Book,* p. 126.
[4]McMurtrie, *The Book,* p. 127.

surprising to know that today one newspaper company can have over 1 million subscribers.

The printing press has also made industry more efficient. Take, for example, the basic contract. If every contract, whether financial or otherwise, had to be handwritten, where would the consistency and efficiency in management be? There was a method created through the printing press that allowed procedures and policies of a company to be equally delivered to all employees, customers, vendors, and so on.

We have the printing press and Gutenberg's application of it as a major factor propelling us into the future. We no longer live in the Industrial Revolution, nor are we a major industrial nation any longer. We are postindustrial and are currently building an information society to live in. Because of this fact, information must be easily obtainable. We are approaching the point of being able to put our hands on any information whenever, wherever, and however we desire. This process was initiated by that primitive printing press hundreds of years ago. As each year goes by, printed information becomes more easily obtainable, thanks to a process begun with Gutenberg and his press.

SOCIAL

It may seem that in the middle of the fourteenth century society was in a state of darkness and despair. But to the contrary, it was a time of intellectual activity. Culminating forces in human affairs that had been slowed during the Dark Ages gave rise to the great rebirth of knowledge, the Renaissance. The Great Schism had brought about deterioration of churches, abbeys, and schools and a great falling off in the devotion, enthusiasm, and discipline of the clergy.[5] Minds were once more becoming eager and inquisitive, and an era of exploration dawned as new opportunities were sought for advancement, restoration, and artistry.

Giving rise to a reborn society, the printing press had given the people an opportunity to restructure society, to regain the power to visualize known and familiar facts into new relations, and to apply them to new uses. For example, the textile industry adapted inking techniques to the textile printing industry.[6] Printing shops expanded the industry to include paper mills, metal clerks, pressmen, inkers, engravers, delivery boys, book binders, and many others. It is no great wonder that this early communication tool was so vital to the survival and very existence of fifteenth-century society. The printing press gave rise to the concept of freedom to think, talk, and of course print the truth, no matter what the consequences. We can see the rise of a socialistic bartering system in which one can find businesses growing rich as they cornered the market on goods that had been printed. Not only did the business structure grow, but also the religious society.[7] Regaining trust in the Church by means of the translated Bible, which had been printed in several languages, was a major step in the restructuring of religion during this period, giving rise to new Church ideologies and denominations. The printing press was spreading its cul-

[5]Irving B. Simon, *The Story of Printing* (Hudson, NY: Harvey House, 1965), p. 20.

[6]McMurtrie, *The Book,* p. 131.

[7]McMurtrie, *The Book,* p. 186ff.

tural shock to the far corners of the earth. One could experience the excitement as hundreds of people sought to spread knowledge and carry the art of printing throughout the world. Their accomplishment was the satisfaction and livelihood of the printing press today. Society was ready for a change, yet it had no concept of what was ahead in terms of the invention and the legacy of the creation of movable type and the printing press.

The social implications of the printing press are far too numerous to document here because they reached every aspect of life at the time of its birth, giving rise to a new social structure, perhaps creating the upper, middle, and lower classes as we know them, or perhaps culturally feeding a starving humanity that might have degenerated in superstition, restlessness, and boredom. One might say that society was awakened to a concept of applying existing knowledge in a new fashion, giving way to the great Industrial Revolution as a result of the spread of new or rediscovered ideas through the print media. The printing press allowed society to apply itself, to learn about other cultures, and to become specialized in the ideology of learning and experimenting. It is from this ideology that a powerful influence, to change and develop, would grow and prosper to the point of actually restructuring the social and economic scale. It is this restructuring effect of the printing press that has made this particular invention so vital to the beginning of humanity's striving for excellence in the areas of invention, education, and knowledge. Socially, the printing press has left a mark of opportunities for humankind to advance into the realms of the unreachable goal and to achieve a society that works together, using existing knowledge in an attempt to advance the conditions of a society in which the only road is progress.

TECHNOLOGY

Our technology has moved us into an information society. The fact that the percentage of gross national product attributable to heavy industry is ever declining while our unemployment percentages are dropping as well is evidence that we have yet to find a way to measure what our nation is really producing. Using industrial output as a measure of prosperity is an archaic approach, developed in the industrial era and no longer appropriately applicable to conditions in this country. The United States produces information and the methods previously used to determine the gross national product cannot be used to determine the amount of information produced and sold in an information society. We can produce information in any way, shape, or form desired. We are also printing it faster every day. It is obvious that the printing press with movable type that Gutenberg designed has evolved into the computer that is so much a part of our lives. Tactical support specialists in the military can load the memory banks of their computers with so much data that page printers will send streaming arcs of printed pages across the room on command, and another single command can destroy the sum total of that information instantly.

Not only is information speedily produced, it is also miniaturized. Microfiche methods have transformed hundreds of volumes of magazines into hundreds of five-inch-by-eight-inch cards. This allows libraries to hold more information without the need to build larger buildings to house the incredible number of books and other printed matter available today. It is conceivable that libraries will become smaller and yet be capable of holding more information due to miniaturization techniques in use today. In the future, a library the size of a bedroom could hold the information of ten of today's libraries.

The ability to print information has been an important and necessary factor of society since the invention of movable type by Gutenberg. It is becoming more important with each passing day. Printing information has allowed one person to build on top of another person's ideas and thus propel the technological revolution forward at an ever-increasing rate. For the information society we live in and are developing, printing is an essential resource. The ability to print is taken for granted for the most part, but without it society as we know it would soon come to a screeching halt. Douglas McMurtrie said it in no uncertain terms:

> In the cultural history of mankind, there is no event even approaching in importance the invention of printing with movable type. It would require an extensive volume to set forth even in outline the far-reaching effects of this invention in every field of human enterprise and experience, or to describe its results in the liberation of the human spirit from the fetters of ignorance and superstition.[8]

The future of the printing press is speed and miniaturization. It will continue to be the basis for an information society. When engineers design a faster printer, future engineers will read their published documentation and use it only to build still-faster printers. The evolution continues. . . .

BIBLIOGRAPHY

Fisher, Leonard Everett, *The Printers.* Lexington, NY: Franklin Watts, 1965.

McMurtrie, Douglas C., *The Book: The Story of Printing and Bookmaking* (3rd ed.). New York: Dorset, 1989.

Simon, Irving B., *The Story of Printing.* Hudson, NY: Harvey House, 1966.

[8]McMurtrie, *The Book,* p. 136.

CHAPTER 3

CREATIVITY AND INNOVATION: THE CRITICAL LINK

There was once a village, located deep in a hidden valley in Bhutan, in the Himalayas. And this village was located at the very base of a high mountain that loomed over it like a gigantic tiger's tooth, its wall sheer and unscalable, its white-capped summit beyond the reach of any traveler. It was, in fact, known as the Mother of Mountains, the greatest monolith in all the world, and the villagers were proud of its name.

And to this village there came a traveler, a wandering pilgrim of great age and great wisdom. And he stayed a while with the people of the village, visiting with this family or that, exchanging his tales of far-off places and strange sights for food and shelter.

When he had been with the villagers for three days, the pilgrim stayed with a merchant, a man named Mahjidi, who dealt with all of the merchants of the south and enjoyed the company of strangers more than most. And as they ate, the pilgrim turned to his host and asked, "Mahjidi, I have studied the great wall of stone that you call the Mother of Mountains, and I have seen what appears to be a long staircase cut into its rock face, yet it leads to nowhere. Can you tell me of it?"

"Ah," said Mahjidi, warming to the opportunity to spin tales of his own, "that is a sad and wonderful story indeed. It begins many years ago, when the elders of our village first encountered merchants from the south. They envied these merchants and their homelands, rich with the wealth of the world and filled with palaces and great fertile plains, with histories of conquest and stories of adventure.

"And in their wisdom, the village leaders determined one spring day that it would be a valuable thing to be able to reach the peak of the Mother of Mountains and look out over the world from its very roof. The towns of far-off kingdoms must surely be visible from such a height. The sight of all creation spread before them would be a treasure well worth the effort to obtain. The village, they were sure, would become a powerful place and the object of many a pilgrimage, if only they could reach those far-off heights. Then the world would envy the village, rather than the village them.

"And so, to this end, the elders brought the people together to begin the great task of reaching the summit of the great mountain. One person in five was recruited for the task, all their needs to be seen to by the others while these selected few attempted the impossible feat of conquering the Mother of Mountains.

"Through the spring and into the summer the selected villagers strived to reach their goal. They labored long and hard, constructing scaffolding to raise them to the small crevice that ran through the sheer cliff of the mountain's base, some three hundred meters above the village's tallest tower. And from there, they began the laborious task of cutting stone steps into the living rock, weaving their way along the crack.

"It was a good plan, they all agreed. It was a plan worthy of their efforts and likely to bring them success. But it was a plan doomed to failure. By late summer, they had barely reached the ledge and begun the steps. By early fall, they had expended the energies of one-fifth of the village on the project, inflicting hardship on all and injury to many.

"'We must press on,' the village leaders would say. 'We must conquer this mountain, and we shall.'

"And press on they did. Through the winter months, they braved the cold to mount their attack on the crevice and slowly extended the long staircase toward their goal. Villagers took turns with the work, no longer relying on only the strongest among them. Everyone was brought into the campaign, each taking his place on the face of the mountain, each enduring the cold, and the danger, and the pain of the labor.

"And in the spring, when the warm winds began to blow from the east and the sun rose high into the sky once more, the villagers found themselves more than halfway along the crevice, having extended their steps up the wall of the mountain by another five hundred meters. Many had fallen along the way. Many more had exhausted their will on the side of the mountain, yet none had given up the quest.

"Then, late in the summer of the second year, the workers on the mountain made a terrible discovery. The crevice had ended. The shelf in the sheer mountain wall had come to an end, and they had barely begun their journey toward the summit. Above the crevice, the mountain mocked them. The glass-smooth surface defied them to continue.

"Even the elders of the community lost heart. The quest was abandoned. The work was ended. And so, to this day, the mountain stands inviolate, the hand-cut staircase to nowhere being the only evidence of the vain hopes of the village. As I have said, it is a sad tale. We never reached the summit and never knew what glories await the first to do so."

"It is a sadder tale than you believe," said the pilgrim.

"How so?"

"Because the Mother of Mountains is not the roof of the world. And the sight of a thousand kingdoms awaits no one upon reaching its summit. There is only the sky and the beauty of a hundred mountains, aglitter in the white snow that never fades away. To the south, more mountains and hills and finally rolling plains greet the eye, with rivers coursing like silver ribbons toward the sea. It is but a small piece of the creation that is the world. To the north, the land of frozen giants, mountains far greater than this, is all that can be seen. It is beautiful, but far from the wonder your forefathers perceived it to be, and hardly worth the lives and misery of so many people."

The merchant, Mahjidi, eyed his guest suspiciously. "And how would you be knowing this?"

"Because I have been there."

"No man can scale those heights!" snapped Mahjidi.

"And no man need do so. The far side of the Mother of Mountains is a gentle slope that requires many days to cross, but it kills no one, and it eventually leads to the summit. Has no one thought to look to the far side of the mountain?"

"No one," answered the merchant.

The pilgrim nodded. "A mountain has many faces, Mahjidi, as does all else in life. Sometimes it is advisable to look upon those faces in order to understand the mountain."

It would seem that if the village elders in this tale had taken the time to view their problem differently, they might have saved themselves a great deal of time and misery. But what has a mountain to do with technology? The tale is instructive for our purposes in that it illustrates a key element in the creative process by which our technological world has come about, that is, the need to see things in a new light. Like the mountain, problems in society have many faces. Unfortunately, it is sometimes difficult to see the hidden ones, the ones that may hold the key to the problem's solution, and so the problem may remain unsolved for some time. If a problem seems to have no solution, it is often not because there is no solution but rather because those investigating the problem cannot see that solution *from their individual perspectives*. Indeed, there is a need to see things in a new light.

To be creative, to be able to develop new machines, new forms of technology, or new ways of doing things, it is necessary to escape from old modes of thinking. It is necessary to see things in a new light, to view them from a slightly different perspective, or to *hold them in a different context* than the traditional one. What really happened to the villagers in the story is that their perspective on the issue was too narrow to see the true solution to achieving the goal. They were so intent on conquering the sheer wall of the mountain that it never occurred to them to look for a solution on some other face of the mountain. To succeed, they needed a change in perspective.

But where does this change come from? Are we all equally capable of making this cognitive switch in our approach when the need arises? What of the tried-and-true, traditional ways of doing things? Are they without value to us? Should we simply scrap what we have spent so many centuries learning to use for our benefit and charge headlong into viewing things from a different perspective? Or should we cling to the tried-and-true, already tested ideas of the past?

Actually, we do both. An investigation of an unaltered contextual view of reality is an extremely useful thing for humanity, as has been asserted in the first chapter. It is the way we create our technology. And, as we have seen in the second chapter, this does not necessarily mean that we are going to scrap all that has come before in favor of the new. The key word is still *appropriate*. The issue to be dealt with in this chapter is not whether to innovate but rather to ascertain what affects this creative element in human activity and how we can use that knowledge to find new answers to new and old questions.

THE ROOTS OF CREATIVITY

Creativity can be defined as the ability to combine a number of factors to achieve a solution to a problem or to make an artifact that is both novel and useful. It is the ability to rearrange known facts into new patterns in order to develop new constructs useful in accomplishing what needs to be done.

Exactly how creativity works and how this rearrangement of facts takes place in the mind has been the subject of research and supposition for thousands of years. The subject of *a priori knowledge* addresses the concept as an alternative to the existence of prior knowledge without experience. Modern researchers seek its roots and a better understanding of the mechanism by which it operates. How the human mind consciously and unconsciously enumerates and evaluates the millions of possible combinations of factors that may contribute to the solution of a problem and the creation of new artifacts is a vast, fertile field for in-depth investigation. The creative process may be slow and methodical, following a linear process, such as is characteristic of the scientific method, or it may be instantaneous, relying on brilliant flashes of insight to bring about an answer. Neither approach is any less creative than the other. Both methods require a new way of thinking.

Creative thinking, the ability to find unique solutions, is something that is learned as much as it is natural. It is *experiential* in nature. The way a person is taught to view life and the experiences one has greatly affect the degree of creativity that person will achieve. Problem-solving abilities are greatly enhanced by experience. The more opportunities a person has to seek solutions, the greater the likelihood that the individual will improve her or his ability to find solutions. A person who has always done the same job the same way without any outside input on alternative methods will be less likely, on the average, to develop unique solutions within the context of that job than someone who has experienced a wide range of different situations and different approaches to doing the job. There is simply a wider range of possibilities available in the second person's mind from which solutions can be formulated.

MOTIVATION

Risk and Reward

Creativity is related to the concept of risk in that risk acts as a deterrent to behaving in a creative manner. The fear of experiencing loss can reduce the desire to be creative. With each attempt at new solutions, one may risk a number of losses, ranging from a loss of personal esteem and social acceptance to a loss of monetary position or even life itself. The first successful manned aircraft was an exceptionally creative thing. Those who achieved it were aware of the risks involved to life and limb and opted to try it anyway. Risk prevented many other casual thinkers from putting their money, and their lives, to the test to prove their ideas. Risk is an important element in the process. It is one of the reasons that the proverb about necessity being the mother of invention is so easily believed. Since there is always risk involved with being creative, it is not until there is a great need (necessity) that people become aware of the problem and are moved to solve it. Often the risk of *not* solving the problem must assume equal proportions to the risk of seeking the solution before the creative process is sparked.

Reward operates similarly as a factor in creativity. It is the reverse side of the risk coin. Reward offers a carrot for those who would seek creative solutions, motivating them through the promise of a better life, more wealth, more efficient use of resources, or a feeling of accomplishment, to name a few. It represents the search for a desirable positive result just as risk represents the avoidance of an undesirable negative result of not being innovative.

Together these two motivations make up what Freud would describe as the "pleasure principle" of motivation, involving a seeking of pleasure and avoidance of pain by human beings. Technology and creativity certainly classify as means of achieving that "idic" goal.

Hierarchy of Needs

A much more useful and succinct survey of motivational factors present in the creative process is offered by Abraham Maslow in his *hierarchy of needs*. Maslow sought to determine what motivates people to behave in the way that they do and, beyond that, to determine what makes a person balanced, successful, and "happy." Through his research, he developed an inventory of needs that must be satisfied in humankind if a person is to achieve a happy, successful, balanced life. This inventory of needs is arranged in a hierarchical structure, ranging from the strongest and most basic motivations on the bottom of the hierarchy, to the weakest yet most advanced goals toward the top. The Maslowian system has been found useful and appears in several different forms throughout motivational literature as a basic explanation of what determines people's behavior.

Maslow divides motives for actions into five need categories. Starting with the most basic and strongest, they are *physiological, safety, social, self-esteem,* and *self-actualization*. Each category represents a specific type of need in humankind.

Physiological needs are survival needs. They include most bodily needs such as freedom from thirst and hunger, the need for shelter, the need to continue the species (sexual drives), and other bodily functions.

Safety needs are also survival needs. They represent the need not only to be sated, warm, and healthy, but also to ensure that these conditions will continue to exist in the future. Safety needs involve security and protection from harm, either emotional, mental, or physical.

Social needs, or *belonging needs*, as they are also called, refer to the needs to belong to and be part of a group or society. Human beings are social animals, gregarious in nature, and require the presence of others for their well-being. The need to become part of a group is a survival mechanism no less important to humankind than to members of a lion pride or baboon troop, where cooperation and joint effort ensure survival in a hostile environment. In humankind, this need translates itself into the need to offer and receive affection, acceptance, friendship, and a general feeling of belonging to some group. It is from the drive to fulfill this need that the human race divides itself into tribes.

Self-esteem needs involve internal rather than external satisfactions. Self-esteem has little to do with the opinions of others, being based on a personal feeling about one's self. It is considered a higher order need involving one's self-respect, autonomy, independence from the control of others, and the feeling that one is achieving. Self-esteem needs translate into such external needs as status in the group, recognition for achievements, and attention from others. With self-esteem the person seeks not only to belong to the group but also to stand out in that group, achieving as high a position within the group as can be accomplished. This is also a basic survival need in humankind, acting as an internal driving force to push people toward their limits.

Self-actualization needs represent the highest-order needs that a person has. They represent the need of a person to strive to be all that she or he can be at any given point in time. Self-actualization involves striving to reach one's full potential, to fulfill one's highest aspirations, and to "grow" as an individual to become what one is capable of being. It is the farthest removed of the needs from the instinctual, animal base of the human psyche.

Consider a primitive human being who is thrust into a wilderness area of North America. This primitive person, alone and naked in the world, first seeks to satisfy the most basic of human needs, the physiological. Our primitive—we'll call him Og—immediately goes about the business of satisfying needs to feel sated by first locating the nearest available stream, from which he drinks, and then running to kill a wild beast, say, a small mountain lion, which he then proceeds to eat. Once he is no longer hungry and thirsty, he finds himself a nice cave by the stream and settles in, having most of his basic needs satisfied.

Now it is time for him to consider the next level of needs, safety. It suddenly dawns on him that in killing the mountain lion he has put himself in great peril, particularly since mountain lions tend to hunt in pairs. Although he had previously been controlled by his great hunger, that now seems secondary. Once he is full, safety becomes important. He proceeds to build a fire to fend off any possible attack from his victim's mate, and he checks around to be certain that there is a sufficient supply of food (nuts, berries, roots, game coming to the nearby stream to drink) to keep him satisfied in the future. When this is accomplished, not only has he sated his physiological needs, but he feels safe as well.

This is all very well and good, but he now notices a definite lack of companionship. His cave is comfortable, the stream and local flora and fauna are available to satisfy his basic needs, but there are no other people about, which threatens his need to belong. After all, Og is a very belonging being.

To fulfill this need, he takes a hike to the other side of the valley where he has noticed a small group of wandering nut gatherers and invites them to move into the caves next door. They do so, and he now has a group to belong to. So much for social needs.

What about self-esteem? How does Og satisfy his need to stand out in the group? This is no problem for our enterprising primitive at all. Og simply rips a tree from the ground by its roots, walks over to the biggest, meanest guy in the tribe, and bashes him with it six or seven times, proclaiming himself chief and inquiring whether there are any objections that might exist among other tribal members. Obviously, after a display like that, there are none.

Now Og not only belongs to the group, but he stands out in the group. He is chief, leader, and "head honcho." Everyone respects and looks up to him. Everyone listens to his advice and agrees with him. He gets the choicest bits of meat and accepts the responsibility of leading the group. His self-esteem is assured.

But this is still not enough for Og. He has almost everything he could want, but there's still something missing. He feels "unfulfilled." He feels that he isn't getting the opportunity to fully express who and what he is. As a result, he wanders off to the back of the cave and indulges in his favorite hobby, cave painting. For hours on end, Og mixes pigment and makes brushes and paddles for applying it to the walls, sketching and coloring in fantastic panoramas the hunts and other events from the tribe's life. Now he is balanced, happy, and successful to his full capacity, all his needs being met.

There is still one element of the hierarchical structure of need fulfillment to be taken into account. Maslow tells us that if a lower level need is threatened, the upper-level needs seem less important and no longer receive attention. It is only after Og was sated that he decided to look out for his safety. It was only after he had become socialized and was a member of the group that he decided to assert himself and become chief. And the desire to express who and what he was did not become a major factor until all of the other needs in the hierarchy were satisfied.

What happens if there is a threat to a lower need? What if the tribe comes to Og and says, "Listen here, Og. We've been thinking that you sit around back here painting walls while we do all the work. We thought we'd pick a new chief and kick you out." Og responds by bashing a few heads, of course. Or, if they've all got clubs and he doesn't, he may just resign in order to preserve his membership in the group.

As soon as a lower need was threatened, the upper needs were put on the back shelf. This is the manner in which most people in society behave.

People operate on a principle of achieving needs and protecting themselves from the loss of achieved need fulfillment. Yet the most highly creative people operate at a level of self-actualization. Does this mean that the only creative people are the ones that are totally balanced? Fortunately for humanity, the answer is no. Needs are never satisfied totally. What we seek to do instead is *satisfice* our needs, that is, we determine minimum levels of acceptable fulfillment for our needs and work toward the satisfaction of those needs until that minimum acceptable level of achievement is reached. No one is ever completely safe. There is always some risk involved in being alive. If people wait until they *completely* satisfy the need to be safe before working on any higher needs, they would never get any farther than that point. No one ever has a total feeling of belonging. There are too many differences among individuals to ensure complete acceptance and understanding from others. If complete acceptance was a criteria for working on *self*-acceptance, no one would ever get as far as considering the possibility.

Satisficing allows us to work on a need until it no longer pulls at us more strongly than another need. We then work on the one that does pull on us most strongly. Without the satisficing compromise, there would never be any great painters, composers, engineers, scientists, poets, military leaders, presidents, doctors, or practitioners of any other profession that demands the participation of the practitioner on the basis of excellence to the limits of one's abilities.

Any of these needs can motivate a person to be creative. Yet the drive for excellence for its own sake, the self-esteem and self-actualization needs, seem to produce more creative solutions than do the risk-oriented lower needs.

In all of the need hierarchy, there is either a motivation of protecting what one has (fear response) or striving to obtain what one does not yet have. Higher order needs are reached based on the desire to achieve and to accomplish, whereas the downward slide into working on lower order needs is a fear response designed to protect against loss. It is through progress that one can achieve higher order needs. It is through fear (threat of loss) that one is forced to concentrate on lower order needs and shelve the desire for higher order need achievement.

Creativity can be viewed as the result of (a) innate ability, (b) experiential learning, and (c) motivation. This third factor, motivation, can be viewed as being based on either fear (the need to protect against loss) or achievement (the desire to achieve some purpose or gain something not presently extant). Further, these motivations of fear and achievement can be viewed according to Maslow's hierarchy of needs as a matter of an individual seeking to satisfy one or more of the classifications of needs in the hierarchy, either by protecting lower level needs already satisficed or by seeking to achieve need satisfaction farther along in the hierarchy. This is in relation to the creative nature of the individual. Yet to be considered are the social conditions surrounding the individual and the effect these social conditions have on the performance of individuals in reference to their creativity, or the performance of the society as a whole in reference to its creativity. To understand this, we must make a comparison.

CULTURAL IMPETUS: A COMPARISON OF ORIENTAL AND OCCIDENTAL APPROACHES

To illustrate the ways in which cultural and social patterns can affect the degree of technological innovation and the kinds of technological innovation that take place, a comparison of Eastern and Western (Oriental and Occidental) technology serves well. There are a number of differences in the ways in which Eastern and Western cultures formulate the individual and societal approaches to the subject of change in general and technological change in particular. There are three major factors to be considered: *cultural restriction, linear thinking,* and *philosophical point of view.* All three of these factors are capable of limiting or expanding the possibilities available to a culture.

Cultural Restrictions

Western culture began exerting its influence on China in the seventeenth century. Yet long before that influence began, the Chinese culture experienced a limited but truly advanced technological development. The list of important inventions first developed in China is impressive.

China is credited with the first use of block printing, and the invention of paper, a necessary adjunct to the use of such blocks. Indeed, the first known Chinese dictionary was created circa 1100 B.C. In addition, the Chinese invented silk, the art of weaving, and the first development and use of astronomical instruments, which took place prior to the erection of England's famous Stonehenge, a Neolithic arrangement of boulders designed to predict important astronomical events, including the winter and summer solstices and the vernal and autumnal equinoxes by which the early Britains could note the passing of the seasons. While European Britains were piling their boulders together, the Chinese were refining their measurements of the heavens.

The Chinese were also responsible for the first use of the horizontal loom, the spinning wheel, the waterwheel, the first mechanical clock, gunpowder, porcelain, and the concept of "coastal naval defense," "rockets," and "hand grenades." This is an impressive list. The Chinese culture was an extremely inventive one, possibly the most inventive prior to the Renaissance in Western Europe. The culture was prime for a technological revolution with the possibility of using their inventiveness to great advantage. Yet the technological revolution in China never came about. It would be more than a thousand years before the initial concepts would be rediscovered or borrowed and put to practical use.

The Western European countries, in contrast, were extremely slow to develop an inventive tradition, plodding through centuries of stagnation and extreme restriction in the availability of knowledge before the blooming of Western technology began at the end of the Middle Ages. The Industrial Revolution of Europe occurred only recently from the historical point of view, well after the founding of the United States. What caused the Western culture to succeed in taking advantage of scientific knowledge and China to be unable to make the same transition?

The Chinese culture is an old one. The first known formal government under imperial control was that of Emperor Huang-Ti, who is known to have established the first Chinese kingdom in 2697 B.C. The first dynasty, the Hsia, was founded as early as 2500 B.C., and there was already a considerable ancient cultural tradition in place at that time. The society was organized and strictly structured under a long succession of emperors whose elaborate and extensive administrative networks held total control over the population. The society was rigid and traditional. Social position was strictly maintained, placing the citizenry in distinct classes across which there was very little mobility. Under such a tradition, any change was seen as a threat to the stability of the society as a whole and the government in particular. A citizen achieved the status of engineer or scientist not by virtue of ability but by virtue of social rank. An individual became a scientist because the father was a scientist. And his son would take on the position after him by virtue of the same reasoning. So tightly controlled was the structure of the society that the population was solidly locked into doing things in the traditional way. Change was considered a thing to avoid. The very religion of the country, based on ancestor worship, was an expression of the people's veneration of the past, past ideals, and maintenance of the status quo. China was considered to be perfect, while the remainder of the world was viewed as barbaric, backward, and uncivilized. To change a perfect social structure would be unthinkable.

In Europe the conditions were diametrically opposed to this stable, tightly controlled Oriental structure. Prior to the rise of economic mercantilism and intellectual freedom, Europe was a collection of petty kingdoms ruled by feudal lords and loosely controlled by kings. Political boundaries changed frequently. The squandering of resources by an expanding population was staggering, and the political control of the Church, the closest functioning body politic approximating stability, was being challenged from within and from without. By the fifteenth century, with the expulsion of the Moors from Europe, the discovery of the New World, the final perfection of the printing press in 1454, and the dominance of European countries in the Mediterranean following the Battle of Lepanto in 1571, Europe was ripe for its own explosion of inventive development. The difference lies in the attitude of the people and the freedom of the people to act. Change among the European cultures had been constant since the beginning of the Renaissance. The people were becoming acclimated to the possibilities of change. When

Gutenberg began publishing in 1454, that single event struck the death knell of stagnation in Europe. Knowledge was the key to technological change in Europe, knowledge and its dissemination among the general population. With the relative freedom to improve one's lot offered by the existence of an already entrenched middle class of independent and mobile workers in the trades and the guilds, a burgeoning of scientific inquiry and practical application was inevitable. It was this freedom to act in combination with the availability of information that allowed the Western cultures to move inexorably toward the Industrial Revolution of the eighteenth century.

This is the first step in the process by which Western linearity leads to the explosion of Renaissance knowledge and, eventually, to the Industrial Revolution. Two other factors contributed significantly to the entire paradigm of development; the Protestant Reformation in the 16th and 17th centuries and the ages of Reason and Enlightenment of the 17th and 18th centuries. Each of these provided additional fuel for the movement toward technological and scientific revolutions.

In the case of the Reformation, the Protestant movement led inexorably toward the development of individualism and countered the lock-step acceptance of "Mother Church" in determining the nature of reality. Protestants openly questioned the Church's authority. In doing so, they presented a new view of the world that removed the separation of God from the individual, by eliminating the intercession of clerics to achieve understanding and salvation. No longer was it necessary to depend on priests to know and understand the Word of God, because increased literacy had led to the ability among the laity to discover and consider religious teachings on their own. Suddenly, individuality and free thinking became important. This lead to more original thoughts on not only religious subjects, but also on the nature of reality in general. Rationalism, not blind dogma, emerged as the philosophy of the day, as is personified in the work of both Rene Descartes and John Locke, who taught reason, logic, tolerance and utilitarian practicality, not dogmatic adherence to the mystical. Because the Protestant forces forged a new understanding of the relationship between God and man, one that was more direct and less magical, there emerged a new image of humanity; one in which *the individual was as important as the religious or political body*, and one in which *happiness in the here and now can be pursued and gained through personal effort*. Suddenly, the individual had importance and was free to think step-by-step through their problems.

When people began to make their own choices, i.e., to think for themselves, the entire social structure of Europe changed. Everything from government structures to social norms were redefined, which spawned revolutions, economic expansion, and a further flourishing of the arts into new and different directions. The move toward secularity in society fractionalized the formerly coherent elements of traditional European society; the state, the church, and the people.

Similarly, during the Age of Reason and Age of Enlightenment, there was a shift to rationalism in thought and a great desire to use a rational, structured approach to studying the world and discovering the way things really were. During the Renaissance, individuals who had the money and the opportunity for an education were able to excel and expand their knowledge. During the Ages of Reason and Enlightenment, this became a mass movement, open to all and eminently available, thanks to the new rational approach to knowledge. It was not just an artistic movement or a religious one. It was a movement that permeated all aspects of life. In religion, it replaced blind dogmatism with logical and rational discussions of the nature of God. In politics, it replaced blind loyalty to a dictator with a system more and more based on individual choice. In social structures it led to the demise of planned marriages and introduced the concept of romantic love, in which individual choice and desire held sway over duty and social structure. In education, it flowered into the modern university, teaching not only philosophical skills but practical ones as well. Enlightenment meant both self discovery and the discovery of provable, demonstrable principles upon which to base decisions. By the time of the Industrial Revolution, around 1825, the progressive process of individualistic, intellectual journeying had led to the modern world.

Linear Thinking

Another major difference between Eastern and Western cultures lies in the manner in which Occidental and Oriental people cognate. The belief among Westerners that Orientals are inscrutable, their ideas unfathomable, and their actions unpredictable and illogical is no more valid than the curiosity and amazement that the Oriental holds for the ideas and actions of the Occidental. Each culture experiences this feeling of paradox when dealing with the logic of the other. The fact that an Oriental sees nothing illogical or paradoxical about a logically Oriental proposition whereas an Occidental may, and vice versa, is testimony to the differences in the approaches of the two cultures.

Western logic is linear in nature. It moves through a sequence of defined statements that leads to conclusions about the world. Statements of logical analysis in Western cultures are bound together by their cause-and-effect relationship. Each statement infers the next. Facts are combined into syllogistic combinations designed to prove that if one event occurs or if one set of circumstances exists, then another event or set of conclusions must "logically" follow. One moves step by step from conclusion to conclusion, each one building on the preceding one in a web of well-founded proofs toward the ultimate conclusion being sought.

For instance, *if* I drop the ball from the tower, *then* it will fall toward the earth. *If* the ball falls toward the earth *and* there is someone standing beneath it, *then* the ball will hit him or her. *If* these events occur *and* the ball is four inches in diameter and constructed of iron *and* the tower is forty feet tall, *then* the person who is struck will be seriously injured. *Therefore, if* I drop the ball under these conditions, *then* a person will be seriously injured.

This is the type of linear (syllogistic) logic experienced in Western culture. And it is reinforced to a high degree by the written language. Occidental writing and books are constructed to proceed step by step (word by word, sentence by sentence, paragraph by paragraph, chapter by chapter, and so forth) from an initial statement toward a final statement. The movement is physically left to right, front to back (relatively speaking), and top to bottom. The reader must follow a linear sequence from beginning to end that reinforces the ideas of cause and effect, temporal linearity, and beginning-to-end direction. Any idea or concept that does not neatly fit into this type of logical progression tends to be less easily internalized by Occidentals. Indeed, the very use of the word *progression* indicates the Western predisposition toward viewing the world as a *forward movement* of events (cause and effect) through time. Every action has a reaction, and each reaction becomes an action, creating another reaction.

The Oriental mode of thinking is quite different. In China, particularly during the imperial period, the logical process followed a different path—all elements were seen as merely parts of the whole. This holistic viewpoint, creating the image of the universe as a huge, single entity rather than as a collection of discrete parts, predisposes cognition toward an acceptance of an event or condition without regard to its resultant effect on other events or conditions. The Chinese language itself, with its abundance of characters and highly specialized nuances of thought, reflects this. Chinese characters are pictographs, each one representing a complete thought. Although each character actually contains (a) a literal meaning, (b) a symbolic meaning, and (c) a sound value, the original purpose of each pictograph was to convey a single meaning and to represent a single thought or concept. Adding pictographic elements can create new sounds and generate more complex ideas, and the language is capable of amazingly complex and entertaining multilevel puns. It is this capacity that also makes Chinese difficult to translate accurately, as at times the pun is the true essence of the passage. The written language itself is fundamentally a pictographic system, and this limits the language's ability to convey the same cause-and-effect procession innate to a Western alphabet. Occidental alphabetic characters do not have conceptual meanings until they are joined with other characters. The only exceptions are special cases of characters coming to symbolize specific concepts, such as *I* as an expression of reference to one's self, or the Greek letter π (pi), used mathematically in the formula for determining the area of a circle. The twenty-six letters of the English alphabet, for example, may change both their own sound value and the meaning of the word they are part of strictly by changing their linear relationship to one another.

The result of this linearity in Western cultures is that whenever there is a new condition, a new discovery, or a new application of known physical laws, one of the questions that follows is, "What's next?" There is always a "What's next?" awaiting the discoverer, and it is a natural consequence of a linear approach to cognition to ask it. For this reason, Western cultures tend to carry advances to their logical conclusion. There is always one more application. There is always one more idea or use inherent in what was previously done with a principle. The sequence progresses until the value of the next step is less than the cost incurred by taking it.

Oriental cultures do not do this. Their holistic approach tends to incorporate a new discovery into the whole *as it is.* Each invention or technological device is seen as a solution to a particular problem and not as the first step in a long series of applications. The Chinese simply did not consider continually asking "What's next?" to be a valid motivator.

Philosophical Point of View

The holistic thinking of Eastern cultures is further exemplified by the philosophical approach developed in the Oriental world. In China, two fundamental principles were applied to the sciences, including astrology, medicine, alchemy, and other Chinese intellectual activities. The first of these, the idea of *yin* and *yang,* presents the universe in terms of a fundamental dichotomy of opposites. Rather than the good–evil dichotomy used in Western culture, in which good must triumph over evil, yin and yang see nature as merely seeking a balance, all chaos coming from an absence of this balance, and a movement toward balance as the way to serenity in nature and in humanity. Balance is the key in the process. Opposites are seen as neither good nor bad, but simply as opposite ends of the scale. Some of the major dichotomies are heaven and earth, male and female, action and passivity, and so forth. Technological change was seen as a means of reachieving balance when there was none. If nature appeared to be in balance, then there was no need to change. This Oriental point of view brings home once more the concept of the preference for the status quo and the veneration of the traditional rather than change.

The second concept that so influenced the Oriental approach to science and technology was the principle of the *five elements.* All reality was classified in terms of five elements, and all spatial relationships and temporal relationships were understood in terms of these five elements. The five elements were *fire, wood, earth, metal,* and *water.* This is somewhat similar to the Greek classification of matter into the elements of earth, fire, air, and water. To the Chinese way of thinking, the efficacy of these five substances as universally elemental in the makeup of physical reality was absolute and inviolable. There was no thought of changing the concept or altering this approach to the understanding of the real world.

All science and all events were described in terms of the five elements and in terms of the balance of the yin and the yang. The Chinese adhered to this theory, unlike Western scientists, who put aside the traditional Greek approach of four elements in favor of alternative theories that more closely fit observation. The Chinese explained observed irregularities as examples of the fact that, upon occasion, even the universe goes astray. Astrologers particularly displayed this refusal to accept evidence. As an example, the five elements were thought to be associated with the five planets. In the face of the discovery of Uranus, the Chinese astrologers simply ignored the inconsistency.

In the West we see a striking contrast, specifically during the time of the Renaissance and subsequent political, social, and scientific upheavals. In Europe, with the spread of available literature for the masses and an increasing degree of literacy among the influential, the opportunity for exposure to new ideas arose. Coupled with a cause-and-effect orientation, the population began to question the traditional, to look for new explanations and possibilities, and to turn their attention outward toward the largely unknown and unexamined world around them. The whole idea of academics changed. Papers were no longer written, they were "published," that is, made public, a phenomenon that would have seem ludicrous and without value before the print media had come into general use.

Westerners were willing to look for new solutions, to ask new questions, and to seek information just for the love of knowledge itself. They were relatively unfettered by political restrictions, the information was more available than ever before, and the desire to profit from new technology, coupled with audacity (a "marriage" proven by the success of the rising merchant middle class), combined to push a "next-step" mentality in the West.

The Chinese were eminently practical in their thinking. An invention had a purpose. A discovery was a discovery, and it represented a new piece of information that fit into the whole of the fabric of the universe, or it was ignored as a fluke. The culture was inventive, it was organized, and it was highly developed, but the very structure of the logic, philosophy, and social order of the Oriental culture kept it from taking advantage in the long term of knowledge as it developed. In contrast, the immediate reaction to a discovery in Western cultures was, What can it be used for, and what does it mean? What is the next step? What logically follows? It is this linear aspect of Western thinking that gave birth to the Industrial Revolution in Europe once all of the physical and intellectual requirements were met. A similar revolution didn't occur in the East until much later, with the introduction of linearity into Oriental thought. It should be noted that in the twentieth century, this condition has reversed itself. The industrialization of Japan and the strides made by China in recent years are evidence of the revolution in thought and physical conditions wrought by the "what next" linearity in the modern Oriental approach.

THE SCIENTIFIC METHOD

No better single example of linear, logical Western cognition exists than the *scientific method.* It exemplifies the step-by-step, proof-oriented approach to the cause-and-effect world that the Western mind embraces. And yet, surprisingly, it begins not in knowledge but in faith.

The scientific method depends for its development on the faith of the investigator at the outset of any scientific investigation. The first action or first step in the process involves developing a hypothesis, an assumption of a real-world phenomenon, and then accepting that hypothesis tentatively, pending proof by experimentation or investigation. The scientist is, in essence, taking the truth of the hypothesis on faith when the investigative process begins.

This is a necessary preamble to scientific inquiry. In the face of limited knowledge, certain assumptions must be made. Based on these assumptions, the scientist then proceeds to test them, thus proving or disproving their validity. Few phenomena are more linear than this scientific approach of observation, hypothesizing, and testing. We *begin* with faith in the Western approach, tentatively accepting observation in conjunction with the proven information available to us. However, rather than stopping there, as the Oriental might do in fitting all observations into a well-ordered belief system, the Westerner *tests* the hypothesis before declaring it to be a true representation of the nature of the world. A key factor in Western science is that there are no absolutes, that "scientific laws" are under scrutiny and are accepted if not disproved, and that we are free to give up a preconception about the makeup of the real world if a better explanation comes along.

Most representations of the scientific method list a set of five steps through which the investigator discovers the truth of a problem. They are as follows:

1. *Defining the problem.* It is first necessary to have a clear idea of exactly what the problem is in order to choose a direction of investigation and properly formulate later experimental techniques.
2. *Observing the evidence.* An accumulation of information is necessary as an initial step in investigation to ensure that all previous work on the problem is known to the investigator and that she or he clearly understands the nature of the problem under investigation.
3. *Forming the hypothesis.* The investigator uses intuition and logical thinking to discover perceived patterns of behavior from the preliminary data and draw tentative conclusions from those patterns. This is a primary creative step by which a new way of viewing the nature of the reality under investigation takes place.
4. *Experimenting.* A method of testing is designed to either validate or deny the truth of the hypothesis. The design of the experiment itself is a crucially creative process. The nature of the experiment will automatically exclude and include what can be discovered by its results.
5. *Formalizing the theory.* The results of the testing of the original hypothesis and conclusions are made available to the public (published) for the scrutiny of the investigative community as a whole and to use to further investigate the subject.

Again, note the step-by-step linearity of the process. The scientific method lends itself to creativity and innovation both in the sciences and in the implementation of known concepts in practical technological application. To form a hypothesis about a problem requires creativity. To design an experiment to prove or disprove a hypothesis requires an innovative approach. To view science and the scientific method as not creative because of their dogged requirement of proof before acceptance is to miss the point. True creation of useful technology requires this rigid approach, but within an orientation that admits to the possibility of change and that encourages the usefulness of innovative methodology to improve efficiency and the quality of life in the society. Change for its own sake, as we will discuss later, is an equal confusion of issues, but change for the betterment of the population is valid, natural, and desirable.

On the negative side, the scientific method by its nature may pose as great a threat to innovation in the society as that of restrictive social norms. The five-element approach of the Chinese is no more than a given point of view regarding the nature of the structure of the universe. That is equally true of the accepted "modern" viewpoint. Modern scientific theory represents a different viewpoint, but it is still only a viewpoint. Westerners use the scientific method to discover new facts within their worldview. Each hypothesis is built on the basis of observations and beliefs that support that accepted worldview. Because of this, their ideas tend to support the traditional body of knowledge and what it expresses. It is only with great difficulty that the major paradigms of a culture are changed. It is shortsighted to believe that the use of the scientific method alone can bring about a discovery of unique and final answers to questions of science. Too many external influences serve to taint the investigator's objectivity. Social, political, and economic motivations sway opinion and prejudice choice of experiment. It is safer to agree with accepted, dominant scientific beliefs than it is to break radically new ground. The kinds of conclusions we reach as a result of our observations tend to follow tradition. People have an incredible capacity to pigeonhole experiences into neat packages that fit in with their present view of the world. Radical departures are often ignored or rejected as "illogical" or "unrealistic." Anyone who has seriously considered the informational content of the general theory of relativity will tell you that it defies common sense. That is one of the reasons that its acceptance was so very long in coming, even in the face of experimental proof.

A radically different scientific theory can only be accepted if there is a change in the worldview of the scientific establishment. People must be willing to change their ideas about a subject and be open to the possibility of an alternative interpretation's validity. This does not come easily.

Herein lies the main problem. Experimentation tends to follow accepted theory. Experiments are committed to theorizing, instrumentalizing, conceptualizing, and developing methods of problem solving that are bounded by the present paradigm. Without a willingness to diverge from the generally accepted paradigm, the resolution of problems whose solutions lie outside society's present realm of knowledge is difficult if not impossible. Many a discovery has been made in ignorance of accepted scientific beliefs, which, if the accepted beliefs had been known, would not have been possible. Yet it is only by changing its worldview that a society can progress. If the answers to major scientific problems lay within the present belief systems, they would have already been solved. The scientific method can be a valuable tool, and, just as with any other tool, it can also be misused. The question is not one of supporting or destroying the present paradigm. It is a matter of discovering the truth of our physical environment and how that truth can be used to society's betterment. As long as this is the key element of research, the scientific method itself will maintain its usefulness.

CONCLUSION

Technologizing is a creative process, and the nature of creativity and innovation by which technological progress takes place is affected by the culture and social structure within which that change is occurring. The manner of cognition, the sociopolitico-economic structure, and the philosophical makeup of the population can all either detract from or encourage creativity and technological development. The direction that scientific inquiry takes and the manner in which a culture puts scientific knowledge to practical use depend on the *belief systems of the culture,* the *opportunity* and the *ability to technologize,* the *motivation* to do so, and, most importantly, the *freedom to question* the established concepts of the culture. Without these elements, technological progress does not take place.

The creativity and innovation discussed in this chapter is useless and, in fact, will probably not occur unless there is some reason for a person to undertake it. Admittedly, the tendency to imagine and be creative is innate and, for most of us, cannot be avoided; however, turning those creative impulses into usable artifacts, methods, and structures requires a reason for doing so. I may have the intellectual capacity to "invent" Archimedes' screw for lifting water for irrigation, but unless I need to do so, I probably won't bother. In other words, there must be some form of *utility* connected with the creation. Some perceived value must be attached to it.

As soon as we begin to talk about utility and value, we enter the realm of economics. From the point of view of the archaeologist as well as that of the economist, economics is merely the study of all those ways a society goes about the process of providing itself with the goods and services necessary for its survival. It consists of production and distribution of valuable goods—that is, goods that have utility for the consumer—so that they can be consumed. In terms of the development and use of technology to both create these goods and services and consume technological goods and services, capitalism has been shown time and again to be the most efficient and effective method of doing this. If we are to understand the role of technology in our lives, we must investigate its relationship to economics in general and, more specifically, how it plays out in our daily lives. Chapter 4 is a discussion of this process.

KEY TERMS

Balance
Belonging Needs
Creativity
Hierarchy of Needs
Innovation
Linear Thinking
Abraham Maslow
Motivation
Occidental Thinking
Oriental Thinking
Paradigm
Physiological Needs
Risk
Self-Actualization
Social Needs

REVIEW QUESTIONS

1. What is creativity?
2. What is innovation, and how is it related to creativity?
3. What is the role of innovation in the development of new technologies?
4. How does Oriental thinking differ from Occidental thinking?
5. How does the scientific method foster innovation and the creation of technology?
6. Is the scientific method necessary for creativity and innovation to take place?

ESSAY QUESTIONS

1. Leonardo Da Vinci is considered to be one of the most creative and prolific figures of the Renaissance. How are the creative principles discussed in Chapter 3 reflected in his inventions, and how are they reflected in his art?
2. Choose a technological field, such as architecture, civil engineering, or electronics, and discuss it in terms of being a science and then in terms of being an art.
3. In your opinion, is the creation of technology merely an element of the Maslowian proposition of need satisfaction, or is there more to the process than merely genetic predisposition?
4. Is the creative nature of people in our culture increasing, decreasing, or staying the same? What do you think are the causes of our current condition?
5. Of the various motivations and impetus discussed in Chapter 3, what do you consider most important to the development of new technologies? Use examples to explain your answer.

THOUGHT AND PROCESS

1. The following suggestions were presented by students during a brainstorming session designed to release their creative bent in determining the solution to an industrial problem. The problem given them involved making a useful product from a relatively useless byproduct, in this case, coffee grounds. Scan the list and choose an example that you consider to be logical and one that you consider illogical. For each of these choices, develop a rough plan of how to produce and market the product you have chosen. It may surprise you how logical some illogical ideas can be when thought of in a creative manner.

Possible products from used coffee grounds include the following: (a) fertilizer, (b) records, (c) chewing gum, (d) ashtrays, (e) concrete filler, (f) weaker coffee, (g) products with a coffee aroma, (h) dye, (i) pigment for paint, (j) pet food, (k) an agent for sopping up oil spills, (l) soap, (m) an abrasive, (n) a protein substitute, (o) hog food, (p) fuel—after fermentation, (q) disposable garbage cans, (r) particle board, (s) fire extinguishers, (t) a song, (u) ballast, (v) insulation.

2. Test your own creativity. For best results, it is suggested that you try this question before reading the next one. For those of you who have just resisted that temptation, this should be a most effective exercise. Place a single piece of paper in front of you on a desk or table and place a pencil beside it. With no other materials to work with, *be creative!*

3. If you had a problem with the preceding question, it may be a result of programming. Many people have the belief that they are not creative and therefore are incapable of doing something creative. To bypass such programming, try doing question 2 again with the following substituted instruction: Make something. After all, anyone can make *something*, even if it is only a mess!

4. Innovation is nothing more than finding a better way to do something. Choose an everyday job that you dislike but must perform and be innovative in discovering a new way to accomplish the same task. Creating an artifact is not a stipulation of this exercise, yet if you can incorporate such a physical construct into the process, you will gain more understanding.

5. Go sit under a tree. How do you view the world around you? Do you analyze it piece by piece or view it as a whole? Try to follow your thought processes as you look around and take in your environment. Whether you find yourself to be linear or holistic in your approach to viewing the world around you, take a few minutes and try it from the opposite point of view. Now try this with other areas of your life and see how it increases your ability to be creative, innovative, and able to solve problems.

CHAPTER 4

ECONOMICS AND CULTURAL IMPETUS

One of the most potent determinants of technology among the subsystems of our sociopolitico-economic system is in the area of economics and economic structure. Technology is an integral part of economics, being one of the chief causes of success and failure not only for individual enterprises but for the economic system as a whole. What is it about technology that affords it such a major role in the performance of the economy? Is it really such an essential part of the economy's functioning? What would be the effects of less or more technology from the economic point of view? These are some of the questions that we explore in this chapter.

CAPITALISM: THE TRADITIONAL APPROACH

To understand the place of technology in the economy, it is necessary to first explore the nature of the economy as a whole, with particular emphasis placed on technological progress and its resultant effects in the economy.

Economics can be defined as the study of how societies choose to use scarce productive resources for the production of goods and services and how they choose to distribute those goods and services to the general population for their consumption. Inherent in this definition are the primary elements of any economic structure, specifically, *production, distribution,* and *consumption.* In brief, economics studies how goods are produced and distributed and indicates that the purpose of all this activity is the consumption of the produced goods. (*Consumption* is the utilization of goods and services to produce satisfaction in the user. It is not within the scope of this book to delve into an understanding of exactly what is meant by "producing satisfaction." Since the concept of satisfaction is such a personal, subjective experience, that is one particular "bucket of worms" that the author gladly passes by.)

The decisions that are made regarding how the production and distribution are to take place are as numerous as the economies that exist today. We shall concentrate on the modern economic approach used in the United States, which is a "mixed" economy.

The roots of the modern American system of production and distribution lie in capitalism, a theory of economic structure first introduced by Adam Smith in his *An Inquiry into the Nature and Causes of the Wealth of Nations.* Published in 1776, Smith presented a basic explanation of the economic structure of "modern" nations and how that structure functioned.

Characteristics

To explain the how and wherefore of economies, Smith started with certain assumptions about how people behave in the market and how the market structure performs its major functions. These initial premises were true in 1776 for the environment within which Smith operated and with certain modifications are largely true today. We discuss five of his assumptions in the following subsections.

Private property

Smith stated that one of the characteristics of a successful and naturally functioning economy was adherence to the principle of private property. According to Smith, individuals hold ownership of the means of production in the form of *natural resources* (virgin land with its original fertility and mineral deposits), *labor* (the potential and real contributions of people to the production of goods and services through work, which involves both physical and mental effort), and *capital* (all manufactured productive resources such as plant and equipment, machinery, and improvements made to land that render it more suitable for the production of goods and services). He further stated that the property owners have the right to do what they will with that property, whether it be to use it (consumption), sell it, rent it, or do nothing (save it). As such, the owners of property and other productive resources are free to determine to what use that property will be put and what goods and services, if any, they will be used to produce.

The principle of self-interest

Smith expressed his assumption that all individuals operating within the economy carry on purposeful activity designed to further their own self-interest. Smith saw economic action as a means of bettering one's position in life and as only having that purpose. The idea of self-interest, Smith contended, may seem somewhat cynical and negative on first inspection, but upon deeper reflection, it can be seen not as something to be criticized but as something that is a fact of life. Technically, Smith noted, there are no actions, even altruistic ones, that are not done for the purpose of self-interest.

This "idic" Freudian approach to economics is nothing more than a statement of the pleasure principle, which holds that individuals tend to seek pleasure and avoid pain. Far from being in opposition to the religious and philosophical beliefs of Smith's era (he was living in a Scottish environment during the age of Puritanism in Great Britain, when the Protestant ethic was dominant), the concept is quite a valid one. As presented, Smith merely reminded the reader that the desire to improve one's economic welfare (to be better off) is the reason people interact in the market and involve themselves in the economic processes. Why would someone want to work if they received no reward for their efforts? What would be the motivation to create new products or increase the efficiency of one's job performance if the result was not of some benefit to the person? Even those who enter relatively less profitable professions receive benefits, either to the best of their personal ability or as a combination of monetary reward and personal satisfaction. The motivation for doing anything is to improve one's quality of life. This desire for self-interest in the economic sense translates into the concept of the *profit motive,* that is, the concept that the reason for people involving themselves in economic activities is that they believe that they can receive a profit (when revenues exceed costs) by doing so and, therefore, improve their economic position. In fact, it is assumed that the economically rational person always seeks to maximize his or her profit.

Competition

The definition of *economics* offered at the beginning of this section dealt with the choices available to individuals regarding production and distribution of goods and services. The necessity of choice concerning what will be produced and what will not be produced reflects the fact that resources are *scarce.* There is not enough land, labor, and capital available to satisfy *all* objective desires, and we must therefore choose which economic desires will be taken care of and which will not be taken care of. As individual firms and individual consumers, all with different ideas about what is and is not an important use of resources, we must *compete* for the right to use what resources there are. The functioning of the market system is such that this competition works to our advantage by forcing the society to produce only that which is most desirable and only in the most efficient way; to distribute it to the largest number of people; and, in the process, to maximize the amount of profit we are able to accumulate. Paradoxically, even those who fail in this "survival of the fittest" sort of competitive mentality end up economically better off as a result of the process. However, to see this more clearly, it is necessary to deal with the concept of "competition" in conjunction with another principal characteristic of capitalism—consumer sovereignty.

Consumer sovereignty

Who makes the decisions in a capitalistic system? Who decides what will be produced? Who decides who will receive the manufactured goods and in what quantity? Who decides what prices will be and in what quantities goods and services will exist? According to Adam Smith, it is the consumer who makes all of these choices, and it is because of this that the market system is self-sufficient, self-regulating, and self-correcting. The scenario that results in this conclusion develops in the following way.

In an economy made up of many buyers and many sellers, each of whom holds a relatively small part of control over what is bought and sold, producers are at the mercy of the market. They have no control over the prices or over their own individual success, and they cannot individually affect the activities of the market. They are operating in what is known as a *perfectly competitive market.* Suppose a company is producing pencils and is in competition with many thousands of other pencil manufacturers. Such a manufacturer must compete with other pencil makers for the right to scarce resources such as pencil lead, rubber for erasers, and wood for the casement into which the pencil lead is fitted. The pencil makers also compete for labor, for machinery and equipment, and for the money necessary to set up and run a business. In the product market, our pencil manufacturer is faced with the problem of selling the company's product to the public, something that all the other pencil manufacturers are also trying to do. In the face of such competition, how can the pencil manufacturer get ahead? How can the company's economic welfare be improved to the greatest degree by maximizing profits? Since profits are the difference between total revenues, that is, whatever is brought in from the sale of pencils, and total costs, or whatever is expended on resources to produce the pencils, the manufacturer must either have a higher price (to make revenues higher) or a lower cost (to make total costs lower) or both so that the difference between total costs and total revenues is as great as it can be. (Total profits equal total revenues minus total costs.)

Case 1: Raise Price By raising the price, the pencil company should reap higher total revenues, thus raising profits if costs remain the same. However, since consumers have so many choices as to where they can purchase pencils, they would be foolish to buy at the higher price when so many other companies are selling the identical product at a lower price. *Conclusion:* Our pencil manufacturing company loses *all* of its business when it raises the price and will make *no* profits rather than more profits. Thus it cannot raise the price and survive.

Case 2: Lower Price This seems like a viable solution. If the company loses all of its business to other companies when it raises its price, perhaps it could take all of its competitors' business by lowering its price and thus increasing total profits by selling more items. In this case, however, our pencil manufacturer is faced with a new problem. As soon as the company lowers its price, its competition will see the threat to its business and will drop its own price to meet our company's price. The result is that everyone is selling about the same number of pencils as they did before, but because the costs are the same and the price is lower than before, profits are less. Hence the pencil manufacturing company will not improve conditions by lowering the price any more than it will by raising it. It is stuck with the price that the market offers, that is, the price that *consumers* are willing to pay. *Conclusion:* Consumers set the prices in capitalistic markets, not manufacturers.

Case 3: Increase Production What about just making and selling more pencils? If the manufacturing company can do that, it can make more profits without changing the price. Can it do this? It cannot do so easily. To sell more than the company's equal share of the pencils than are demanded, the manufacturer must have a product that people want to buy *more* than they want to buy someone else's product. That is, the pencils must be better in that they must present more utility to the public. The manufacturer will strive to make a pencil that is as good as possible in order to encourage the public to buy it. Unfortunately, so will all of the competitors, with the result that *all* of the pencils that are available in the marketplace will be of the highest possible quality. If one company has a better product, the other companies lose customers, and they are not about to let that happen. If they do, they go out of business and are no longer part of the market. *Conclusion:* Consumers determine how much of a product is going to be demanded in the marketplace, and they further force manufacturers to produce the highest quality product that can be made, given present levels of technology and present levels of expected customer utility.

Case 4: Reduce Costs Another possibility is open to the pencil manufacturing company if it wants to maximize profits. It can lower costs and thus increase the differential between total costs and total revenues. This is done by using every technological and human relations skill at its disposal to minimize production costs—by searching for better production methods, by buying resources at the lowest possible cost, and by generally keeping expenses to a minimum—to maximize the efficiency of the operation and to waste nothing. This tactic works to maximize profits, but, unfortunately, it works equally well for the other manufacturers with whom it is in competition. All the pencil manufacturers strive to be as efficient as possible in methods of operation. The result is that the manufacturer must maximize efficiency just to stay in business. *Conclusion:* In order to maximize profits, competition forces a company to produce the highest quality goods it can, at the lowest possible cost, to sell them at the lowest possible price that allows a profit, and to make them available to the largest number of people.

Consumer sovereignty exists not only in the product market where final goods and services are bought and sold, but it is equally evident in the factor markets, where natural resources, labor, and capital are purchased.

Earlier it was mentioned that these same manufacturers who were trying to maximize profits in a product market were also in competition for productive resources. To get a share of those resources, they must be willing to pay a high price for the right to their use. The factor market is an auction from the point of view of the manufacturer. It consists of bidding against all other producers for the right to resources. Only the manufacturer who is willing to pay the highest price will be able to share in those resources.

Where does that money come from? It comes from profits generated in the product market. If a company is unsuccessful in selling its products, it will not have funds to buy raw materials for the production of more goods and services. Only successful sellers are capable of being successful buyers in the factor market. And how does the economy decide what products are bought and what products are not bought? *Consumers* do it.

Each time a product is bought, the consumer has voted for the right of that product to exist. Each time a consumer refrains from buying a product, he or she has chosen to vote "no" to the right of that product to exist. Only the products with the most votes (those that are sold the most, creating the most revenues for their producers) are allowed to exist. "No" votes result in no money, which results in no product. *Conclusion:* Because of competition and consumer sovereignty, the market creates the largest number of goods of the highest quality for the lowest price and makes them available to the most people, with the final consumer having the final say in what will be produced, how it will be produced, who will produce it, and to whom it will be distributed. The market system is therefore democratic, self-perpetuating, self-regulating, and self-correcting in nature.

Capital investment

Adam Smith stressed the importance of investing in the means of production. One of his clearest messages in the *Wealth of Nations* was the belief that *economic power stems from the ability to produce.*[1] It is only because of a nation's capacity to produce that it has economic strength, and the maintenance of that ability ensures a viable economy. He further felt that investment should be encouraged and that it is through an expansion of the capital base, that is, the level of investment in the means of production (plant, equipment, and so forth), that an economy can experience economic growth and an improvement in prosperity. As will be illustrated later, it is the validity of this conviction that brings us to the actual value of technology in the economic structure.

Laws of Supply and Demand

The importance of these two primary laws of market operations lies in the fact that they are central to both predicting the actions of suppliers and consumers in interaction and in explaining why prices in the market reach the levels they do and why they tend to maintain those levels in the short run. Simply, the laws may be stated as follows: The *law of demand* states that there is an inverse relation between the price of a good or service and the

[1]Adam Smith, *An Inquiry into the Nature and Causes of the Wealth of Nations,* ed. Edwin Canaan (New York: Random House, 1937), Passim.

amount of that good or service that will be demanded in the marketplace. The *law of supply* states that, in general, there is a direct relationship between the price of a good or service and the amount of that good or service that will be supplied by producers for sale in the marketplace.

These laws of economic behavior are an expression of common sense, based on the observation of consumer behavior and producer behavior in market situations. They are, to a considerable degree, self-evident. As buyers and sellers, virtually everyone who has interacted in the marketplace has a personal experience of their validity.

Law of demand

The law of demand states that if the price of a good rises, the public will demand fewer units of that good because fewer people are willing to pay that higher price, because those who do purchase the product purchase less of it or because of both of these factors. Hence, people ask, "How much?" before deciding to buy. Price is a determining factor.

A simple illustration of the law of demand is seen in graphing the relationship between changes in price and changes in demand (see Figure 4-1.) Note the downward slope of the demand curve indicating the inverse relationship between the price of the product and demand for the product in question.

Law of supply

The law of supply is similar in that it expresses another facet of market behavior, that is, the determining nature of price in deciding how much producers are willing to produce. As we indicated above, individuals seek to

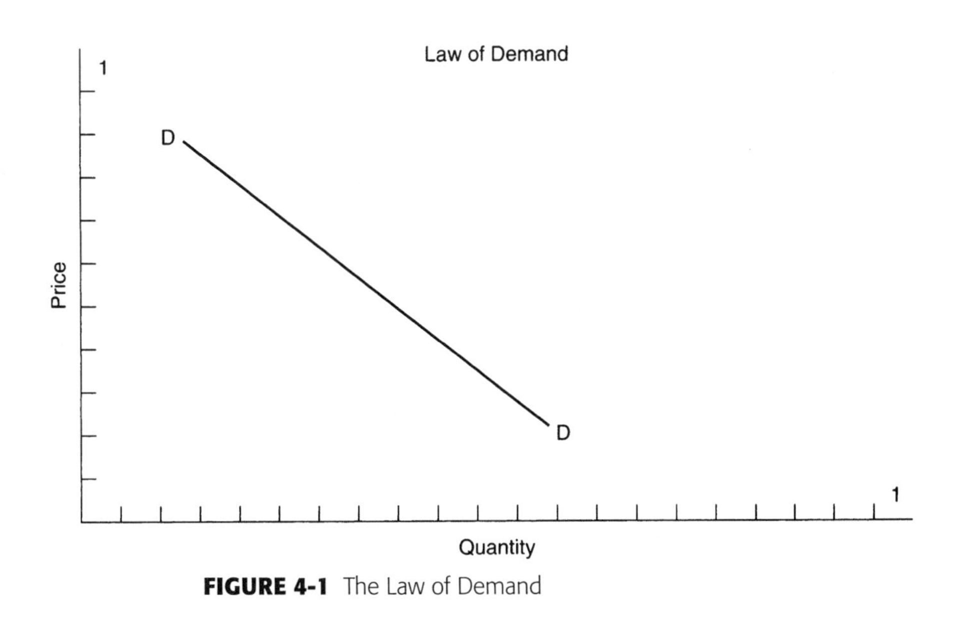

FIGURE 4-1 The Law of Demand

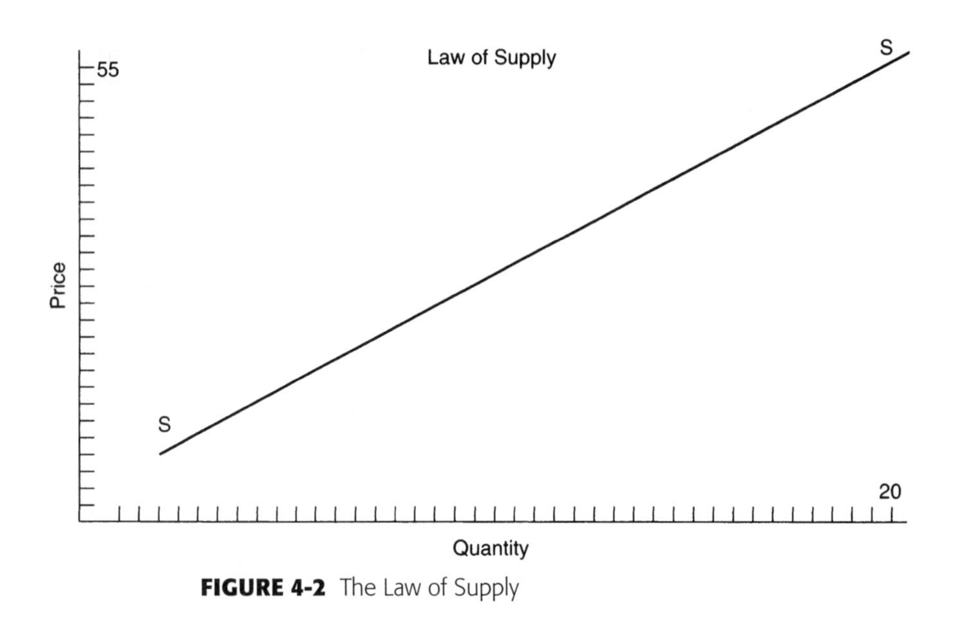

FIGURE 4-2 The Law of Supply

maximize profits in their production of goods and services. The greater the profit that can be gained by producing a good or service, the greater the probability that a producer will be willing to produce and offer that item in the marketplace. Hence, if the price of a good rises, producers will make more of it, as it represents an increase in the contribution each unit sold will make to their economic welfare.

Figure 4-2 illustrates the direct relationship between price and quantity demanded. The supply curve slopes upward to the right, indicating an accompanying rise in quantity supplied for each increase in unit price.

In the case of the law of supply, it should be noted that there are some exceptions. The wording of the law of supply begins "In general. . . . " This qualifying phrase is included because of three cases in which the law does not hold: the case of *fixed supply*, of *fixed price*, and of *economies of scale*.

The case of fixed supply is illustrated in Figure 4-3. Here we are faced with a market in which, no matter what the demand for the product or what price consumers are willing to pay for the right to possess that product, the quantity of the product available remains the same. There is no relationship between price and quantity supplied in such a case. Examples of this phenomenon include such things as the *Mona Lisa,* the Hope Diamond, or the quantity of moon rocks available on Earth. With each of these products, no matter how much the price changes, the supply is fixed and constant. (It should be here noted, however, that this can be a short-term phenomenon. In the case of the moon rocks, if the price were to go high enough, someone would mount another expedition to recover more. The price must exceed the cost of the journey for this to happen, however, which is not likely in the foreseeable future with the present levels of technology.)

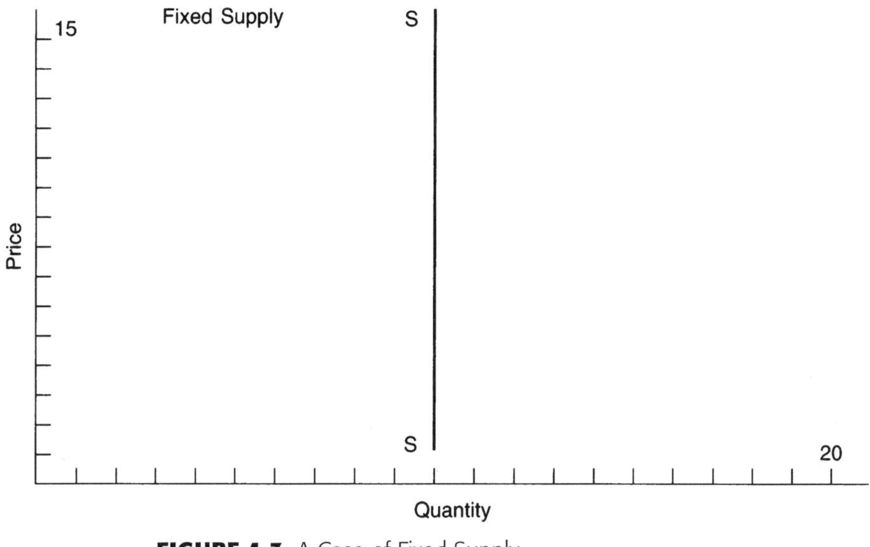

FIGURE 4-3 A Case of Fixed Supply

The second exception, the case of fixed price, is also indicative of a market in which there is no relationship between price and quantity supplied. In this case, however, it is price that remains constant, not fluctuating at all over a wide range of quantities of product supplied. The supply line becomes a horizontal line (see Figure 4-4), since quantity is free to vary as much as it wishes while price stubbornly refuses to budge. Price controls imposed by a government or a cartel are an example of this. If the price of

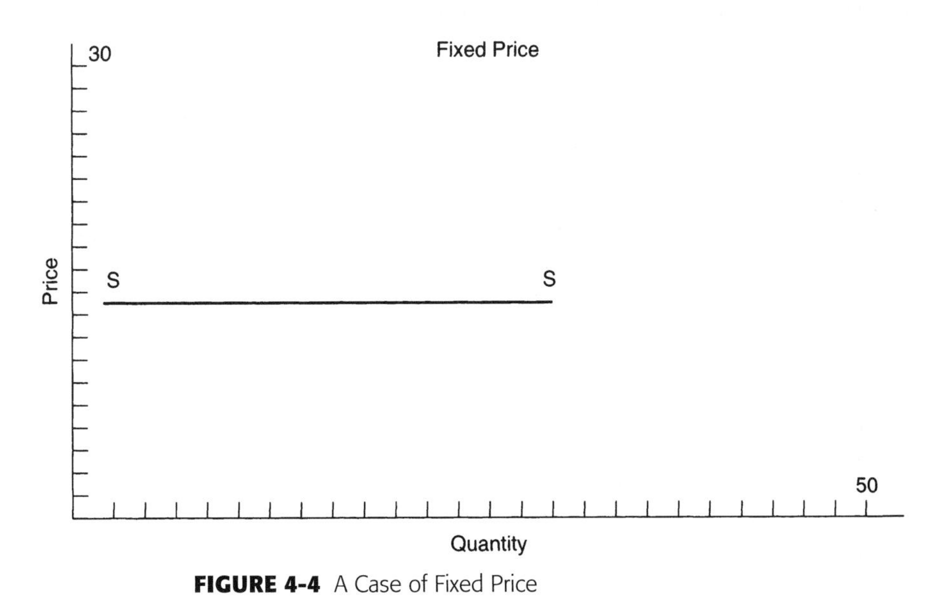

FIGURE 4-4 A Case of Fixed Price

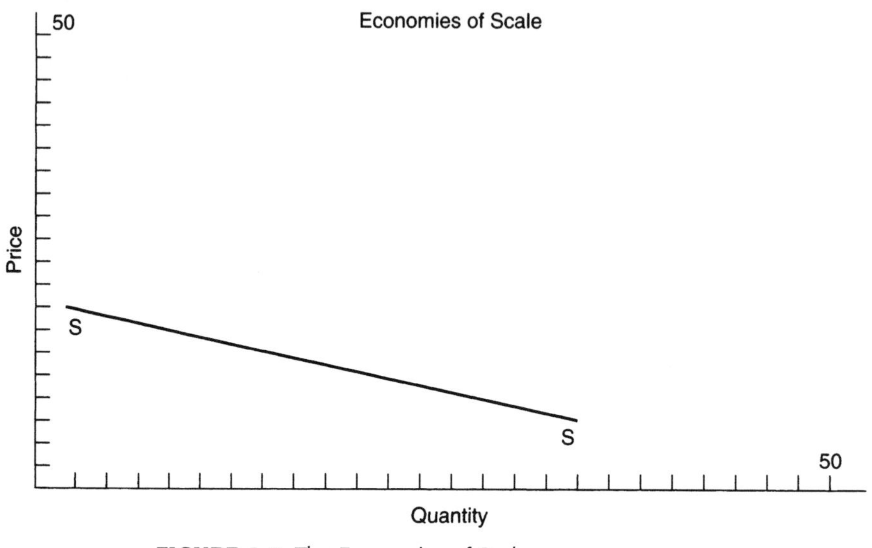

FIGURE 4-5 The Economies of Scale

gasoline or the price of airline tickets is fixed, computed by a predetermined method independent of demand for the service, it is a case of fixed price and results in a breakdown of the law of supply.

The third exception, that of economies of scale, is a bit more complicated. The supply curve in this case slopes downward to the right rather than upward as we would expect (see Figure 4-5). It *seems* to indicate a situation in which as the price of the good drops, producers are willing to make more of the item. Such an idea is contradictory to everything we know about producer behavior. To illustrate, it would seem to indicate that if the going price of washing cars declined from $7.50 per car to $2.50 per car, more people would be interested in doing it! How can this be?

The fallacy lies in the nature of what is taking place in the market. Rather than describing changes in supply in response to a change in price, the graph describes the relationship between changes in *cost* and changes in price.

As the number of units produced increases, manufacturers find that their overall costs tend to decline. This is due to an increase in the efficiency of plant and equipment as production rises. As an example, consider the information provided in Table 4-1.

If our pencil manufacturing company purchases a pencil making machine at a cost of $100,000, that amount of money is added to its costs of production. If it makes one pencil with the machine, that pencil will cost it the price of the raw materials, the cost of one worker to push one button one time, and *$100,000 in machinery cost per pencil*! To make any profit at all, the pencil would have to be sold for a price in excess of $100,000! However, if the machine has a capacity of producing 1 million pencils before it must be replaced, and if used to do so, the pencil company has now greatly

Fixed Costs for Machinery	Units Produced	Fixed Costs per Unit	Left for Profit @ $5.00
$100,000	1	$100,000	$99,995)
100,000	5	20,000	(19,975)
100,000	10	10,000	(9,950)
100,000	100	1,000	(995)
100,000	1,000	100	(95)
100,000	50,000	2	3
100,000	100,000	1	4
100,000	1,000,000	0.10	4.90

TABLE 4-1 Dilution of Fixed Cost as Production Rises

reduced its machinery cost from $100,000 for one pencil (a cost of $100,000 per pencil) to $100,000 for 1 million pencils, or a cost of $0.10 per pencil. In order to achieve this improved level of efficiency, all that is required is an increase in output. This means that as a manufacturer increases production, the *cost per unit* declines, resulting in higher profits.

This is all very well and good, but why would our pencil manufacturer want to lower the price if this was true? Why not just keep the price the same and get more profits per pencil sold as costs drop per pencil produced? The reason is that *in order to sell all of those extra pencils, the manufacturer will be forced to reduce the price.* The law of demand states that to increase demand for a product, it is necessary to reduce the price. So what does the company do? It lowers the price as it increases production and still makes more profit as long as the costs per unit decrease faster than the price per unit decreases. Hence, the peculiar downward-sloping nature of the supply curve.

Market Equilibrium

The laws of supply and demand, as an expression of consumer behavior and producer behavior, offer an explanation of how quantities of goods made available in the marketplace and demanded by consumers differ with a change in price. The curves represent the collection of price–quantity combinations one would expect to encounter in the marketplace. However, in and of themselves, they do not tell us what *will* be produced and what *will* be purchased by consumers in the marketplace, nor what the actual price will be. It is necessary to view the behavior of both consumer and producer in concert to discover the actual price and the quantity bought and sold. This is called *equilibrium.*

Equilibrium exists in the marketplace when buyers and sellers agree on both the price of goods bought and sold and on the quantity of goods produced and sold at that price. If agreement is not reached, there is no *prevailing price* in the market, making demand and supply uncertain. The *equilibrium price* is defined as the price at which suppliers and consumers agree to exchange an identical quantity of goods or services and agree to go on exchanging that quantity at the same price. As in Figure 4-6, if the supply and demand curves are constructed on the same chart, it can be seen that the same combination of price and quantity exists on both the supply curve and the demand curve at only one point, that is, where the two curves cross. It is the only set of (x, y) coordinates (quantity and price) shared by both consumer behavior and producer behavior patterns. Graphically, this is the point of equilibrium, where the supply and demand curves intersect. The price prevailing at this point is known as the equilibrium price, and the quantity demanded at this price, the equilibrium quantity, is the quantity that will exactly match consumer demands with producer desires. Any other price will create instability in the market.

Suppose that for some unknown reason the price was actually higher in the marketplace than the equilibrium price. If this were the case, there would be a difference between the quantity demanded (at a higher price, demand would drop) and the quantity supplied in the marketplace (at a higher price, the law of supply dictates that there would be more produced). The resulting difference would mean there is more product available than there is demand for, and suppliers would have leftover supplies after they had sold all that they could in the market. This excess product costs them money. They have purchased resources to produce these goods, they have their capital tied up in them, and they must pay for the warehousing of the produced items while they are still in the suppliers' possession. The suppliers are very interested in

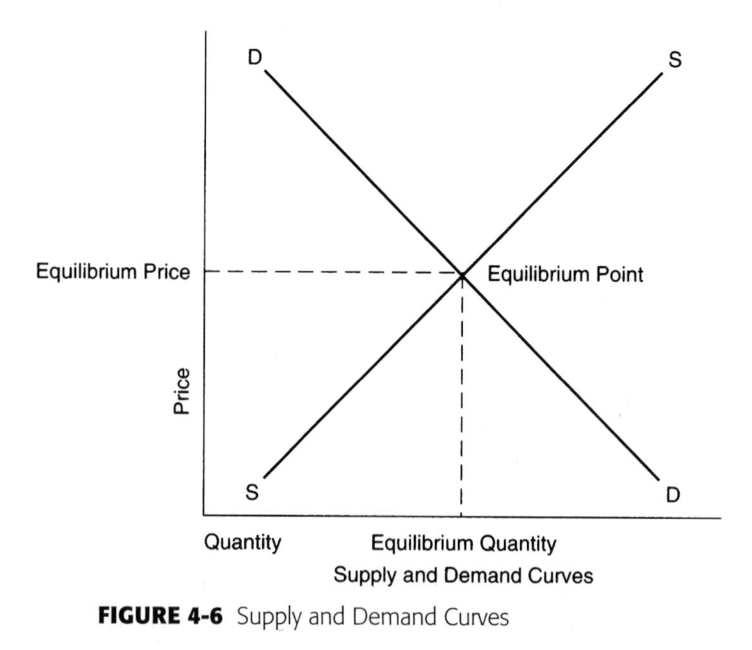

FIGURE 4-6 Supply and Demand Curves

getting rid of the surplus, and the only way to do that is to sell it, which means convincing the public (by lowering the price) to buy more than they have of the item.

As this process takes place, the dropping price makes goods more attractive to the buyer, who is now willing to buy more; whereas suppliers, no longer able to command such a high price for their goods in the marketplace, are less interested in producing the goods. Thus supply drops toward the equilibrium quantity level. Eventually, equilibrium price is again reached, and quantity supplied once more equals quantity demanded, resulting in minimum waste and maximum efficiency.

Result 1: If the price in the marketplace is higher than the equilibrium price, there will be surplus product, which forces producers to lower the price, thus disposing of merchandise as demand rises, forcing price back toward equilibrium.

What if the price is below equilibrium price? At a price lower than equilibrium level, manufacturers will not be so ready to produce, since they cannot make as much money for their efforts as they can at a higher price, and they will cut back production. Hence, supply drops. At the same time, the lower price will appear very attractive to consumers, who will demand more of the product than at the higher price, and demand will rise. This results in *shortage* in the market, a situation in which demand for the product is higher than the supply.

Under such circumstances, consumers find themselves seeking a relatively limited supply of goods, and they must *compete* with one another for the right to own those goods. Such competition on the part of consumers is called an *auction.* The consumers bid up the price until one person wants the item enough to pay a higher price than anyone else is willing to pay. Bidders drop out as the price rises; at the same time, the higher price results in suppliers supplying more of the product for consumption. Demand drops and supply rises as price rises, until, at equilibrium, demand and supply are once more equal.

Result 2: If the price is below equilibrium price, a *shortage* of goods results, forcing a rise in prices through the auction process until equilibrium price is again reached, supply is exactly equal to consumer demand, and there is no waste of productive efforts.

One of the major benefits of a capitalistic free market, according to Adam Smith, is exactly this tendency of the market to always seek equilibrium. It means that the market system is self-correcting, that it is not necessary to control it, to second-guess it, or to force it to perform in an efficient manner; this is something that happens naturally.

Economic Trade-off and the Production Possibility Curve

In *Wealth of Nations,* Smith points out that resources are often scarce. There are not enough of all productive resources to fulfill all subjective desires for goods and services. Although this may not be strictly true of all resources (the problem may be realistically viewed as one of distribution rather than true scarcity, in that a person residing on the shores of Lake Superior may argue against the idea of there being a shortage of water, whereas a citizen of Egypt or Libya would not hesitate to agree), the economy behaves as if there really is scarcity, and behavior is, after all, what we work with in economics. It can be assumed, then, that we have effective scarcity, and as a result, we are forced to make choices among the various uses to which we can put our resources. Every hour of labor put into producing automobiles is an hour not put into the production of water buckets. Every ton of steel used in the production of railroad cars is a ton of steel not dedicated to the production of skyscrapers, or subways, or M-1 tanks, or battleships, or any of the other choices available. Not only are we forced to make such choices, either overtly or through the market system, but we are forced to pay a price for the decisions we make. This concept of the *opportunity cost* is the foundation of the phenomenon of economic trade-off and the resulting production possibility curves.

An *opportunity cost* is the value of the alternative *not* chosen. It is the value of what was given up in order to do something else instead. If a person has the opportunity to either go into the restaurant business at an expected profit of $20,000 the first year, or into the tire business at an expected profit of $32,000 the first year, but is unable to do both, then a decision must be made. The economically rational human being is expected to seek maximization of profits and would therefore choose to enter the tire business and earn a profit of $32,000. An accountant would consider the profit of the tire business to be exactly $32,000, the amount earned after expenses the first year. But to the economist, this approach would ignore the cost of giving up the opportunity to make $20,000 by entering the restaurant business. Thus the economist would consider the profit made *because of choosing to go into the tire business* as only $12,000, since the individual would have made $20,000 no matter which business he or she chose to enter. He or she simply would absorb a loss of $20,000 resulting from foregoing the opportunity to enter the restaurant business, and the *economic profit* would be $12,000, the difference between the total accounting profit and the opportunity cost.

Translating this concept into a general case concerning economic structures, each time goods are used for one purpose, it costs us in that we are forced to pay an opportunity cost for not using the resources for the second-most profitable purpose. We *trade* the right to use it for one thing in favor of the right to use it for another. This *economic trade-off* (a sacrifice that must be made to obtain something) is inherent in each economic decision.

Economic trade-off can easily be demonstrated if we consider the case of an economy that is forced to choose levels of two goods that compete with each other for some or all of their inputs.

Imagine a society whose production capacity is limited to making either bicycles or canned soup. It can either put all of its productive capacity into making bicycles, or it can put all of its productive capacity into making canned soup, or it can make various combinations of bicycles and canned soup. Each of these cases involves a trade-off. The production of nothing but bicycles results in no canned soup being produced, and vice versa. If we decide to produce *some* canned soup, we must give up *some* bicycles to do so, as every case of canned soup requires resources that would have been used to make bicycles, thus reducing the number of bicycles that can be produced. By plotting all of the various combinations of bicycle production and canned soup production that are possible, we create what is called the *production possibility curve* (see Figure 4-7).

Note that any combination of production that lies on the production possibility curve itself (known as the *production possibility frontier*) represents a case of full utilization of productive capacity. It defines the *maximum* combinations of output of the two products, given that all resources are fully employed. It points out that if we want to produce an amount of canned soup, *x*, then the maximum number of bicycles that can be produced at the same time will be an amount, *y*, that corresponds to the point *x, y* on the curve, given the present level of resources and a constant technology. Any combination of bicycles and canned soup lying beneath the production possibility frontier is possible but represents less than maximum use of resources. Any combination outside the production possibility frontier is a combination that is impossible as it is beyond the capacity of the society to produce. The *production possibility frontier* is the limiting expression of possible production combinations of the two goods due to economic trade-off in the economy.

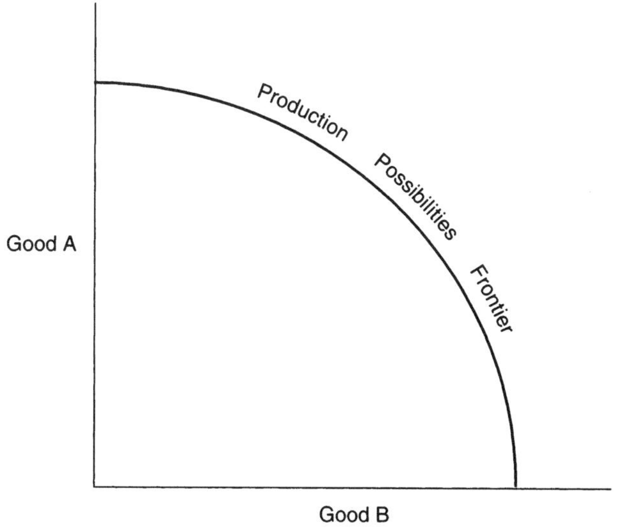

FIGURE 4-7 Production Possibilities Curve

Although there are no societies in the world today whose productive expertise is limited to the production of either bicycles or canned soup, the concept of economic trade-off and the production possibility curve is still quite a valid one and, as we see shortly, one that is intimately involved with the problems and benefits of technology in the society.

Capital Formation

Smith stresses the importance of capital formation throughout his exposition of capitalism. Indeed, the name *capitalism* itself stems from his assertion that investment in the means of production is essential for a successful economy. He states that unlike the assertions of the mercantilists who viewed economic power as resulting from amassed physical wealth, true economic wealth stems from the ability of the society to produce goods and services. It is this ability that brings economic prosperity, growth, wealth, an increase in economic welfare, and all of the other positives of a capitalistic system. Investment, he says, is the key. Without an investment in the ability of the society to produce, machinery and equipment would soon wear out, and there would be no replacements; production would thus stop. The society would stagnate and slowly begin to drift backward toward less productive times. Efficiency in the use of resources and the availability of productive ability keeps the economy going. Investment is essential.

To a large degree, all of these characteristics and principles are as true today as they were in 1776, yet we find some major differences in classical capitalistic theory and modern capitalistic approaches. One obvious example is the tremendous contributions to the economic system made by government at all levels, especially at the national level. Smith stated that the only legitimate role of government in economic affairs was to protect property rights and that its role should be limited to just that. We do not seem to have followed this precept too closely.

Why? What is different now? Why do we find the traditional capitalistic theories so inadequate to explain modern economic phenomena? Primarily because of an occurrence that took place after Smith's theory was formulated, an occurrence that totally changed the complexion of economic structures. What changed everything so dramatically was the Industrial Revolution.

CAPITALISM AND TECHNOLOGY: THE MODERN APPROACH

In the industrialized societies of the twentieth century, the classical approach to economics presented by Adam Smith is no longer adequate to explain what takes place in the market. We no longer live in the simplistic agrarian society that surrounded Smith in eighteenth-century Scotland; all that was put to rest by the advent of the Industrial Revolution and the rise of industrial states. The level of industrialization necessary to create and maintain an industrialized nation is far beyond anything that Adam Smith had envisioned. Progress is more rapid; the range and availability of goods is many times greater; and the productivity of the worker due to improvements in technology, methodology, and communications has exceeded that of Smith's time so completely that his approach is simply inadequate.

This is not to say that the Smithian approach to economics was incorrect or that it is now useless. Because of the industrialization of a considerable portion of the planet, circumstances have changed so much that his theory no longer reflects actual conditions.

The effect of the Industrial Revolution and its consequent explosion of technological change is similar to the effect of quantum theory on traditional physics. Sir Isaac Newton successfully predicted the results of a range of combined events through his classical physical laws. His laws of gravitation, action and reaction, and so forth were sufficient to explain observed phenomena in his time. However, as the phenomena studied by physicists changed, and specifically as the phenomena increasingly involved particles traveling close to the speed of light, the results of the observations diverged from what the classical laws said should be happening. It was necessary to wait until the beginning of the present century for Einstein, Bohr, and others to formulate new laws before the observations could be fully understood. The Newtonian approach to physics was not wrong; it was simply inadequate to explain phenomena occurring at relativistic speeds.

Theory was not able to closely approximate observation again until the advent of major breakthroughs in economic theory, such as the Keynesian approach, developed by John Maynard Keynes. In this century alone, economic thought has shifted from a patchwork of classical theory to the Keynesian approach, then to the emergence of the Monetarist school, and finally to what is known as the "modern economic approach." The adherents to the latter approach do not seek to decide whether Keynesians, Monetarists, or other theorists are correct in their assumptions, choosing instead to concentrate on a more pragmatic approach and use whichever theory works at a given point in time.

Technology and Efficiency

In the classical approach, it was indicated that individual manufacturers compete for the dollars of the consumer by producing as good a product as they possibly can. This is still the case. However, there are now other forms of market systems in operation in our society besides the traditional highly competitive free market. There are *oligopolies,* markets made up of large manufacturers of goods, each of whom is able to command a significant percentage of the market. Industrialization allows such a market to exist.

Why would a market exist with a limited number of producers? What happened to the large number of small producers, each with a small part of the market? They have given way in many areas of endeavor because of the insatiable desire on the part of the economic system as a whole to operate more efficiently. *Efficiency* is the ability to operate with a minimum of waste, effort, and expenditure of resources. In industry, that often translates into bigness.

Oligopolies have come about because of the fact that as industrialization takes place, efficiency is created through large-scale operations. Economies of scale dictate that to produce a good product at a low price, it is necessary to make many of them, at least until the process becomes too large and cumbersome and the benefits received from bigness begin to decline. Since, as previously noted, industrialization seeks bigness in order to achieve efficiency, if a market is limited by the size of the consumer sector of the economy, that is, the number of consumers and the number of units each wishes to consume, the resulting production profile shows that only a few large companies are needed to satisfy all demand for a product. The market still operates competitively, with each company doing its utmost to achieve maximum production at minimum cost to create the best product it can and sell it to the largest number of people at the lowest price. Yet this is now achieved through a small number of relatively large companies in whom is concentrated the lion's share of industrial capacity and not in a large number of small companies operating without benefits of economies of scale. We call this phenomenon *industrial concentration,* and the greater the degree of concentration, the more highly oligopolized that particular industry is said to be.

At the retail end of the economic process, we have seen in recent years another burgeoning oligopolistic system in the consolidation of department store chains through buyouts and takeovers, and through the rise of such "hypermart" chains as Wal-Mart, where the value of streamlined central control functions and the immense savings benefits from mass purchasing, distribution, and marketing have created an evolutionary pattern similar to that of heavy industry earlier in the history of the U.S. economy. Thus we have a small number of automobile companies, a small number of steel companies, a limited number of major computer manufacturers, a few large electronics retailers, and so forth.

This does not rule out large markets. In some cases, bigness can lead to complacency. When that happens, inefficiency is created. The economic system, being a self-correcting system seeking balance, has an answer for this. When the imbalance becomes great enough, new producers are encouraged to enter the market, albeit an expensive process, and the inefficient producers are forced to return to a competitive status or lose business. This was classically illustrated by the influx of foreign automobiles into the United States from foreign producers who were able to produce at a lower cost and compete favorably for consumer dollars, in spite of protective tariffs and stiff resistance from domestic producers. The American auto industry found itself in a position of having to increase efficiency to compete.

How does a company increase efficiency other than through economies of scale? Again, it does so through the use of technology. Technology is what allows bigness in the first place, but that is not enough. Industries are constantly seeking new methods and better ways of handling production problems, and they are constantly seeking to develop new products for the consumer. Technology is a means of doing both. Money spent on the research and development of products and on improving production methodology is as much a part of the attempt to compete as it was in Smith's time. Only the highest quality and most useful products at the lowest possible prices are able to survive in the market for any length of time. And that means improvements in technology.

Economic systems, particularly capitalistic economic systems, are reciprocal in nature. They offer a return based on value. That is, people are recompensed for the value they contribute to the economy. How much are people paid for their services? It is generally dependent on how useful their work or service happens to be. The more valuable the input into the productive process, the more recompense they are able to command in the marketplace. Their value is, of course, in proportion to the availability of workers doing the same type of work. Scarcity alone does not produce value. Buggy whip lacers, for example, are extremely rare in our society, but that is not a guarantee of a high price for their services, as the service being offered is not particularly valuable to the society in itself. In contrast, an electrical engineer in a world of electrical gizmos, gadgets, and goodies is a valuable resource, and if there is a limited supply of these skilled professionals, the value of each one is relatively high. What is valuable commands value in return for the right to use it.

In addition to scarcity, productivity creates a high price for the services of workers. The more productive one is, the more valuable one becomes. Likewise, productivity is created through technology. In an industrialized society, which is dependent on the efficient use of knowledge, machinery, equipment, and so forth, it is through technology that we maximize the value received from the production process. The only reason for technology, as indicated in Chapter 1, is to create artifacts based on natural law for the purpose of doing something more efficiently and with less effort and expense. Labor becomes more productive through the addition of technology to the productive process, and the wielding of that technology in its efforts to create goods and services.

If an economy is to produce goods and services efficiently, it must have the technology to do so. Technology becomes a limiting factor on the capacity of a society to create economic welfare for its members. This brings us to technology and the production possibility frontier.

Technology and the Production Possibility Frontier

Societies make many choices, and the economic trade-off of each contributes to the nature of the society. As an example, the political form associated with a particular economic construct is dependent in part on the choice that a nation makes between creating and distributing goods through the private sector (industry) and creating and distributing them through the public sector (government).

Any good may be either a public good (provided through the governmental structure) or a private good (provided by private industry and made available to the public through the market system). There is nothing inherently "public" or "private" about any given good or service. This is strictly a matter of choice. The deciding factor is whether the society feels that a particular good can be more efficiently (more economically) provided through one sector of the economy or another. Private armies, for instance, are not unheard of, but they are not nearly as efficient and economical on a large scale as a public defense system organized and paid for by the society through the public sector. Housing is generally created and supplied by industry in the private market, but when it is necessary to build houses for the good of the society and the private industrial sector is unable or unwilling to do so, the government (public sector) is not averse to stepping in and creating that housing through the process of eminent domain and law. This is simply a *choice* that is made by the society.

Another choice that the society makes is whether to invest productive capacity in the ability to produce (building capital inputs such as machinery and equipment, new knowledge through research, and so forth) or to create consumer goods (e.g., shoes, ships, sealing wax, and electronic computers). This choice determines, among other things, the rate at which an economy grows and the economic welfare of a society over time.

Even as Smith said, it is necessary to invest in the means of production. A society must set aside part of its productive capacity for use in producing capital if it wishes to have the means necessary to supply that society with goods and services. The choice is how much to set aside.

Investment in technology is part of that decision. Technology dictates the total productive capacity of the society along with other factors, such as the availability of resources. How technology is handled affects economic growth. A society can choose to invest any portion of its wealth in capital, and the more it invests in capital, the greater the future capacity to produce all combinations of goods and services. It is through this investment that the production possibility curve can be extended. If productive capacity is increased to more than compensate for depreciation and depletion, then the result is *more* capacity to produce than before. Therefore the production possibility frontier moves outward. Since the area beyond the production possibility frontier is defined as an area of impossibility, by extending the curve, we decrease what we cannot do and increase what we can. Economic welfare rises with an increase in the availability of goods and services, and economic growth is experienced.

Determinants of Economic Growth

Economic growth takes place in accordance with a mix of determining factors. As these factors change, so does the ability of an economy to grow. The major accepted determinants of growth include the following:

1. *Population:* This factor includes the size and nature of the population, as well as demographics such as age, sex, health, location, and total numbers.
2. *State of the arts:* This determinant of growth includes the type of technology available (state of the arts), the availability of that technology, and the applicability of it to the production and distribution of goods and services.
3. *Growth of knowledge:* The degree to which knowledge increases in the society among individuals and the highest level of knowledge obtained by the society as a whole, how generally knowledge is dispersed through education, how specific it is to content (the nature of knowledge), and how easily exchanged and communicated that knowledge is, are important factors in what constitutes the intellectual base of a society and the rate at which new knowledge can be assimilated and put to use.
4. *Available resources:* This fourth major determinant of growth includes the type and quantity of resources available; the rate at which resources are used; the rate at which various resources, if any, are replaced; whether replaceable resources are an important factor in the economy of the society; and the expectation for future increases in resource availability.
5. *Rate of capitalization:* The fifth major determinant of growth is the rate at which the society is willing to capitalize or invest in productive capacity, within the limits of the other factors mentioned here. The capitalization rate is as much a measure of attitude toward economic progress and growth by the members of the economic community as it is a measure of the rate of change in productive capacity itself.

Three of these five determinants of economic growth—state of the arts, growth of knowledge, and the rate of capitalization—are directly related to technology itself. The state of the arts is certainly a matter of technological development in the society. The rate of capitalization determines how much of society's resources are going to be devoted to improving and using technology for the benefit of the society. Technology as artifacts *is* the investment of the society in capital. The application of what is known to create efficiency in the productive process and to produce an increased availability of goods is central to economic welfare.

The third factor, the growth of knowledge, is generally viewed as the most significant source of long-term economic growth. It is through increasing our understanding of the physical world and then educating the population in that knowledge that we are best able to deal with and manipulate the physical world to society's benefit. It is this that makes education such a critical factor in the future of the country, and it is this that creates the opportunity for a wide range of people to experience the benefits of improvements in technology. Technology results from manipulating nature, an ability that stems from our understanding of its behavior and one that results in an increase in knowledge.

Unlike resources, the quality and quantity of technology and knowledge can be increased in a society. As resources decline in abundance and quality, innovative use of the remaining resources increases the importance of technology as a means of maintaining or improving the state of the economy. Efficiency in the face of diminishing natural resources and a growing population that demands more and more goods can only take place through ever-increasing amounts of capital, knowledge, and technological application of that knowledge.

Population, too, has a link to technology. The ability to gain new knowledge and manipulate that knowledge to the benefit of society through the creative process to form technological artifacts is a function of the availability of intelligence in the society itself. Assuming that it requires a certain level of intelligence to manipulate new information effectively, the size of the population can be critical.

Genius, as an example, is measured in percentages, reflecting the top 2 or 3 percent of the population in the ability to solve problems. Assuming that this is a valid measure, the number of geniuses available in a given society is proportional to the size of the population. A population of 1 million people would be expected to yield an average of 20,000 people who qualify as geniuses (2 percent). If the population reaches 10 million, that number jumps to 200,000 citizens with genius capabilities. The larger the population, it can be argued, the larger the pool of genius intelligence on which the society can draw. We must also note, however, that the larger the population, the greater the number and extent of societal problems to be handled.

As for natural resources, there are only two possibilities available to us for solving the problems of limited, rapidly depleting supplies of raw materials. In the face of these decreasing resources, we must either increase the efficiency with which we use what we have, thus increasing the amount and quality of goods and services yielded from a given amount of input, or we must search out new supplies to support our present level of consumption and increased future desire for consumption. In either case, technology is necessary to accomplish the goal.

Efficiency, as already noted, requires improved methodology. In the case of finding new supplies of raw materials, consider the parallels between the discovery of the New World and the exploration of near space. When the New World was discovered in the fifteenth century, the European countries increased their supplies of raw materials dramatically. In a time when the denuding of the huge European forests had all but destroyed available supplies of wood, the New World was found to have abundant supplies readily at hand. A single huge forest stretched from Canada to Florida and from the East Coast to the far side of the Mississippi River and beyond. Gold and silver, gems, and iron and coal deposits were all available in huge quantities.

Mercantilism, which dictated that economic power stemmed from physical wealth, flourished with the availability of a virgin world to take that wealth from, during what has come to be known as the "rape of the New World." Later, however, the capitalistic possibilities were realized fully, and colonies developed to take advantage of these increases in raw materials. It was no accident that most of the British navy during the heyday of British colonization was built from American naval stores. Eventually, the process led to the emergence of the world's greatest industrial powers.

Similarly, we exist in an era of decreasing resources. Yet just beyond our reach, in a sea of velvet void, there float incalculable supplies of nickel, iron, and other minerals just waiting to be scooped up and processed by some enterprising group of people. The only factor that has kept this process from occurring to date is the expense involved in comparison with the profits that would be incurred from the enterprise. When costs drop low enough or demand for those resources rises high enough, a profit potential will result, and just as any economically rational human being would be expected to do, entrepreneurs will venture forth in search of greater economic welfare for themselves and their people. The potential exists for such an enterprise in the near future, and the results could be the birth of economies undreamed of before this time.

Technology and Economic Welfare

Economic welfare refers to the quality of life that exists in a society. This is not necessarily synonymous with economic output, nor is it dependent on economic growth per se.

Suppose you were living in a small, agricultural nation with a low level of technology and a large but relatively poor population. Suppose also that the basis of dietary staple in your country is a hypothetical tuber called "blivet." Nearly every meal is centered around some form of blivet, which is chiefly ground and cooked to form a brown porridge with a flavor similar to cardboard, and which almost no one likes to eat. Now suppose that improvements in agriculture are introduced into the country that result in a massive net increase in agricultural output, due to trade agreements, some fortuitous investment from outside the country, and an extremely enlightened government. The increase in output represents growth in the economy with a rise in the gross national product (GNP) from a level of 100 million cabas (the caba is your monetary unit) to a level of 200 million cabas. What a wonderful turn of events! In just one year, your economy has doubled! The country should rejoice in its good fortune!

Or should it? Before the rise, blivet represented 60 percent of the GNP of your country. Every family in the country could afford three bowls of blivet porridge per day, a very adequate diet. Now, with a doubling of production, each family finds that it can have *six* bowls of blivet per day! Is this progress? Are six bowls per day of sticky brown porridge with the consistency of cardboard an improvement over three? This is a highly subjective question.

It appears that the economic welfare of a country may be equally dependent on the *quality* of products consumed as it is on the *quantity*. It is the quality of life we are establishing, not just the quantity of goods.

In this regard, technology contributes not only to the efficiency with which a nation is able to produce but also to the variety of goods that are available. Consumers have more to choose from in their efforts to satisfy consumption desires if the level of technology allows for a wider selection of consumption items. And through this process, technology contributes heavily to the economic welfare (quality of life) of the members of the society.

Technology and Negative Externalities

It is inequitable to view technology in society only from the perspective of economic benefits. A balanced treatment of the subject must also note ways in which the economic structure suffers as a result of technology. The side effects of technological industrialization are not always pleasant. Industry's use of technology, if done without regard for the total fabric of the world system within which it exists, can lead to the polluting of rivers and streams; the saturation of the atmosphere with deadly substances; the destruction of delicately balanced ecosystems; and the decrease in living conditions through crowding, tension, pressure, and a host of physical and mental disorders in the people that technology is designed to help. Collectively, these negative side effects are referred to as *negative externalities,* and even with the homeostatic tendencies of the species, it may not always be possible to resist economic progress enough to catch these deleterious effects. In truth, it may be a price we think we are willing to pay.

One negative externality is the cost that must be absorbed by some third party outside the exchange process of the consumers and producers, as a result of the exchange process taking place in the market. Yet it should be understood that in the case of technology, it is the *application* of technology and not the concept of technology itself that causes the problem. It is the way in which we *handle* technology that creates difficulties.

For example, a farm products company locates on the banks of a stream in a rural setting, close to several small towns with a readily available supply of labor and in close proximity to the farming communities to whom it wishes to sell. Being a good businessperson, the owner of the factory endeavors to produce a high-quality product at a reasonable cost and, as part of that effort, chooses to dump low-level toxic byproducts into the local stream. The alternative waste disposal process would be so expensive that the price to customers would dramatically increase. In good faith, the company is seeking to keep costs down and to produce a useful product, which it then sells to wholesalers and farmers in the surrounding area. The farmers benefit from a farm products plant being so close at hand, the wholesalers benefit from the easy availability of goods to sell, and the manufacturer benefits from lower production and distribution costs. The consumer of the vegetables and other produce benefits from better quality food at a reasonable cost. Everyone seems to win.

Or do they? What about the waste flowing into the local stream? If the stream can handle the increase in toxic materials, everything is fine. However, if it cannot, there are negative externalities. Suddenly the manufacturer is no longer seen as a firm being maximally efficient by minimizing costs but rather as a firm that is passing costs on to a third party. Downstream, the farmers may begin to experience problems with their livestock that drink from the stream, having to deal with low milk production and even illness and death of productive animals. Towns farther downstream may experience increases in purification expenses in their efforts to draw drinking water from the water source, purify it, and distribute it to community residences. Plant life may suffer from the chemicals in irrigation water, resulting in erosion, loss of productivity in fields (requiring, by the way, a greater amount of fertilizers and other farm products to maintain yields), or even serious flooding in the area due to heavy losses of plant life. All of these negative consequences have to be paid for by someone. The cost of production did not go away, it was merely *transferred* to the community as a whole.

Does the farm products company really get away with anything? Are its customers really free to purchase the farm products at an artificially low price? Economic systems are reciprocal in matters of externalities as in other matters. The consumer and the manufacturer pay for the disposal of the toxic wastes, but in deceptively different ways.

Taxes in the community pay for water purification. If the farm products company is located in the community, it shares in that tax burden. Its customers also share in it, which reduces the amount of money they have available to pay for farm products, thus reducing the company's sales. Farmers suffer losses in crops and livestock, further reducing their ability to buy from the manufacturer. Loss of prestige and esteem may plague the firm if the negative externality becomes known. The general quality of life in the community may decline. Lawsuits may result. The employees and principals of the company further suffer from higher taxes and a lower quality of life. In many small ways, the cost of the dumping of waste is paid for, and paid for by the firm creating the problem. As with the other factors in the economy, technology creates a trade-off through negative externalities, exacting an opportunity cost from the society.

HUMAN OBJECTIFICATION: THE PRICE OF CAPITALISTIC INDUSTRIALISM

As with anything else, capitalism has its downside. From a purely sociological point of view, that downside is personified by the tendency of companies—and indeed the whole society—to objectify an individual to the machinations of the economic process. In a capitalistic system, the gathering of productive resources, including labor, to create an efficient and productive commercial enterprise has a tendency to create a view of those productive resources as nothing more than elements in a process. For natural resources and capital (raw materials, machinery, and equipment) that works well, but labor involves human beings. There is a propensity merely to view them as "cogs in the wheel," which objectifies the individuals involved in the process. This view did not take long to develop after the British Industrial Revolution, but became codified in the works of Samuel Taylor, the father of scientific management, and the Gilbreths, the original efficiency experts. By viewing subject workers in terms of performance alone, they helped to remove the humanistic aspects of the labor force from consideration.

A firm that refers to workers as "productive units" illustrates the point well. There is an entirely different understanding of what one is dealing with when one refers to "productive units" rather than workers. The former is far less personal than the latter. Rearranging or eliminating productive units is different than transferring or firing people. A firm that refers to customers as "profit units"objectifies their customer interaction in order to ensure the company's livelihood.

Objectification goes with the machine-model approach to industry, and even in our post-industrial information age economy, this tendency prevails, as we have done little more than change the type of technology with which we are involved. The fact that we are no longer relying on large factories and mass production of finished goods to fuel the economy does not mean that the technological process is bypassed. On the contrary, we are more involved in technology and the transfer of technology to the general population than ever. With capitalism we undoubtedly find ourselves involved in big business, which means large scale thinking, and which reduces the tendency to think of the individual. In spite of the ability of all these technological gadgets, goodies, and gizmos to increase our individual mobility and freedom, the very concept of mass production and distribution breeds a reduction in the consideration of the individual over the whole.

If we mass produce, we seek to design for the masses, not for the individual. If we serve a large public, then the problems and needs of any individual customer tend to diminish in importance as long as the statistics show that we have provided the most goods for the largest number of people. Thus, we rely on the input of statistical data, gained increasingly through true data processing and delving into the desires of a consumer *group* rather than the individual consumer for market decisions, and on efficiency studies of production rates and cost controls to make decisions on plant size and work force profile.

There was a time when workers could go to work for a major corporation and establish a career that would last until their retirement, if they continued to perform at an acceptable level. There was tremendous loyalty among employees to their companies. There was high resistance to changing jobs mid-career, but this is no longer the case. It is not the result of objectification alone; the objectification process has been in place since the Industrial Revolution. It is equally the result of the increase in the rate of change in our society and in our technology, which requires people and firms to reinvent themselves on an ongoing basis. The casualty of all this is again the individual. Employees no longer feel strong ties to the firms for which they work. Their job security is not in the firm but in their skill set, and they are as likely to quit and move to another job as the firm is to "downsize" them out of a position. Effectively, what has occurred is that employees have taken their cue from employers and have objectified their employers in response to their own changing needs. The personal relationships that once existed between firm and worker has been blurred at best, and destroyed at worst. The processes and causes of this phenomenon are well known, but the indirect costs are not often noted. Those costs include less consideration for individual circumstances among workers, less security, a greater tendency toward 'looking out for number one,' a reduction in loyalty to any group—social or professional, increased anxiety, decreased security, and a general reduction in cooperation beyond what is absolutely necessary to get the job done. One needs only to think about the level of personal loyalty between themselves and the company for which they work to understand this shift.

Social critics of the capitalistic system often point out this objectification as a primary difficulty of using this economic scheme, yet the increased production and efficiency in the process often overrides the objections. The truth of the matter lies somewhere in between. A balance exists between the personalization of small enterprise and the objectification of large firms in the capitalistic system. What everyone must remember is that whereas the mass production of industrialization objectifies employees, the freedom of the capitalistic system allows for anyone who has the desire, the drive, and the ingenuity to do so, to independently become successful. The beauty of the capitalistic system is that one person having a large slice of the pie does not prohibit someone else from having as large a slice. That is, the wealth of one individual does not eliminate the possibility of wealth for anyone else. As economic growth takes place through free market systems, *the pie just gets bigger!*

Our culture causes people to forget this, and look for their security, rather than take a risk to embrace opportunities. The questions become, "How much of the objectification exists because of the nature of the process itself? How much is the result of cultural belief systems that lock individuals into the view?" If one does not wish to be a cogwheel and realizes it is not necessary to be one, they then have the freedom to individualize their own success. Just as technology has led to mass production, it allows individuals to seek their own route to economic well-being. It allows them to compete in the marketplace with much larger and much wealthier players, and succeed.

The error in thinking that people often make about the process of objectification is to assume that the only goals of a culture are economic. To an economist, all decisions are based on the capacity to produce and distribute goods, and in the process to maximize profits. This alone objectifies the individual. There is no room for individuality in the equation. It is all *process.*

From the point of view of a sociologist, the purpose of economics is to increase the economic welfare of the people. It is a process that relegates economics to a supportive role in the achievement of cultural goals. Both viewpoints are incorrect, and paradoxically, both viewpoints are correct. That is, we are faced with an apparently dialectic argument in which a synthesis of the two opposing positions presents us with an answer. This is not an either–or situation. As for technology, it is neutral. It can be used for both the freedom of the individual and the development of strong economic structures. Which is important depends on who is viewing the process. In truth, both are important, and successful capitalistic economies will always take the individual into account; including that individual's needs and how to fulfill them, both as consumers and as producers. Any successful firm should understand the necessity of treating both employees and customers as individuals and that the way to satisfy their own economic goals is to satisfy the goals of those individuals.

CONCLUSION

Technology affects the economic structure of a culture through a number of relationships that exist among social content, cultural characteristics, and the nature of the technology itself. Through its capacity to increase efficiency and raise the level of output of a culture, technology creates the opportunity for the society to grow and develop. Innovation is an essential element in the creation (production) and distribution of goods and services to the population so that they can be consumed and create satisfaction. The very concept of economic welfare, that is, how well off the population is in terms of the quality of the life that it leads, is linked with the capacity of the economic structure to supply these goods and services and to continue to do so at ever-increasing levels of sophistication and variety. Technological innovation depends on the scientific capabilities of the society itself; the quality and the extensive nature of the society's educational system; and the percentage of the total productive wealth that is invested in the development of research and new methodology, new processes, and new products and services.

On the negative side, for every technological innovation and for every advance in the economic utilization to which a society puts its natural resources, labor force, and capital base, there is an inherent cost that the society must pay. Because of the natural law of reciprocity, a culture cannot avoid paying this price. If the effects are not handled logically and with forethought by the society, the payment may be higher than necessary and may actually reduce the level of economic welfare through destruction of the environment, overuse of limited resources, or other unforeseen results of technological activity. For this reason, technology and its role in economic structure should be taken into account in its development and utilization *before* negative externalities occur beyond an acceptable level and beyond the capacity of the society to absorb them.

So we see that technology is supported by and supports an economic system that distributes goods and services to the population for consumption. Yet why are some technological goods developed and produced rather than others? Why is it that some ideas were developed long ago and then left dormant for hundreds of years before they really began to make a major impact on the lives and societies that eventually used them? Why is it that if we can do something, we do not, and then we suddenly decide at a later date that we will? Is it merely a matter of all of the elements having to come together at the right time, or is there some more fundamental process at work? In other words, why do inventions, processes, and technological breakthroughs happen when they do, often at what seems like just the right moment?

In the next chapter, we look at why inventiveness happens as it does and how we seem "magically" to know when an idea's time has come.

KEY TERMS

Capitalism
Competition
Economic Trade-Off
Economic Welfare
Growth of Knowledge
Law of Demand
Law of Supply
Market Equilibrium
Negative Externalities
Production Possibility Curve
Self-Interest
State of the Arts

REVIEW QUESTIONS

1. What is the relationship between capitalism and technology?
2. What are the five determinants of economic growth?
3. What is the production possibility frontier, and how is it related to technology?
4. Define negative externalities and its relationship to technology in industry.
5. How does technology help to create the economic system in which we live?

ESSAY QUESTIONS

1. Chapter 4 emphasized the importance of economic considerations in the development of technology. Are there powerful motivators,

other than economic ones, that lead to major technological breakthroughs? If so, what are they? If not, why not?

2. According to the text, growth of knowledge is the single-most important factor contributing to economic growth. Do you find this to be true? What other factors, included or not included in the text, do you find to be equally important?
3. The presentation on economics tends to view growth of knowledge as a positive force to the economic and social betterment of the culture, yet there are equally compelling observations suggesting that it serves to objectify the individual and classify him or her as merely a cogwheel in the machine of progress. Discuss these two views and develop a synthesis that engenders both points of view in order to explain our actual experience in life.
4. How is the cause of individuality and freedom at odds with the concept of cooperative economic effort through the use of technology and division of labor? How is it aligned with cooperative economic effort?

THOUGHT AND PROCESS

1. What are some of the ways in which technology has changed the work experience for your family in the past ten years? Consider both positive and negative consequences of technological change as they have affected you or other working members of your family.

2. Make a list of technological changes that are presently under scrutiny for their detrimental effects on the environment, on the safety of the members of society, or on the economic well-being of one or more groups of workers in our society. Include some less obvious examples affecting only small parts of the population at present.

3. Assume that through innovative technological change, the costs of building houses were to be reduced by 50 percent in the next six months. Speculate about some of the probable economic changes that would take place as a result of the new technology.

4. What changes in the economy are likely to be the result of a general use of industrial robots? What changes have already come about?

5. If a single input into the economic process (natural resources, labor, capital, technology, entrepreneurship) had to be limited to its present level, which would have the most devastating effect on the development of society? Why?

The decline and fall of the Roman Empire is a subject that has been studied and debated for hundreds of years. What is important about the Empire in terms of our present study is the high degree of technology and cultural organization achieved by the Romans and how, through careful planning, decentralized control, and the effective use of mechanical technology, they were able to achieve a cultural sophistication unparalleled in the ancient world and, with the exception of China, unequaled until the Industrial Revolution in Europe.

We tend to think of the ancient world as rather provincial and backward, with limited technological development and little organizational structure. However, as you read the following report, you will find that that assessment is untrue. Notice the relationship between the technology of the era and the social structure of the culture. Note how each supports the existence of the other and how each created the other in the form it maintained.

The Historical Perspective of Roman Engineering and City Planning

CITY PLANNING

The *Encarta 98 Multimedia Encyclopedia* defines city planning as the unified development of cities and their environs. It is further explained that for most of its history, city planning primarily focuses on the regulation of land use and the physical arrangement of city structures, as guided by architectural, engineering, and land-development criteria. Greek and Roman eras are often associated with unconventional, highly evolved examples of city planning. Were the Romans masters of city planning, architecture, engineering, and government from their humble beginnings?

Most of our research will concentrate on the Romans' well-developed cities and not the experimentation and appropriation of the many technologies used to reach this level. We will cover three distinct aspects of Roman engineering and city planning. These aspects will be based on the period of development for Roman technology. The first will be considered a precursor to city planning, followed by the steps of implementing the planning technologies, and ending with the technologies that have been utilized by future generations. We will be as factual as our research allows, but in the words of French Emperor from 1804–1814, Napoleon Bonaparte, "What is history but a fable agreed upon?"[1]

[1]Libraryspot Quotations, http://www.libraryspot.com/quotations.htm.

HISTORY OF ROME

According to Roman mythology, the history of Rome began in 753 BC when a basket, containing twin infants Romulus and Remus, floated down the Tiber River and washed ashore at a place near seven hills. Their uncle, Amulius, had set them adrift to prevent them from interfering with his aspirations of becoming king. Their mother was the daughter of a local king, and their father was the god of war, Mars. A she-wolf nursed the infants until a shepherd and his wife adopted them. When the twins reached maturity they killed Amulius and established a city on the seven hills. During a petty quarrel Romulus killed Remus and became the first king of Rome.[2]

The myth depicts Rome's humble beginnings. Farmers capitalizing on the water supply of the Tiber River inhabited Rome. Rome was definitely not the empire it would be in the future. The farmers, Latins from the Latium plains by the Tiber River, were struggling to survive amongst the other settlers. Over the years the Greeks, Etruscans, Samnites, Sabines, and Umbrians conquered Rome and subsequently Rome conquered all the other settlers. Throughout these warring times, the Romans began to collect artifacts from the other societies. Although Rome lent its name to the civilization, the city was only a small piece of the Roman Empire.[3]

According to the Latin calendar found at Anzio, the "Parilia-Roma Condita" or Festival of Pales-Foundation of Rome, is the 21st of April. Although many different years have been cited for the birth of Rome, 753 BC has been adopted. It has been difficult to deduce the vast, rich history of early Rome, as well as conflicting views of lore throughout the next twelve centuries, since there have been very few definitive historical works. Livy's *The Early History of Rome* reads more like an embellished account or fictional tale, but it is often cited as an explicit depiction of early Rome.[4]

Following the humble beginnings Rome had become a wealthy and flourishing city, by the standards of 600 BC under a foreign monarchy. Around 500 BC the monarchy was overthrown by the citizens and formed into a republic. Over the next few hundred years, the continual wars left Rome as the only great power in the Mediterranean world. The immensa Romanae pacis maiestas, boundless majesty of the Roman peace, consisted of an area about two-thirds the size of the continental United States with about half of its population. This vast empire assimilated and diffused the cultural heritage of Greek, Oriental, Semitic, and Western European people. The extensive economic, political, and social trepidation that accompanies such an expansion led to the eventual downfall of the Roman Empire. By 476 AD the western part of the empire had been overrun by barbarians, later evolving into the nations of modern Europe. The Eastern Empire remained the most powerful and civilized state in the world until it was conquered by the Turks in 1453 AD.[5]

[2]*Encarta 98 Multimedia Encyclopedia,* 1998 ed., s.v. "Rome." CD-ROM.

[3]Susan Smith, *Ancient Rome* (London: Dorling Kindersley, 1995), pp. 16–19.

[4]Donald R. Dudley, *The Civilization of Rome* (New York: Meridian, 1993), pp. 9–11.

[5]Dudley, *The Civilization of Rome*, p. 9.

CHAOS OR DETERMINATE RANDOMNESS

Chaos, in some mythologies, is the formless and disordered condition of matter from which the universe evolved.[6] Chaos can be defined as "a state of utter disorder without structure of any kind, the exact opposite of systemic structure".[7] The hypothesis involving chaos theory is further defined by determinate randomness. Determinate randomness can be described as "any phenomenon that has multiple outcomes dependent on random events which can be determined mathematically, given a set of actual events".[8] Early Roman city planning was more of a random collection of artifacts pieced together as a perceived need arose. Early Rome was not an example of the distinct city planning expertise the Romans would soon display. But development had to start somewhere. The I Ching states this concept best, "Before the beginning of great brilliance, there must be chaos. Before a brilliant person begins something great, they must look foolish in a crowd".[9]

EARLY DISORDER

In the beginning, the city of Rome was comprised of a muddle of huts in a complicated maze of alleys. The city grew without any organization or overall planning. The people of early Rome were actually thankful for the narrow, winding alleys since they kept the "unhealthy winds" of sewage and rotting food from traveling throughout the city. Most of the streets of Old Rome were less than sixteen feet wide. The alleys were flanked by insulae, modern-looking brick apartment houses that filled an entire block, which made them dark and gloomy. Ancient cities did not have zoning laws. As a result all types of buildings were mixed together with no semblance of order. There was a conglomeration of hovels, insulae, mansions, taverns, temples, warehouses, and workshops. Julius Caesar had to order wheeled vehicles to only travel the city streets at night, with certain exceptions, to help alleviate the dense traffic jams.[10]

The technology at this time was of an experimental nature. Many different concepts were tried, cognizant or not, on city development. The Romans had an intense desire to enhance their chances of survival against the other tribes, or nations, of people attempting to overrun Rome. This survival trait spurred a profound need to exercise their adaptation skills by use of artifacts.[11] Max Frisch, a German author, looks at this adaptation through the use of technology in a more negative aspect. He states, "Technology is a way of organizing the universe so that man doesn't have to experience it".[12] We view technology as a means to survival of the fittest, through any methods necessary to achieve this.

[6]*Webster's Dictionary of the English Language* (New York: Modern Promotions/Publishers), p. 50.

[7]Paul A. Alcorn, *Social Issues in Technology: A Format for Investigation*, second edition (New Jersey: Prentice Hall, 1997), p. 163.

[8]Alcorn, *Social Issues in Technology*, p. 165.

[9]Libraryspot Quotations, http://www.libraryspot.com/quotations.htm.

[10]L. Sprague De Camp, *The Ancient Engineers* (New York: Ballantine Books, 1963), pp. 188–189.

[11]Alcorn, *Social Issues in Technology*, pp. 6, 8, 13.

[12]Libraryspot Quotations, http://www.libraryspot.com/quotations.htm.

Their first method of survival was to fortify their towns against their enemies. The city walls were instrumental in the protection of the city. To be effective the walls had to be at least 30 feet high and 15 feet thick. The wall was actually comprised of two walls both constructed out of stone. The inner wall had to have 30 yards of clearance for the movement of troops while the outer wall needed a clearance of 200 yards to monitor the outlying area. The outer wall was built to extend 30 feet below ground level to ensure no one could tunnel underneath. The top of the outer wall contained crenellations, which are alternating high and low sections, so the soldiers could launch their weapons over the low sections and were protected behind the high sections. Between the two walls was a layer of rock and dirt. The inner wall was higher than the outer wall to thwart rocks and arrows from making it into the city during an attack.[13]

ROMAN ENGINEERING

Roman engineering was primarily intensive utilization of simple principles, with an abundance of inexpensive labor, raw materials, and time. Prisoners-of-war, slaves, and soldiers required to assist in construction provided the cheap labor. Rome had access to a copious supply of brick, stone, and timber. Roman engineering consisted mainly of civil engineering. This included the building of roads, bridges, public buildings, and other permanent structures. Modern writers often note that Romans contributed practically nothing to pure science. "To the scientific habit of mind which has made our present attempt at civilization possible and is rapidly making it impossible, no Roman ever contributed anything."[14]

The Romans spent an inordinate amount of resources on the development of public works. They built roads, harbors, aqueducts, temples, forums, town halls, arenas, baths, and sewers. Magnates, governors, and even the emperors believed they gained veneration and eminence by presenting the common people with some useful or entertaining public project. Many of these works were destined to become more revered and ornate than the last public undertaking. Some public bathhouses were so extravagant in their decorations they were mistaken for temples or mansions.[15]

MAY I BORROW A CUP OF TECHNOLOGY?

The Romans rarely developed unique engineering processes and ideas. They were masterful in the art of borrowing aspects of design that worked well with other cultures and improved on them. Their ideas came from the Etruscans, Greeks, and Asians who came to Rome voluntarily or as captives. The early Roman house was built on a plan developed by the Etruscans. The house, or domus, of an affluent early Roman bourgeois contained about twelve rooms encircling an atrium. The atrium had a square hole in the roof to allow sunlight in and to let the rain fall through to a cistern. The Romans added a second court to their homes when the Greeks influenced their building style. The peristylum or

[13]David Macaulay, *City: A Story of Roman Planning and Construction* (Boston: Houghton Mifflin Company, 1974), p. 33.

[14]DeCamp, *The Ancient Engineers,* pp. 172–173.

[15]DeCamp, *The Ancient Engineers*, p. 172.

peristyle, from the Greek words meaning "surrounded by columns," were used for planting flower gardens. The indigent peasants continued to live in the staple one-room huts. The opulent gentry built large county homes, called villas.[16]

Many innovations were actually reinvented several times throughout the ancient world. The number of people who were engineers and inventors was extremely small by today's standards. Since there was no such thing as patent protection, inventors kept their projects secret so the ideas could not be stolen. Some innovations were lost when the principal inventor died without leaving a description of the work. Still other inventions were lost when cities were ransacked, destroying the only working model and the inventor.

The Romans capitalized on the invention of central heating. A Roman businessman, Gaius Sergius Orata, was successful in raising fish and oysters for the market. He got the idea, probably from the sweat baths of Baiae, which were heated by volcanic steam, to keep his supply flourishing through the winter by heating his tanks. He built new tanks above the ground, utilizing a fire under the tank to circulate hot air and warm the water. He decided to market his invention to equip country homes with balnae pensiles, or "raised bathrooms." The rooms were heated by allowing the hot air to circulate in ducts under the bathrooms. Orata's invention, called hypocaustum from the Greek words for "under" and "burning," was later developed to heat an entire house.[17]

Wooden shutters had been extensively used to close windows due to a lack of a better method. During Hellenistic times many translucent artifacts were experimented with for windowpanes. Some of the materials tried by the builders included oiled cloth, sheepskin, mica, horn, and gypsum shaved down to a thin pane. Many of these had drawbacks that made them impractical for mass usage. Romans adapted the art of glass-making from the Egyptians and improved upon the techniques. This made it possible to make windowpanes out of glass that were clear enough to see through. They also developed the ability to blow glass, thereby lowering the cost of glass drinking vessels so even the common people could own glass. The classes did have the distinction of clear glass vessels for the rich and opaque glass vessels for the poor.[18]

During these times of extensive building, emulating existing technologies, and attempting to have a lavish lifestyle, the Romans began to understand the importance of organization and city planning. A. A. Milne, famed author of the Winnie-the-Pooh series states, "One of the advantages of being disorderly is that one is constantly making exciting discoveries."[19]

The Romans developed aqueducts to ensure a constant water supply to the city, sewer systems to rid the city of waste, bath houses for the citizens to bathe, public bathrooms to ensure human waste was not dumped in the streets, roadways to travel between towns, and huge public buildings for gatherings. Many of the technologies employed for building these structures were advanced methods adapted from other cultures.[20]

[16]DeCamp, *The Ancient Engineers,* pp. 173–176.

[17]DeCamp, *The Ancient Engineers,* pp. 180–181.

[18]DeCamp, *The Ancient Engineers*, p. 178.

[19]Libraryspot Quotations, http://www.libraryspot.com/quotations.htm.

[20]Smith, *Ancient Rome*, pp. 84–93.

The Roman culture had many problems trying to overcome homeostasis throughout their time of technical growth. *Homeo,* meaning *the same*, and *stasis*, which pertains to equilibrium, refers to maintaining a status quo and being resistant to change.[21] The ancient rulers usually agreed with the outcry from their subjects concerning technological advances. There was a belief that people's livelihood would be threatened by the adoption of these inventions. During the reign of Emperor Vespasianus, "An engineer offered to haul some huge columns up to the Capitol at moderate expense by a simple mechanical contrivance, but Vespasian declined his services: 'I must always ensure,' he said, 'that the working classes earn enough money to buy themselves food.' Nevertheless, he paid the engineer a very handsome fee."[22]

ADVANCEMENTS THROUGH DESTRUCTION

The survival of ancient communities depended directly on manpower. Roman boys were needed to plow, cultivate crops, build, and fight when the community was attacked. Daughters, at first, were thought to be a superfluity. The early Romans did not regard human life as sacred. They routinely killed any weak, malformed, diseased, subnormal, or surpluses of healthy children to control the growth of population. This gruesome practice had a precept to obtain the concurrence of five neighbors in sentencing a child to death. In fact, the father retained the power of life or death over his children until his own death, when the eldest son would take on the family responsibilities.[23]

It was the Roman tradition, in the times of the Roman Republic, to require the parents to educate all the children they had decided to bring up in the world (*ut omnes liberos suseptos educarent necesse est*). Much of the information concerning training children comes from the writings of the Elder Cato (234–149 B.C.), known as the Censor. The children learned reading, writing, arithmetic, law, public affairs, and Roman traditions. The activities of throwing spears, fighting in armor, riding horses, boxing, enduring heat and cold, swimming in rapid rivers, running, jumping, and fencing were pursued not to keep fit but rather in preparation to fight on the battlefield in defense of their city or attacking a foe.[24]

War was an integral aspect of the ancient world. Many civilizations were trying to establish themselves within common territories. The citizens of Rome had to defend themselves to survive. By learning warring technologies and strategies, Rome became a strong Empire. They became the aggressors, attacking many different cities and peoples. Through the destruction of cities and conquering of people came enlightenment of city planning and cultural erudition. The Romans learned how other cultures built water supplies, heating, sewage, building structures, and many other technologies by emulating what they saw in foreign cities, taking over the cities, and improving on many ideas.

It was also very important to connect their new cities, conquered cities, and established cities together for a trade route and ensure the quick movement of the military, if necessary. Therefore, it was imperative that the Romans establish a strong infrastructure

[21]Alcorn, *Social Issues in Technology*, p. 20.

[22]DeCamp, *The Ancient Engineers*, pp. 178–179.

[23]F.R. Cowell, *Life in Ancient Rome* (New York: The Berkley Publishing Group, 1976), pp. 35–37.

[24]Cowell, *Life in Ancient Rome*, pp. 37–38.

of roads to connect these cities. The Roman Army built most of the roads with the use of slave labor. Many of the architects and laborers for many of the projects, such as public buildings, roads, apartment houses, shops, city walls, bridges, aqueducts, and sewers were supplied by outside cultures through slave labor, prisoners of war, and others looking for work. Standardization of city layouts was a vital basis of city planning. The Roman Army often traveled from city to city to defend, build, or settle in the city. Consequently the preliminary city planning goals dealt with ensuring similarity in the structure so traveling members of the Army could navigate the confines of the city.[25]

The legionaries made good use of sound engineering techniques that are still practiced today. When the Roman armies ended a march and set up a fortified camp, they utilized a standard, square plan that remained identical regardless of the terrain. The Greek army, for comparison, adapted the formulation of their camp to the specific ground, while the Roman army altered the ground to construct their camp. This practice was followed regardless of the amount of excavating that had to be done. These techniques transferred over to public building. Roads were built to remain as straight as possible to ensure a quick march of the army. This entailed the removal of small hills, building of bridges, and the filling of small trenches to flatten the terrain.[26]

INNOVATORS OF CITY PLANNING

Julius Caesar was the first emperor to attempt to organize Rome in a systematic way. Although he did not live to see the results of the works, which included building a canal in the Tiber River, his adopted son and successor, Augustus, carried on his strategy. He attempted to transform Rome into a suitable capital for the Roman Empire. Augustus and other noblemen provided Rome with many fine public buildings, baths, theaters, temples, and warehouses.[27]

The early Roman theory of government believed that the chief magistrates were expected to build public works for the masses. To fund these endeavors, they used cash from the sale of public lands, riches acquired from foreign wars, or their own amassed fortunes. Some of the ferocious foreign conflicts and barbarous looting of the provinces by Roman leaders were probably the result of acquiring funds needed to build the public works they felt obligated to furnish. These communal enterprises were to be presented to the Roman people for their use and delectation for free. The only restitution a benefactor would hope to receive was recognition for the endeavor. This included having statues erected to commemorate them and poets reciting panegyrics to give an impression of magnificence, pride, and self-esteem, which for a Roman nobleman was more important than life itself.[28]

The construction projects completed by Augustus, which included restoration of old buildings, provided ample employment, but the lack of an overall city plan resulted in unsafe and unsanitary conditions. His most important contributions included the reorganization of city administration and organization of fire brigades. In 7 B.C. he divided Rome

[25]Macaulay, *City: A Story of Roman Planning and Construction*, pp. 26–28.

[26]DeCamp, *The Ancient Engineers*, pp. 199–200.

[27]*Encyclopaedia Britannica,* 1996 ed., s.v. "Rome History: Rome of Antiquity." CD-ROM.

[28]DeCamp, *The Ancient Engineers*, p. 212.

into 14 *regiones*, wards, and the wards into *vici,* precincts. Each of these subdivisions had officials who performed both administrative and religious functions. The *vigiles*, professional firemen, also fulfilled minor police duties, particularly at night.[29]

Nero has been credited with introducing the most modern ideas for town planning. After the great fire in A.D. 64, Nero built new streets and his spectacular Golden House. He encouraged private citizens to construct commodious, fireproof homes and apartment buildings with better access to the public water supply. Other emperors in the late 1st and early 2nd centuries added grandiose imperial forums, temples, arches, baths, and stadiums. Trajan's Forum consisted of a complex of buildings and courtyards, his market with tiers of shops and majestic market hall was one of the preeminent achievements of city planning in Rome.[30]

TECHNOLOGICAL ADVANCES BREED COMPLEXITY

Although the ancient Roman civilization is nothing like modern society, there still were systems that had to be governed. In relation to Roman predecessors, Rome was larger, more complex, with a greater degree of mechanization, and required a more expeditious speed of action that warranted the need for automatic controls.[31] This is evidenced by the need to regulate and charge for water usage. This was a difficult concept for the Romans to grasp since water was a natural resource. They had to fabricate a *calix*, or standard nozzle, using a *quinaria*, which was a length of bronze pipe 1¼ digits (0.728 modern inch) in diameter and 12 digits (8.75 inches) long to connect the user's pipeline to the distributing tank. Using a crude computation to calculate the proportion of the cross-sectional area of their nozzles, the user would be charged a fee. As with any society, people devised ways of defeating the system. This included illegally tapping into water supplies, installing oversized or additional nozzles, or even stealing water directly from the aqueducts.[32]

Rome had established more complex building structures, government, technological advances, law, social structure, and normal daily life. In attempting to make their lives easier and less convoluted, they succeeded in complicating their life with new procedures and needs. The best analogy for this situation was from an article in the *Atlanta Journal* concerning the growing complexities in law. It stated "The 10 Commandments contained 297 words. The Bill of Rights is stated in 463 words. Lincoln's Gettysburg Address contains 266 words. A recent federal directive to regulate the price of cabbage contains 26,911 words."[33]

[29]*Encyclopaedia Britannica,* 1996 ed., s.v. "Rome." CD-ROM.

[30]*Encyclopaedia Britannica,* 1996 ed., s.v. "Rome." CD-ROM.

[31]Alcorn, *Social Issues in Technology*, pp. 114–115.

[32]DeCamp, *The Ancient Engineers*, pp. 212–213.

[33]Libraryspot Quotations, http://www.libraryspot.com/quotations.htm.

Urban Life in Rome

City and country depended on each other: the city concentrated people, experiences, and monuments of Roman greatness; the country supplied food, water, and fresh air.

NEIGHBORHOODS

A typical Roman neighborhood consisted of landmarks, statues, buildings, with various colors and textures to see; sounds of construction and chariots, shouts and conversations in Latin, Greek, and several barbarian languages to hear; the fragrance of bakeries and food shops and the smell of smoke, sewage, sweat, and urine; jostling crowds to touch; and a kinetic sense of movement through many types of urban environments.

The tenants of damp, smoky, dark apartments could lean out windows and talk to neighbors across the street, or they could go down into the streets and squares. Along with a glimpse of the sky and a breath of fresh air, they would find a crowded scene populated by peddlers selling water, cooked beans, sausages, or salt fish. There were also vagabonds charming snakes or reciting poetry and begging priests of some type of religion.

Open public spaces had been an important part of the Roman scene and offered casual concerts, for business and pleasure. In the park on the Capitoline, a politician could work the crowd as it promenades there, enjoying the view and shopping among the displays of silk and jewelry. There were also opportunities for romantic meetings which were no longer quiet as whole bands of young men serenaded desirable women and at the Circus Maximus or the Subura, where the streetwalkers walked.

SUBURBS

For all its symbolic value, the wall built in the fourth century BC could not contain the city in its centuries of growth. By the Augustan period, the city had grown in such a ragged fashion and was so bound up with the surrounding countryside that it gave the onlooker the impression of a city stretching out into infinity. In fact, Roman laws regularly included the city and the mile outside within the urban jurisdiction.

Along the outer edge of this suburban area, the rich bought small farms and built private villas and gardens on them, creating a broad green belt that sometimes threatened to choke the city and prevent its outward growth. To the masters of these villas, however, a semirural environment was available within easy reach of the city's Forum, temples, and markets.

As if to demonstrate architecturally that life in the country was the reverse of life in the city, the parts of villas were arranged, from the point of view of a town house, backwards. Instead of being closed in from the urban street, they were wide open to the surrounding landscape. The atrium, freed from its city role as a semipublic reception room, was set back as a private living room. The peristyle was in front, where people entered. The best-known example of such a suburban villa is the Villa of the Mysteries located a short distance north of the Herculaneum Gate in Pompeii and built early in the second century BC.

SOCIAL LIFE IN THE CITY

As Rome grew to become the largest, most important city in the world, it became more cosmopolitan and sophisticated in its diversions, occupations, distractions, and pleasures. Its inhabitants were constantly assaulted and enticed by the variety of stimuli, and while some observers of the urban scene were delighted by the city and celebrated its diversity, others disapproved of its decadence.

Roman cities tended to conduct their activities outdoors with parades and promenades, trials, political meetings, and plays. Dinner at a restaurant or at home was always more pleasant under a trellis than a ceiling. Even though some activities had to take place under a roof, the Roman architects seemed determined to devise ways to paint their walls with such realistic vistas or build their roofs with such wide spans that the indoor scene became an almost acceptable substitute for the outdoors.

BATHS

The bath, with its pleasant distraction from the cares and squalor of ordinary life, became a way of life for the Romans. It was the setting in which they washed themselves, took their exercise, spent their leisure, were exposed to art and cultural programs, made business and political contracts, and conducted their social activities. Heating brought a change in the mystique of the bath; it became a more comfortable place, and people tended to spend more of their waking hours there.

For a minimal admission fee, all this luxury was available for anyone who wanted to enjoy it. In these villas of the poor, the commoner found opportunities for exercise, swimming, steam baths, saunas, and poetry readings.

Customers had a wide variety of exercises to choose from such as lifting weights, playing ball, and swimming. There were a variety of spectator activities as well, from spirits at the stadium track to literary readings in the hemicycles. Service personnel were available to sell cakes, sausages, and bath oil. In the changing-room, slaves guarded their master's clothes. In corridors underground, other slaves tended furnaces and washed towels.

DINNER PARTIES

Among the Romans, as in most cultures, dining was an opportunity to express common interests and consolidate social cohesion through the shared space of the dining room and the shared food and drink. Every house of any pretension had at least one dining room, and the more elaborate houses had several. The dining room was called a triclinium (three-couch place) because couches for diners were normally arranged in rows around three sides of the room. A dining room used in the summer should face the north, and one for spring and autumn should face east, so that the sun would warm it and have it ready for an evening's socializing. A winter dining room, however, should be relatively enclosed, and decorated simply, in a way that will not be spoiled by smoke from the braziers and oil lamps.

RESTAURANTS AND TAVERNS

A wealthy person who lived in a private domus or on the ground floor of a luxury insula was apt to stay at home in the evening. The poorer resident of the city might try to escape from a dank room by seeking out companionship at a caupona[1] or a popina[2]. Here we might find a stand-up snack bar, a room with tables and chairs, and in more elegant establishments, an inner garden with private dining rooms complete with couches. In many of these inns, slaves or freedwomen were available as prostitutes, and eventually the word caupona took on the connotation of a low-life dive.

PUBLIC AND PRIVATE SERVICES

In ancient Rome the interactions between urban government and private residents shaped a very different profile of services from that which is familiar in modern industrialized cities. The problems to be discussed were not necessarily what Romans considered essential issues. Rome's city planning was limited to what resources were at hand to help individuals cope with the day-to-day problems of living in the ancient city of Rome.

FINANCES

To provide public services was generally a responsibility, and an opportunity, of the ruling class, which not only sponsored much of the building in the ancient city, but also established endowments, supplemented the grain supply, or paid for games as a type of public obligation. A problem with depending on the usually self-interested generosity of an upper-class sponsor was the potentially haphazard way in which services were delivered. When the emperors assumed responsibility for construction or other public services, they used their own purse, in much the same way that members of the aristocracy traditionally used their family fortunes to pay for their donations to the public.

Funds belonging to the state treasury flowed into the aerarium in the Temple of Saturnus. Taxes and tolls, produce and rents from public land, generated this revenue. The money was available to aediles and cesors, who used it to construct and maintain public buildings and roads.

POLICE

The state did its best to provide peace and security through its military policy, but during the republic, law enforcement and police protection were in the hands of private individuals. The Twelve Tables make it clear that although the state took responsibility for conducting trials and passing judgement through its magistrates (the praetors), it was the job of the victim or his or her family to apprehend and prosecute the lawbreaker. The only specific law enforcement officers mentioned in the Twelve Tables are the quaestores paricidii, whose duty it was to investigate parricide, the premeditated murder of a free person, the one crime in which the state took an explicit active interest. The Twelve Tables recognized private suits between citizens,

[1]A caupona was a full-service establishment that offered meals and drinks as well as rooms.

[2]A popina, a simpler restaurant, served wine and some hot food.

though often these were arbitrated by some third party. Starting in the second century BC separate tribunals were established to judge crimes against the state, although it still remained the responsibility of private citizens to gather evidence and conduct the prosecution.

WELFARE

In Rome and other cities of Italy, state-sponsored systems of loans to landholders provided interest income to support annual payments to children. These typically provided 16 sesterces a year for a boy if he was legitimate, 14 if he was illegitimate, 12 for a girl if legitimate, 10 if illegitimate. This was a small enough amount to suggest that it be aimed primarily at the poor, to subsidize their basic sustenance or encourage them to have children. In the Roman sense of values, helping the poor and homeless was simply not the traditional way. That kind of charity was an Oriental concept codified by Jews and Christians who became conspicuous and a little suspect in their zeal for taking care of the sick and the poor.

HEALTH

In the Roman tradition, medical care, too, was a private matter. The duties of the father of the household included tending to the sick of his family. When a wife, son, or daughter took sick, the traditional Roman father administered remedies; lower-ranking members of the household would be cared for by slaves.

The image of the physician in Roman literature is negative. He and his female counterpart tended to slaves or ex-slaves. They worked with their hands, and often stooped to dubious methods of attracting attention. Frequently the cures did not work, and visits were stereotyped and parodied: drain blood or apply a dressing. Other sources of medical help included the neighborhood barber, who knew something about herbs and knew how to handle sharp instruments.

Compared with a modern city, sickness was more visible on the streets of ancient Rome, a function of haphazard medical care and the nonexistence of hospitals and rest homes. Inscriptions and literature indicate that the mortality rate among children and young adults was high. Grief was a frequent emotion in ancient Rome.

EDUCATION

Some upper-class families used their trained slaves as teachers; others and the lower classes hired teachers or sent their children out to a ludus, where reading, writing, and rote learning formed the basis of the curriculum and discipline was a proverbial and constant problem. Boys and girls went to school together, in a rented shop, or the portico of a public square or a courtyard.

After the basics were acquired, an upper-class young man would spend several years in the company of his father's friends, observing and helping as they attended the Senate, campaigned for office, spoke in assemblies and law courts, and went on military campaigns. As Roman society become more complex during the late republic, it needed additional forms of education. Slaves, for instance, were sent out to ludi to learn the skills necessary to their jobs as teachers, bookkeepers, secretaries, shoemakers, weavers, physicians, musicians, cooks and carvers. Public libraries first opened under Augustus, and facilities for lectures and readings in the baths provided opportunities for continuing education for large numbers of the city's residents.

City Planning

ELEMENTS OF ROMAN HISTORY

The growth of the Roman Empire did not happen by chance. Rome was one of the most organized and structured cultures of its day. The Roman engineers, architects and military leaders used this organizational structure to boost their efficiency in various elements such as military and civilian life. This structure embraced a popular concept of our current age known as the assembly line, which increases the productivity and efficiency of workers vs. the handmade work done by others.

City planning in the Roman Empire was a very calculated concept. Cities were built to handle only a certain number of people and once this limit was reached, a new city was built. This may sound strange to people living in the sprawling cities of today, but it wasn't as easy in 50 B.C. to build another subdivision when the old one was too small.

Roman cities usually abounded with many public works, as each new emperor and territorial governor tried to impress the people more than the last had done, or simply keep the people happy. These public works included forums, amphitheaters, aqueducts, fountains, and coliseums. Above all "government, law and order were signified in civic buildings."[1]

The first Roman emperor, Augustus (63 B.C.–A.D. 14) appointed General Vituvius, who designed three main ideas that would be incorporated in most of the towns and cities constructed by the Romans. These three ideas were principles of architecture, the rules for the placement of cities, and the rules on how basilicas and forums would be constructed. These rules were designed to "extend the power of the empire"[2] through uniformity of design so that wherever a citizen went they would feel at home, soldiers would know how to traverse cities they had never been in before, and the splendor of these cities would humble or impress the "heathens" who populated the lands that the Romans conquered.

[1]Lecture Outlines, http://wdsroot.ucdavis.edu/dept/1da/courses/classites/30w98/communities/romanfilm.html

[2]Lecture Outlines, http://wdsroot.ucdavis.edu/dept/1da/courses/classites/30w98/communities/romanfilm.html

SITE PREPARATION

The first idea that was generally used was the rule for the placement of cities. When a new city was being proposed a group of soldiers and slaves would travel to a likely site to see if it qualified. Surveyors would scout an area looking for an area that had enough flat ground for the city, but that also had enough slope to the land that it would drain properly and avoid floods. When a site had been found that matched the criteria, a priest was usually called in to examine the livers of small animals like rabbits, pheasants, etc., that lived on the land to check to see if the animals looked healthy or not. If the animals were found to have been healthy, one last check was performed to look for stagnant pools, and if none were found, then the site was confirmed, the gods thanked, and development began.

Once a site was selected the soldiers and slaves who were to be its first citizens and builders started to work. The first task was to construct a base camp, or *castrum*. In order to build this camp they first dug a ditch and built a stockade around a rectangular area. This area wasn't big enough for the city proper, but helped defend the workers until the planning was complete and the official city walls constructed. The following tasks included the building of the main streets that would extend north to south with a corresponding street extending east to west. There was an open area, or *forum*, below where these streets crossed, and this was the area where the soldiers received instructions, food, supplies, and also served as headquarters for the commander.

The site selection was usually done during the spring and summer, during the time the *castrum* was built. Following the initial *castrum*, the tents were gradually replaced by permanent wooden structures during the late summer and fall, and any rivers or streams the soldiers had to cross had temporary wooden boat bridges built over them. While this initial construction was taking place the planners and surveyors were designing the rest of the city over the summer, fall and winter.

ARCHITECTURAL DESIGN ELEMENTS

During the planning stage, the principles of architecture came into play. These principles were *order*, *eurythmy*, *symmetry*, *propriety*, and the *proper orders assigned to temples of the gods. Order* was ensured so that all the individual parts of the city were planned and built in such a way that they combined for maximum strength and effectiveness. *Eurythmy* ensured that no building in the city could be more than twice the height of the street it faced, while the matching layout of the insulae, markets, and other buildings ensured *symmetry. The proper orders assigned to the temples* assured that each temple was built in an approved style Doric, Corinthian, or Ionic, in accordance with prescribed practices. Finally *propriety*, or the "perfection of style which comes when a work is authoritatively constructed on approved principles,"[3] ensured everything was built right the first time.

As mentioned earlier, Roman cities were built for a certain number of people, for purposes of convenience (further discussion will include the building of a Roman city for a population of 50,000 people). When the surveyors were done planning the city they started construction. The main north to south street, *cardo*, and the east to west street, *decumanus*, were both lengthened and widened and the *castrum's* size increased to

[3]Lecture Outlines, http://wdsroot.ucdavis.edu/dept/1da/courses/classites/30w98/communities/romanfilm.html

720 yards long by 620 yards wide. This would allow for a city large enough to hold the number of people as planned.

The next step was to divide the area with "roads into a chessboard pattern"[4] of blocks, approximately eighty yards wide, called *insulae*. A priest controlled a plow drawn by a white cow and bull around where the city walls would be, lifting the plow only where the gates would be built. The plowing by the priest and a thirty-foot wide strip of land called a *pomerium*, where no houses could be built, ensured that the gods would bless the city.

In addition to the walls and streets the designers also planned a new and larger forum which would become the religious and governmental center of the city, and they set aside land for the markets, baths, fountains, and entertainment sections that would keep everyone happy, clean, and fed. The last major design feature was where the aqueduct would come into the city and where the reservoirs holding the collected water would be built.

Although the city planners designed the majority of the city, some areas of design were left to the individual citizens such as the design and construction of residential houses. However, certain ordinances indicated that a house could not be higher than twice the width of the street, and anyone who had a house or shop facing a main street had to construct an overhanging shelter to keep rain and sunshine off the citizens' heads.

Once the city was laid out and planned, the city roads, any necessary bridges, and city walls were built. The streets of the city were designed for people to walk on, "therefore adequate sidewalks were built and strict laws were written to control any movement of carts and chariots, which could endanger the health, and safety of people on the streets."[5] The sidewalks were raised as much as a foot and a half above the level of the streets and stepping stones at intersections ensured people didn't get their feet wet during rainstorms. These stepping stones, plus the fact that many streets were dead ends, ensured that cart and chariot traffic was reduced to a minimum in the city.

WATER AND WASTE DESIGN

When the streets and walls were built, and the aqueduct was under construction, the system for distributing water throughout the city was started. Reservoirs were built to hold all the water coming from the aqueduct, and lead pipes ran from these reservoirs to the fountains, toilets, baths, and houses of the wealthy. During a water shortage these pipes could be turned off, first to the wealthy homes, then the baths, and finally the toilets. The homes of the wealthy received their water from a central water tank that also distributed water to all the nearby houses.

The water distribution system was useless without some way to carry off water and waste. Thus the next thing to be constructed was the sewer system. The sewer system was composed of two *cloacae*, which were tunnels large enough to walk in, connected to smaller tunnels that traversed under the rest of the entire city, channeling waste in the direction of the ground slope out of the city itself.

While all this water and sewage work was underway, the last major idea in construction, the rules on how basilicas and forums would be constructed, came into effect. The

[4]David Macaulay, *City: A Story of Roman Planning and Construction* (New York: Houghton Mifflin Company, 1974), p. 13.

[5]Macaulay, *City: A Story of Roman Planning and Construction*, p. 44.

forum was "paved and covered two entire blocks"[6] consisting of at least four major buildings, the temple to Jupiter, Minerva, and Juno, the *Rostrum, Curia*, and the *Basilica*. The temple to Jupiter, Juno, and Minerva represented a combination of the most powerful, cunning, and intelligent of the Roman gods. The *rostrum* was a raised platform where all the important decrees and news was read. The last two structures, the *Curia*, which was where the senators met, and the *Basilica*, or court of law, were the places where the true power reigned and where all the important people who actually ran the city worked. The forum was surrounded by a two-story colonnade, which contained shops and schools, and was accessible through the triumphal arch, dedicated to Augustus.

SOCIAL ELEMENTS

As the city grew, so did the forum, so it usually wasn't completely finished until after the city had reached its peak population, however this was not the case with the central market. The central market was essentially an open space, surrounded by a two-story colonnade that contained shops and offices like the colonnade surrounding the forum. A statue to Ceres, goddess of agriculture, and a public fountain were the only permanent fixtures in the open market space itself.

Near the market were warehouses, which stored grain, wine and oil. Wine and oil were stored in large jars called amphorae. These amphorae were buried in the ground to keep them fresh, while the grain was kept in sacks on separately ventilated shelves, thus keeping it dry, and also to keep grain dust from causing the building to explode.

As the city grew, the shops for conducting commerce became more numerous. Often shopkeepers of the same type of goods would cluster together, and people would name the street after them, thus you could have a "street of gold, street of bread, etc." Many of the shopkeepers lived above their stores, or in smaller apartments in different parts of the city, while a few who did really well could have their own houses built. There was a variety of shops, barbers, snack bars, jewelers, bakeries, indeed many of the shops you might see in a normal mall today, only on a much smaller scale.

As the city grew, many of the older, small houses and apartments were torn down and larger buildings constructed. As the need for better public facilities increased, older public works, such as smaller bathhouses, were torn down and newer, larger ones built. These larger bathhouses were called thermae.

Thermae usually consisted of three indoor sections devoted to bathing: the *caldarium*, *tepidarium*, and *frigidarium*. The *caldarium* was filled with hot water and used for initial bathing, while most bathers soaked, gossiped and relaxed in the *tepidarium* before jumping into the *frigidarium* right before they got out and got dressed. Another part of the *thermae* was the *palaestra*. The *palaestra* was an open grassy area containing a swimming pool and places to wrestle and exercise, which was also enclosed by a two-story colonnade. There was also an area on the second floor filled with scrolls in case someone wished to read and relax instead of gossiping and soaking.

With the needs of the public health taken care of and business booming, as it should, the last major aspect of city construction, the entertainment center, was focused on. The entertainment center consisted of an amphitheater, also known as a coliseum, and a theater. The

[6]Macaulay, *City: A Story of Roman Planning and Construction*, p. 54.

theater was large, and the amphitheater was colossal, both taking over thirty years combined, in some cases, to build. It is a small wonder then that when both were complete a month-long celebration was held to celebrate.

The amphitheater was an oval- or round-shaped building that could hold up to 20,000 people, and was built in a tiered shape. The floor of the amphitheater was sunken below ground level so that it could be filled with water for mock naval battles, and also featured cages for animals. These animals, along with slaves and gladiators, would sometimes fight to the death to entertain the crowd. The amphitheater also had a covering that could be used to keep the sun off of citizens so they would be more comfortable while watching the shows being staged for them.

The theater was a semicircular shaped building also built in a tiered pattern that could also be covered when the sun was too hot. In addition to this the theater had a wooden frame along the back wall which was used to hang painted backdrops that would make the plays better. Dressing rooms off to the side of the main stage were also included in the theater.

Due to well-laid plans of the city designers, a typical Roman city's water, sewage, food, entertainment, and governing sections were never taxed beyond what they were designed to accommodate, which is a pretty remarkable feat. During later chapters we will discover just how the Romans were able to build all their engineering accomplishments, as well as how the idea of city planning has been used throughout the ages since the times of the Romans.

Tools of the Romans

INTRODUCTION

Today most of us in this world take full advantage of technology. We have computers that can nearly think for us, we have automatic machines that do our heavy lifting, and we have a tool that allows us in some manner to automate almost every task that we do. Is this advancement in technology really to our benefit? Are we better off without the uses of these tools? Personally I do think that the things that are built in today's society are as solid as the things built in the ancient era. How would we react today, if suddenly all of these technological tools were stripped from us? Think about only being able to make do with what nature has given us. I don't think that we would get quite as much work done. People in the ancient times were good at using nature to create buildings, many of which were considered to be beautiful works of art and the Romans were masters of accomplishing these tasks.

In today's society most people have a specific title, whether they are an engineer, architect, construction worker, or surveyor, etc. In the Roman world these activities would have been assigned to one person. By the beginning of the second century A.D. the

Romans had some of the best fabri[1] in the land. The Romans' precision and planning of a project set them apart from many of the world's engineers.

In the days when the Roman Republic was beginning to dominate central Italy, there were two needs that were of top priority. One was to supply the essentials of life to a city that was very large by ancient standards and the second was to secure the conquered territories in Italy. The inspiration of such projects came from within the aristocracy,[2] although the Populus Romanus might be asked to support the proposals, by voting in assemblies. Some of these projects included the building of aqueducts,[3] roads, and bridges, which will be discussed in this portion of our report.

AQUEDUCTS

In order for the Romans to build aqueducts they first had to find water, and they had a unique way of doing this. Just before sunrise they would lie face down on the ground and rest their chins in their hands and take a look over the countryside. If they saw vapor curling up from the ground they knew that there was water. They also studied the soil to determine where the best water was. They found that in clay and fine gravel that the water will be poor in quality and taste; in coarse gravel it will be sweeter and a lot healthier; in tufa and lava it would be plentiful and good.

They also used a bronze bowl to check if the water was up to standards by leaving it in an underground pit overnight and examining it the next morning to see if moisture had condensed in it. The Romans also had to survey the route in which the aqueducts would be built. They used a dioptra[4] to distinguish the difference in height between the source and outfall of the water route. The preliminary work for the aqueducts was trenching in the soft ground. The sides of the trenches would be temporarily supported by the use of timber props. The hardest work of building the aqueducts was the tunneling, in which they would sink a puteus[5] every seventy-one meters and at the rock face they would tunnel forward, while passing back the hewn stone in baskets to be hauled up the shaft. The aqueduct had to be inspected for damages once it was completed and working. Over the years the aqueduct would be repaired or improved to maintain its useful life.

The Romans also took into consideration the condition of the people who depended on the water. They would look at these people and if they were strong and had a good skin complexion, they knew that the water was good.

ROADS

Some of the best roads in Rome were first built for the convenience of rulers. This would explain the reason for the excellent inns alongside these royal roads. Secondly, roads were built for military purposes. The Romans went to great lengths to keep their roads straight and symmetrical. During this time in history there were no surveying instruments such as true-

[1] Skilled craftsmen of the Roman Empire.

[2] The senate of the Roman people.

[3] Roman word that meant *the conductor of water from one place to another.*

[4] A Roman surveyor's leveling instrument.

[5] A shaft that the Romans would use to pass tunneled stone back up in baskets.

scale maps or compasses; the Romans used primary stations or beacons to keep the road straight. Stakes or stones laid out in close intervals were used to mark the distances between the beacons. The roads were layered in the following manner: the statumen, rudus, nucleus, and pavimentum with the statumen being the strong foundation of the roads. The Romans used a cart with an odometer[6] to calculate how far the road had come along.

BRIDGES

Many of the Roman roads would end at the edge of the sea. The Mediterranean occupied a central position in the Roman world. To overcome the small bodies of water that interfered with the Roman roads, the Romans built bridges. The first bridges were built by fastening together a pair of piles one and a half feet thick to verify the depth of the river. These were lowered into the river at an angle and driven in with the use of pile drivers. Beams that were placed above and fitting between the piles supported these. From that point planks were laid between them, surmounted by long poles and covered with hurdles. Other stakes were placed at an angle downstream to secure the rest of the structure and absorb the force of the current.

BUILDING MATERIALS

One of the most important and abundant building materials to the Romans was the cappellacio[7] in addition to the reliance upon mortar and concrete. Stone was one of the most important materials that the Romans used to build with. They used it to lay foundations, build water channels, and surface roads among other things. Tools such as the folding rule, plumb line, calipers, chisels and the trepan were used to shape and mold stones.

It is evident that the Romans had no automated tools to work with, but they created tools from within their natural environment to build with. Nature was technology to the Romans and they used it well.

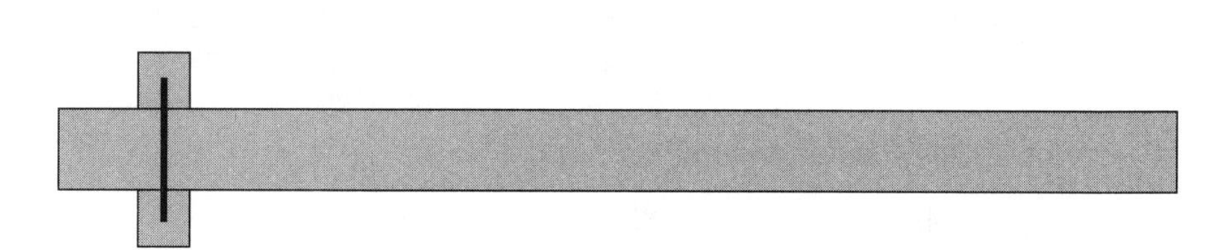

The Future Results of Ancient City Planning

There are many elements to the fascinating Roman way of life in view of their contributions to technology and city planning. The scope of the topic is extremely vast and would in most circumstances require several years of dedicated study. For the purposes of our

[6]An instrument that dropped a small pebble into a bowl after every mile.

[7]A type of soft stone that was formed from the dust of ancient volcanoes.

research project, this section will cover a limited study that involves elements of the evolution of city planning.

EVOLUTION AND STRUCTURE OF CITY PLANNING

King Solomon said that "There is nothing new under the sun,"[1] however when we view the captivating architecture of Washington D.C. or the systematic grid plan of New York City, a person cannot help but wonder if the city planners were ahead of their time. So much thought and craftiness was put into the planning of many modern cities, however, most people fail to realize that these planning concepts have evolved from techniques that have been used for centuries. The truth is that many of the building tools and planning guidelines (such as city grid patterns) we see today have been passed down from the Ancient Romans.

Technology has reproduced itself according to the specific needs of various societies in our diverse world. A lot of technology goes into city planning, but the intricate, strategic organization and structuring is sometimes overlooked as less important when compared to some of the more tangible technological advances. For example, the average person in America can expect to find an operable bathroom in most every restaurant (with few exceptions), but the planning that goes into determining water and sewage capacity according to population is generally not a concern for the person looking to use a bathroom. Their only concern will be that the bathroom (or the technology used to make the bathroom functional) is in working order. However, technology would not be as useful if it were not for strategic planning which is also a key element for the development of a city.

The Romans understood this concept and realized that careful planning was a key to success for the development of their Empire. They were able to look beyond the idea that society had to follow technology and realized that technology could follow society.[2] This has been proven throughout history when we examine the adaptability of the Romans to various regions of the world such as Spain and Israel.

OBJECTIVES OF CITY PLANNING AND ADAPTABILITY

Since technological advances were not the primary concern for the Romans when planning a city, what was? A sound material base and adequate security of the city and the welfare of the people were among the primary concerns of the Romans.[3] Engineers would develop technologies that would help fulfill the main objective of a city plan. The Romans believed that well-planned cities could do more to maintain peace and security than a vast number of military camps.[4] Yet the Romans realized that a secure city also had to be a place where people wanted to live. This is where careful planning by the Roman engineers played a significant role.

[1]Ecclesiastes 1:9.

[2]Paul A. Alcorn, *Social Issues in Technology: A Format for Investigation*, second edition (New Jersey: Prentice Hall, 1997), p. 87 [Homeostatic Reactions].

[3]Edward N. Luttwak, *The Grand Strategy of the Roman Empire* (Baltimore: John Hopkins University Press, 1976), p. 1.

[4]David Macaulay, *City: A Story of Roman Planning and Construction* (Boston: Houghton Mifflin Company, 1974), p. 51.

Engineers not only carefully planned cities, but also created or used technology when unforeseen circumstances arose such as the siege of Masada in A.D. 70–73.[5] At this event the Jewish war had essentially been won by the Romans with the exception of a few hundred resisting Jewish people that held out on a mountain in the Judean desert. Instead of storming the mountain, the Romans reduced the fortress of the Jewish people by implementing engineering skills that included a ramp that would extend to the top of the mountain. This illustration demonstrates the Romans use of a network of elements that could be used to meet specific needs.

The network elements (or systems) that the Romans used for city planning consisted of technology, engineering, integrated diplomacy, military forces and road networks. When used in unison, synergy[6] is created in which the whole is greater than the sum of the parts. These strategic plans would follow civilizations for centuries to come and would be added to, or subtracted from, depending on the needs of a particular society, but the blueprint remained constant as technology reproduced itself in many areas.

Systematic city planning dates back several centuries and the intricate planning techniques used by the ancient Romans have been a foundation from which many modern cities such as London and Paris have been developed. (It is also important to understand how the evolution of technological advances and shifts in the economy from the agricultural age to the information age has caused city developers to modify planning with the progression of time.)

THE IMPORTANCE OF CITY PLANNING

Can you imagine relocating to a newly developed city only to discover that 70% of the buildings and homes flood during heavy rain because the grade of the streets does not slope one inch every ten or twenty feet? Would you become fearful if two thirds of the population have become infected with giardia from the water supply? What would it be like to walk about the streets breathing toxic odors because the city sewage system cannot handle the capacity of the population? These examples help enhance the importance of correct city planning and how it can reinforce or dilute the security of the people. The Roman leaders understood this concept and set standards for city planning to ensure the well being of its citizens, and these concepts would continue to evolve through time to become city planning standards in many parts of the modern world.

The city planners and engineers of ancient Rome took careful consideration to be sure that a proposed city could meet the needs of every individual regardless of their financial status. Adequate space was allotted for houses, shops, squares, and places of religious worship. Other areas of consideration were the number/size of streets, population, water, drainage, and sewer systems in addition to several other factors that could not be overlooked if a city were to function at an optimum level. For this reason military engineers, planners, architects, surveyors, inspectors, and construction specialists would be employed to ensure the successful development of the city infrastructure. Many of these same elements are used when developing modern cities/townships in the twenty-first century.

[5]Luttwak, *The Grand Strategy of the Roman Empire,* pp. 3–4.
[6]Alcorn, *Social Issues in Technology,* p. 156.

TECHNOLOGICAL IMPACT

Technological advances have had a tremendous impact on city planning. City planning in America and in parts of Europe have adapted to endure the changes from an agricultural society that evolved into an industrial revolution that would lead the world to what is now known as the Information Age.

Technological advances of the automobile, airplane and railroad systems would completely change the methods of transportation because people could now access many areas of the world in much shorter periods of time. These advances would change the business industry and how commerce was conducted. It also meant that militaries had to develop new tactics since opposing troops could be deployed with ease to nearly any remote location of the world.

As cities began to grow there was an exponential increase in population. Many people who lived in rural farming communities moved to the more industrialized cities. There was also a flood of immigrants from many parts of Europe that came to the United States to seek greater opportunities for their families. Most of the world would have to compensate for these societal changes and include new elements of infrastructure such as public transportation.

The world today has an information highway that is slowly trying to take control of our daily lives. Technology is a key element of life today in America. However, I personally don't believe that all detrimental possibilities have been considered. Children are being raised on television, video games, and computers, which in my opinion could be leading society down a path of dependency or selective thought processes. The written word found within books, art, culture, and a strong family foundation are proven and successful facets of life that we need to cherish.

BIBLIOGRAPHY

Alcorn, Paul A., *Social Issues in Technology: A Format for Investigation* (2nd ed.). New Jersey: Prentice Hall, 1997.

Cowell, F. R., *Life in Ancient Rome.* New York: The Berkley Publishing Group, 1976.

DeCamp, L. Sprague, *The Ancient Engineers.* New York: Ballantine Books, 1963.

DeSelincourt, Aubrey, trans., *Livy: The Early History of Rome.* New York: Penguin Books, 1971.

Dudley, Donald R., *The Civilization of Rome.* New York: Meridian Press, 1993.

Ellis, Cliff, *History of Planning and City Planning.*

Encarta 98 Multimedia Encyclopedia, 1998 ed., s.v. "Rome." CD-ROM.

Encyclopaedia Britannica, 1996 ed., s.v. "Rome History: Rome of Antiquity." CD-ROM.

Grant, R. M., *Early Christianity and Society: Seven Studies.* New York, 1977.

Hamey, L. A., and Hamey, J. A., *The Roman Engineers.* Cambridge: Cambridge University Press, 1981.

Jones, R. Duncan, *The Economy of the Roman Empire.* Cambridge: Cambridge University Press, 1974.

Kunkel, W., *An Introduction to Roman Legal and Constitutional History.* Oxford: Oxford University Press, 1973.

Libraryspot Quotations, http//www.libraryspot.com/quotations.htm.

Luttwak, Edward N., *The Grand Strategy of the Roman Empire*. Baltimore: Johns Hopkins Press, 1976.

Macaulay, David, *City: A Story of Roman Planning and Construction.* Boston: Houghton Mifflin Company, 1974.

Platner, S. B., and Ashby, T., *Topographical Dictionary of Ancient Rome.* Oxford: Oxford University Press, 1929.

Scarborough, J., *Roman Medicine.* Ithaca, New York, 1956.

Smith, Susan, *Ancient Rome.* London: Dorling Kindersley, 1995.

Van Daman, E. B., *The Building of the Roman Aqueducts.* Washington, DC, 1934.

Webster's Dictionary of the English Language. New York: Modern Promotions/Publishers, 1974.

The Effects of Modernization and Conservation Efforts on the Maasai of Tanzania and Kenya

INTRODUCTION

The Maasai of Kenya and Tanzania represent a pastoral culture spanning more than a thousand years, and are classic representatives of the subsistence strategy. However, in the twentieth century, as a result of restructuring and economically progressive programs instituted by the governments of both nations, the Maasai have found their traditions, their strategy, and their lives threatened in a hodge-podge of well-meaning legislative programs designed to improve their lot and more efficiently utilize the land and resources on which they depend.

In this article, we will investigate the nature of the Maasai culture, its ways of successfully interacting with its environment, and the nearly catastrophic events that have led this proud group of people into conflict with the countries in which they live. To do this, the article begins with a short history of the Maasai, an ethnographic snapshot of the culture as it relates to economics and the subsistence strategies it employs, and an examination of the governmentally instituted programs that have caused the problems. In the end, we will look at the issues that exist and their relevance to the balance of this text.

HISTORY OF THE MAASAI PEOPLE

The Maasai have traditionally occupied a belt of land from the western coast of Africa, through the Rift Valley, to Lake Victoria, and Victoria Falls. This area, known as Maasailand, encompasses a massive area of savannah, or grasslands, that include the Serengeti Plain, the Goi Mountains, the Olduvai Gorge, the Mau Escarpment near Lake Turbanen, and the Northern Highland Forests of Tanzania, as well as the grasslands along the border of Kenya and Tanzania, and the whole of Northern Tanzania along the boarder with Kenya.[1] In Kenya, the principal tribe sharing the land with the Maasai is the Kikuyu, a more sedentary and agriculturally oriented culture.[2]

As pastoralists, the Maasai are herders of cattle. Pastoralism involves a primary dependence on herding and animal husbandry in areas where other forms of technology would be difficult or impossible. The pastoralists are found on grasslands "...Pastoralists secure numerous products from their herds: milk products, fuel, leather goods."[3] Archaeological evidence indicates that there have been pastoralists in the Maasailand area for the past 2300 years.[4] It would appear though that the Maasai as a group have occupied the area only since approximately A.D. 1000.[5]

There are disagreements as to the actual origin of the Maasai and how they came to settle the area, but it is generally agreed that they arrived from the north after a drought that created a mass migration of various peoples. This is supported by the oral traditions and legends of the Maasai themselves. Among the other groups who moved south with them are those tribes now known as the Somali, Borana, and Rendile, who, according to Maasai myth, settled in less desirable land along the way while the Maasai continued on to Kenya and Tanzania.[6]

Linguistic evidence also suggests that the Maasai were a tribal entity in about A.D. 1000, when they separated themselves from the Samburu, a group living in the area of Lake Turkana in Kenya.[7]

According to Blauer, this migration from the North a millennium ago began in Ethiopia. As a racial group, he and others consider them to be a Nilo-Hamitic mix.[8] Yet Gleason stated this to be a misrepresentation. Gleason feels that Maa, the language of the Maasai, belongs to the Chari-Nile family of languages, which is named for the original area of the root language located in the upper Nile Valley, extending westward into the Chari River Basin, almost to Lake Chad. According to his research, several other languages are included in this Nilogic group, specifically, Dinka, Nuer, and Shilluk.[9]

[1]Winnie Mukame, interviews by Paul Alcorn, January–June, 2001.

[2]Larry A. Samovar and Richard E. Porter, *Intercultural Communication: A Reader*, 6th ed. (Belmont: Wadsworth Publishing, 1991).

[3]Johnetta B. Cole, ed., *Anthropology for the Nineties: Introductory Readings* (New York: The Free Press, 1988).

[4]Terrance J. McCabe, Scott Perkin, and Claire Schofield, "Can Conservation and Development Be Coupled among Pastoral People? An Examination of the Maasai of the Ngorongoro Conservation Area, Tanzania," *Human Organization* 51, no. 4 (1992): pp. 353–366.

[5]Etagale Blauer, "Mystique of the Maasai," *The World and I* (March, 1987): pp. 497–513.

[6]Samovar and Porter, *Intercultural Communication*.

[7]McCabe, Perkin, Schofield, "Conservation and Development," pp. 353–366.

[8]Blauer, "Mystique of the Maasai," pp. 497–513.

[9]H. A. Gleason, *An Introduction to Descriptive Linguistics* (New York: Holt, Rinehart, and Winston, 1961).

These languages are also purely Nilotic, though erroneously connected at one time to Asian languages in a group called Nilo-Hamatic, which includes Acholi in Uganda, and Maa and Nandi in Kenya, and Tanzania.[10]

Originally predominant in *those areas* of Kenya and Tanzania which they occupied at the time of European colonization, the Maasai found themselves faced with a changing environment and an entirely new paradigm. Prior to the arrival of Europeans, the Maasai were free to practice their pastoral life which included the raiding of other peoples in order to obtain additional cattle, the primary form of wealth. Raiding, which the Maasai were quite successful at, was not considered dishonorable to the Maasai.

Prior to the colonization period, the Maasai were fierce and relentless warriors when interacting with other tribes in the area, chiefly the Kikuyu, Akamba, and Kalenjin. Through this conflict they were able to infiltrate the territories of the other tribes. They were viewed as fierce warriors who were not to be trusted. Thus, we can account for the violent, warrior image that existed not only in the minds of other tribes, but also in their own minds.[11]

In actuality, the Maasai coexisted with the Kikuyu, Lisii, Kalenim, and Kanbe groups in relative peace. This is because of the symbiotic relationship between the Maasai and the other tribes due to different economic processes within them. The Maasai were primarily herding people who utilized the land for grazing purposes, while the other tribes farmed the more fertile pieces of land in the area which were usually at higher elevations than the grasslands that the Maasai used for grazing. Because of this, no true displacement of other tribes took place and the Maasai, as they traditionally have with these other tribes, traded leather goods in exchange for vegetables to supplement their traditional diet.[12]

With the arrival of Europeans, all this changed. The Maasai signed a treaty with the British ceding lands South of what became the Morebose-Kisumu rail line. In the process, some clans were isolated in the ceded region and moved into northwestern Kenya. Another isolated group continues to occupy lands around Mt. Elgon. This group has been slowly dying out from disease (Ebola and others), intermarriage, and in recent years, conflict with aboriginal groups in the area. The remainder of the Maasai groups adopted a policy of passive resistance, steadfastly resisting any European influences and refusing to change with the rest of the Kenyan and Tanzanian Maasai. As a result, they have been viewed as backward, overly-traditional people. The Kikuyu continue to view them in this way but offer them reluctant respect for their determination and individuality.[13] The Kikuyu, more willing to accept Western influence, adapted to new culture and values after colonization and final independence. It is difficult to determine elements of true Kikuyu culture today, however, because the Maasai rejected Western cultures and values, their culture is far more cohesive and traditional than any of their neighbors' culture.

For the most part, the Maasai have lived, in their grasslands located to the north of the rail line, undisturbed by the changes of the colonial era. In Tanzania, there have been changes through time, but the Maasai of Kenya have a very different history. As with any

[10]Gleason, *Descriptive Linguistics*.

[11]Samovar and Porter, *Intercultural Communication*.

[12]Mukame, interviews.

[13]Samovar and Porter, *Intercultural Communication*.

colonially-influenced community, location affected the degree to which change and adaptation became necessary.

The Maasai living near Nairobi, along the Nairobi-Ansha road, have been more readily exposed to westernization and modernization. In these communities, hospitals, schools, and shopping centers connect the clans with Nairobi. They are also more directly involved in tourism into the Ngorngoro National Preserve. Intermarriage with non-Maasai inhabitants of the area has also created changes in attitude toward education, lifestyle, and diet.

Conversely, the Maasai living in the Maasai Mara plains are still living their traditional lives. It is here that the reluctance to change is highest and the resistance to governmental attempts to modernize is at its strongest. The population still lives in the traditional shelters, called Manyattas, and still practice female circumcision and traditional medicine.

MYTHS, PARADIGMS, AND CUSTOMS

As with any culture, in order to understand the Maasai, it is necessary to understand the way that they view the world and the organization within their lives. Culture and perception work hand in hand. That is, to a large extent, our cultural experiences determine our view of the world. The most important experiences are taught by each generation to the next ones. Therefore, to understand a culture, it is necessary to look at experiences that were passed on to the next generation.[14] In each case, as one examines the content of the Maasai culture, it becomes evident that the primary theme of what is taught and passed down from elder to youth centers on, and is structured by, the pastoral herding of cattle. To the Maasai, cattle are the most important factor in their lives. Without this as a starting point, any attempt to understand their choices and their behavior will be misleading.

The Maasai paradigm has three primary components: (a) Nature, (b) Religion, and (c) Death, all of which influence every individual action and control the behavior of each person's lives. Nature, for instance, is always held in the highest regard. They see themselves as a part of nature, dependent on it and interacting with it rather than controlling it. Nature is seen as the father and mother that supply them with everything they need.

To the Maasai, nature cannot be changed, though if it chooses to do so, it can change itself. That is to say, *people do not control nature, they adapt to it.* This coexistence is followed to the point that no wild animal is ever killed except when it poses a direct threat to the Maasai.[15] Even in the advent of droughts, wildlife is not killed. Lions or leopards are killed only if they take human life, occasionally elephants are killed if *they* kill livestock. Small animals are only ceremonially butchered or strangled, and on rare occasions, a bull is killed ceremonially.[16] They view hunting and cultivation, processes which damage the land, as abhorrent, destructive crimes against nature, and therefore immoral.[17]

The essence of the Maasai's practical day-to-day behavior can be found in their religion. Their religion is a reflection of their understanding of the natural world. Traditional folk tales describe the Maasai decending with cattle through the sky to the earth. They

[14]Samovar and Porter, *Intercultural Communication.*

[15]Samovar and Porter, *Intercultural Communication.*

[16]Mukame, interviews.

[17]Samovar and Porter, *Intercultural Communication.*

also assert that God gave cattle to the Maasai people. Thus, it is acceptable to take cattle from other tribes, as the beasts belong to the Maasai in the first place.[18] Those familiar with the Lakota (Sioux) myths concerning the hunting of buffalo and their place in Lakota life will find in this a reminiscent rationalization for performing an otherwise immoral act.

An alternative tradition indicates that the Maasai originally believed that God gave cattle to all humans, but that the other humans killed their cattle rather than maintaining them as an ongoing resource. For that reason, all cattle that exist in the world must be Maasai cattle originally, and the act of stealing them is merely a matter of returning them to their rightful owners. In either case, the Maasai view cattle raiding as a normal and honorable undertaking.

The Maasai's religion is a monotheistic one, in which their God is called *Engai*, which is the Maa word for sky. The term *Engai* has two distinct purposes and personalities to go with it in a quasi yin–yang dichotomy. The first of these is *Engai Narok*, the black nature of *Engai*, which is benevolent and generous and shows himself in thunder and in rain, an obvious necessity for survival on the Savannah. *Engai*'s other personality is *Engai Nanyokie*, which is the red nature of God and shows itself as lightening. As one would expect from the lightening image, *Engai Nanyokie* is destructive.[19] Thus, the Maasai view of the supernatural centers in natural aspects of a god whose being expresses both benevolence and violence. This duality is reminiscent of other religious concepts beyond simply the yin–yang, including the Christian view of good versus evil, and the Hindu dual nature of God the Destroyer and God the Creator. This is similar to the Hindu dichotomy of Vishnu and Kali. Additionally, it can be viewed as a definition of order versus chaos as described in systems theory.

Engai is found in nature. To the Maasai, God *is* nature and cannot be represented symbolically. For this reason, the Maasai rejected the attempts at the Christianization of the region, as they felt that Christianity viewed God as separate from man.[20]

No priests exist in the Maasai society. The closest they come to priests are secular elders called *Liaboni* (or *Oliabon*—respected one who signs peace treaties). Elders do not either speak for God or preach. Their religious tradition is non-written. It exists only orally, in the form of stories and legends.

To the Maasai, their cattle are central even to their religion, not only because they came to earth with the Maasai, but because they are seen as the determinants of choosing mediators between God and Man, and also among men as well. Politically, they depend on village elders and heads of household who are among the oldest and richest in terms of cattle, wives, and children (in that order) in the village or clan.[21] To the Maasai, herding is the *only* successful livelihood, and thus defines ones life. Anything else is disrespectful to *Engai* and to the Maasai.

Death, the third component of the Maasai ideology, is seen as a reflection of nature, just as religion is. According to Samovar and Porter[22], death is seen as a natural consequence of

[18]Samovar and Porter, *Intercultural Communication*.

[19]Mukame, interviews.

[20]Samovar and Porter, *Intercultural Communication*.

[21]Mukame, interviews.

[22]Samovar and Porter, *Intercultural Communication*.

life, part of a cycle that is ongoing. Their reaction to death is to place a corpse in the wild, to be devoured by hyenas and other beasts, therefore completing the cycle of life. Only the Laiboni, the most revered of the elders, are buried. In either case, the body of the deceased is allowed to return to nature from which it came. Death is a completion and an integral part of the process of being a Maasai.

There are three historical episodes that have influenced the Maasai's perception of themselves, of others, and of events that occur. These are centered in the Creation, as explained in their myths of the beginning of the world, in Fierceness, as exemplified in their attitudes toward other tribes, and in the Reaction to Modernization, which is the focus of our current investigation.[23]

The culture's historical origin, and the fact that they were the only people (per their myths) to make it all the way to Maasailand explains why they feel they are superior to and separate from other tribes. They view themselves as the most fit group, being the only ones to make the entire journey. This breeds self-respect in the Maasai, an almost imperious attitude that marks all non-Maasai as inferior peoples. Further, this pride translates itself into an ethic and level of discipline commensurate with their self-image. To the Maasai, to be a complete person and have respect for one's self, one must have the virtues of obedience, honesty, wisdom, and fairness. Stringent adherence to these principles is part of the reason that the Kikuyu and other tribes often view their Maasai neighbors as steadfast. As Blauer says, "The Maasai do not steal material objects: Theft for them is a separate matter from raiding cattle . . . from this basic belief, an entire culture has grown. The grass that feeds the cattle and ground on which it grows is sacred . . . it is a sacrilege to break the ground for any reason, whether to grow food or to dig for water, or even to bury the dead."[24] Obviously, the burial of the Laiboni is all the more an indication of reverence for these elders, considering their view of nature.

Social organization among the Maasai centers on the family and small extended groups, as is necessary because of their way of life. Children are perceived to be highly valuable. Continuation of the tribe is centered on the children and in the family. "More hands make light work," they would say. Maasai see offspring as an economic investment, as it generally is in agricultural and pastoral societies, where the dependence on manual labor is paramount. This is far different from the view of technologically oriented societies, where children are viewed more as a luxury. They do not return significant wealth to the family coffers as they would if they worked the family farm or herded the group's cattle. Dependence on large families is in direct conflict with Kenya's policy of population control and family planning. No matter how poor a family may be, if a Maasai Elder has many children and grandchildren, he is considered wealthy by the Maasai.

In Maasai communities, there is no difference between children and wives. They are equally loved and cherished. In the market place, if one Maasai greets another, the general form of that greeting is, "Hello. How are ***you*** today? How are your ***cattle*** and how are your ***children***?" Note the order in which the questioner inquires after the welfare of the

[23]Samovar and Porter, *Intercultural Communication*.

[24]Blauer, "Mystique of the Maasai," pp. 497–513.

person greeted. This is an indication of the relative importance of levels of wealth in the Maasai mind.[25]

Because family and age groups are at the center of the community, children are taught in the traditional way by their grandparents rather than undergo any formal education. For men, there are four stages of life: childhood, adolescence (beginning with circumcision), *Moraniship* (warrior status, junior or senior), and Elderhood (junior or senior). For women, there is only childhood, circumcision, and marriage.[26]

Traveling with the warriors (*Morani*), the women build shelters called *Manyattas,* which are arranged in circular groups, facing inward to create what are called *bomas,* with three openings that can be closed up at night to keep the cattle, sheep, and goats enclosed. They are constructed throughout the land and are used only for few months at a time. Then the *Morani* move on to another area when the grazing is used up. Although the practice of migration with the herds has been seriously curtailed by the restructuring programs of Kenya and Tanzania, many groups still cross the border as they have always done (particularly during wet and dry seasons), whenever the cattle require a move to obtain water and grassland.[27]

This Maasai paradigm strongly reflects the value of the group or community as a whole. Sharing is a culturally embedded concept among the Maasai, as is illustrated by the requirements of junior warriors to have, within their warrior village, a close friend with whom they are expected to share everything. Beyond this, general sharing is expected within the community as a whole, to ensure the survival of the people. The Maasai say *"Keng'ar' Imurran ndaiki enye pookin, metaa ore 'Imurran asiinak neitoti Ikulikai,"* which means, *"'Imurran share all their food so that even those without cattle are fed by their fellow age-mates."* This sharing is a matter of personal honor.[28]

Space is even communal as it relates to the grazing of animals. The Maasai see no individual or joint ownership of land, merely territory belonging to all of them. At present, clans have their own areas and boundaries within traditional Maasailand. They are generally respected by others and by other tribes, yet this is more a matter of division of use than it is division of ownership. The land is owned in common by all in the community.[29]

To supplement their herding economy, the Maasai raise other, smaller animals. Within the confines of the Ngorongoro Conservation Area, the Maasai follow a pattern of sedentary life during the wet season in relatively large groups, as they have for nearly a thousand years, then break up into smaller bands and disperse so as not to tax the land during the dry season, when a small area cannot support such a large population.[30] It should be noted that among the Aleut a similar pattern can be observed, villages spring up during the summer for the purpose of trade and finding wives, then disappear with the coming winter as the population spreads out over the arctic tundra and snowfields.

[25]Mukame, interviews.

[26]Samovar and Porter, *Intercultural Communication.*

[27]Blauer, "Mystique of the Maasai," pp. 497–513.

[28]McCabe, Perkin, and Schofield, "Conservation and Development," pp. 353–366.

[29]Samovar and Porter, *Intercultural Communication.*

[30]Peter Rigby, *Cattle, Capitalism, and Class, Ilparakuyo Maasai Transformations* (Philadelphia: Temple University Press 1992).

The Maasai are able to thrive on a herding culture for much the same reason that the Plains tribes of the United States were able to thrive on a culture centered on the buffalo. In each case, the animal at the center of their economic life provided them with whatever they need. From cattle, the Maasai gain food, clothing, housing, medicine, and ceremonial services. It also defines and necessitates their nomadic lifestyle.[31] Their food consists of the milk and blood of the cattle and, in rare ritualistic feasting events, meat. Additionally, they depend on the agricultural communities with which they share the region for grain and vegetables, which they obtain through selling livestock when necessary.[32]

The restricted eating of meat preserves wealth and supports the importance of ceremonial occasions. Meat is also eaten when it is needed for gaining strength, such as in the case of a woman who gives birth or someone who is recovering from an illness. Thus, their normal diet is primarily milk and blood, occasionally meat, and limited amounts of grain. The Maasai live on about 1300 calories a day, as opposed to the average consumption in industrialized countries of nearly 3000 calories. This keeps them amazingly fit and sleek, and Westerners are often struck by their physical well being.

Like most isolated groups, the Maasai are very vulnerable to diseases brought in by outsiders. Their cattle suffered severely from Rinderpest in the 1890s, and smallpox was so extensive among the human population that the Maasai were nearly wiped out.[33]

Finally, in view of the nature of their culture, there is the question of time. The Maasai are extremely patient and relaxed in temporal matters. For them, there is always enough time. This appears to be the result of their passive, pastoral lifestyle.[34] Combined with their rather majestic physical stature, their apparently haughty attitude, and their refusal to give in to western ways, combined with their views on time, presents a picture of the Maasai as a strong, powerful, pastoral culture with deep-seated traditions that are finely tuned to their lifestyle. They are a proud, cohesive group, well versed in the ways of their land and well able to manage a subsistence strategy suited to their environment. Indeed, for more than two thousand years, the Maasai and their forerunners have lived in harmony with their environment. They not only react to it and cooperate with it, but represent a major element of its function. The Maasai and their herds, in concert with the wildlife and natural conditions in which they live, have jointly created a successful and systemically viable ecosystem. They are an integral part of the natural structure of the Maasailand region of Africa and are as responsible for the system's structure as any other element in it. There is a balance in Maasailand that has survived for millennia. The semi-nomadic shifting of herds over the plain as rain and grass shift, the burning of grasslands to improve their fertility and control the devastating tick population that attacks wildlife, and the Maasai's reverence for all living things that excludes all but the most limited taking of life, either domesticated or wild.

Into this environment, the Europeans with their ethnocentric view of what is valuable and their desire to exploit the wealth of a region they consider to be underutilized, came

[31]Samovar and Porter, *Intercultural Communication*.

[32]McCabe, Perkin, and Schofield, "Conservation and Development," p. 353–366.

[33]Blauer, "Mystique of the Maasai," p. 497–513.

[34]Samovar and Porter, *Intercultural Communication*.

to control the land and manipulate it, along with the indigenous population, to their own purposes. *Herein lies the problem.*

RESTRUCTURING AND CONSERVATION

Actually, restructuring can be viewed as having begun with the removal of the Maasai from their most fertile territory early in the colonization experience. This process resulted in the Maasai turning their backs on European interaction. Anything so totally antithetical to their way of life and ethic as dishonorable deception merely served to further relegate all Europeans to a lower category of humanity, unworthy of consideration, and they stoically refused further contact with the newcomers. Yet it was in the postcolonial period, under the new self-government rule, that the true restructuring efforts began, both in Kenya and in Tanzania. In Kenya, the shift took the form of legislation designed to create farms or ranches, areas where the Maasai could take permanent residency, own the land, and have the sole right to its use. Much of the land was marginal or only barely suitable for pastoralization, but this was not seen as a problem to Europeans, because much of the land utilized by the Maasai were marginal lands anyway. Somehow, the fact that their marginal land was supplemented with richer lands during the wet season, or the fact that private ownership of land was a foreign concept to the Maasai, escaped the planners, who were quite determined in their efforts to modernize along European lines.

Until very recently, the most accepted way to carry out the preservation of natural resources in Africa was to bar human habitation in the areas to be preserved. Evangelou points out that the "unexploited productive potential of range areas, regions mainly utilized by largely self-sufficient, non market-oriented pastoral peoples" became increasingly attractive, receiving the attention of national governments.[35] This is a natural approach for national governments to take if they wish to improve the economy and increase productivity. The rich rangeland is seen as perfect for an expansion of market-oriented agriculture and development, and pastoralists are often viewed as inefficient and "in the way." It was with an intention of improving the lot of these simple herding peoples that the process developed.[36]

The group ranch concept as a basis for development in Kenyan Maasailand was consolidated legally in the *Land Act* (Maasai group representatives were included in the formulation of this piece of legislation) and in the *Land Adjudication Act* of the Kenyan Parliament in 1968. The concept was specifically designed to promote forms of development based upon private property, in particular, land. It is through these acts of legislation that the aforementioned ranches were created.

Land set aside for private ownership was often of low or marginal usefulness. New land owners were not willing to let non-owners in the area shift herds as conditions dictated, a necessity for viable pastoral economies. These two facts ensured that not only would the non-owners be ruined, but the owners would not be able to develop any excess meat for sale as in the past. Unfortunately, this was the goal of the two acts to

[35]Phylo Evangelou, *Livestock Development in Kenya's Maasailand: Pastoralists' Transition to a Market Economy* (Boulder: Westview Press, 1984).

[36]McCabe, Perkin, and Schoffield, "Conservation and Development," pp. 353–366.

begin with.[37] Note that the concept of raising cattle for sale is totally antithetical to the Maasai traditional practices (praxes), religion and cultural heritage. European refusal to understand this fact further doomed the process to failure.

The result of this was that people continued to shift herds on a sharing or reciprocal basis after the Land Acts, particularly during the drought in Kenya in 1983. The new laws were simply ignored, a reaction typical of Maasai attitudes toward outsiders. The attempts to exploit the productive potential of range areas in this manner resulted in political confrontation. The implications of the Kenyan efforts were that progress is good, traditionalism is bad, and the Maasai stand in the way of progress. It is a typically ethnocentric argument, creating a dependency on progressive methodology (and in unwanted market economic structure), and the accompanying 'oppression' of the local people in the process of destroying their cultural integrity. This argument is quite typical of Western colonialism in Africa. Imposition of developmental 'progressive' restructuring on the Maasai indicates a conscious or unconscious preference for resource development (cattle and wild animal populations as well as arable farm lands) at the expense of the people.[38]

In Tanzania, a slightly different, though obviously parallel, process occurred. In the case of Tanzania, a national park was created in 1951 and the inhabitants of the area were restricted from the use of fire. In the past, Maasai peoples used fire to renew the fertility of the grassland by reintroducing nutrients, particularly nitrates, through ash into the soil and also to control the tick population and other livestock pests. In 1954, there were further restrictions in which the government prohibited cultivation in the area, a practice not carried out by the Maasai but rather by other groups on whom the Maasai depended for the grain supplementation of their diet. They traded cattle for the grain the other groups supplied. Since then, the Maasai have been forced by circumstances to obtain grain by selling their cattle and then using the money to purchase grain from local suppliers.

This led to outrage, not only on the part of the Maasai but also on the part of other tribal groups affected by the creation of the park. The resulting conflict stemming from this ruling led to governmental separation of the park into two distinct areas; the Serangeti National Park and the Ngorongoro Conservation Area (NCA). As part of this arrangement, the Maasai within the Serangeti National Park were moved out entirely. Most entered the NCA, which not only reduced the amount of land available to the Maasai for their cattle but greatly increased the density of population on the remaining land.

Further, no settlement was allowed within the NCA, though the Maasai were allowed to graze their cattle there under government permit. The idea was to preserve the land for wildlife as much as possible.[39]

In 1977, Kenya and Tanzania closed their common border, which served to separate the Tanzanian Maasai from all contact with the West. This only made the plight of the Maasai within the Ngorongoro Conversation Area worse. The Kenyan Maasai became a tourist attraction, while Tanzanian Maasai remained isolated. As a result, Kenyan Maasai have changed rapidly toward a market-oriented economy while Tanzanian Maasai have not.[40]

[37]Rigby, *Cattle, Capitalism, and Class.*

[38]Rigby, *Cattle, Capitalism, and Class.*

[39]McCabe, Perkin, and Schoffield, "Conservation and Development," pp. 353–366.

[40]Blauer, "Mystique of the Maasai," pp. 497–513.

Some of the laws imposed on the Maasai by the Kenyan and Tanzanian governments have specifically impacted the very essence of the Maasai culture. Hunting in Kenya was banned entirely, and was almost entirely banned in Tanzania, and as a result it became illegal for a Moran to kill a lion in order to demonstrate his bravery and hunting skills. Traditionally, without this ceremonial right of passage, a Moran could not move on to Elderhood and take a wife. The practice is no longer followed, partly because beyond the confines of the Ngorongoro Wildlife Refuge lions are scarce. Today, alternative rituals are employed to signify rights of passage.[41] Beyond this, the cause of the subsequent economic and health problems experienced by the Maasai are the direct result of the restrictions imposed on them in the NCA, particularly as it applies to restrictions on grazing and the banning of agriculture and burning of grasslands.

The NCA authorities see the problem as the lack of good management among the Maasai, a people who, with their predecessors, had been managing Maasailand on their own for over two thousand years.[42]

The consequences are heartbreaking and easily quantified. Since the westernization and conservation efforts have been in force, the size of herds among different clans has varied considerably, showing a lack of the former stability experienced by the Maasai. Whereas some variation through time no doubt existed prior to colonization, it is more extreme and patterned now, and caused by the introduction of diseases, such as Rinderpest, Bovine Pleuropneumonia, and Smallpox, among others.

Economic growth from outside the Maasai culture necessarily means unequal accumulation of stock, which to a traditionally classless semi-nomadic pastoralist society creates confusion, a breakdown of traditional culture and cultural values, and trauma to the internal operations (historical praxis) of the population.

Among the Ilparakuyo Maasai, the variations in herd size exhibited several severe trend shifts, as can be seen in the following table.

Year	Herd Size
1966	10500*
1967	12000
1968	14450
1969	16000*
1974	15000
1975	13000*
1976	12400
1077	12300
1978	10200*

*Nodes of severe trend shifts

TABLE 4-2 Changes in Ilparakuyo Maasai Herd Size from 1966 to 1978

[41]Blauer, "Mystique of the Maasai," pp. 497–513.

[42]McCabe, Perkin, and Schoffield, "Conservation and Development," pp. 353–366.

Changes in livestock units (LSUs) per capita among the Ngorongoro Maasai between 1960 and 1978 further illustrate this point by revealing a shift from traditional cattle herds to more small livestock as an alternative.

Year	LSUs per Capita	Percent of Small Stock in Herd
1960	16.55	38
1966	14.75	42
1970	15.75	39
1974	15.95	56
1978	11.20	63

TABLE 4-3 Change in LSUs per Capita Among Ngorongoro Maasai from 1960 to 1978

Herd size in one Ilparakuyo homestead from 1975 to 1987 yielded these results.[43]

Year	Number of Households	Total Population of Homestead	LSUs per Group	LSUs per Capita
1975	4	91	520	23.37
1978	5	23	380	16.52
1980	6	25	485	19.40
1987	8	—	105	—

TABLE 4-4 Herd Size, Ilparakuyo Homestead, 1975 to 1987

From analysis of this data, several things become abundantly clear. Since the restructuring and conservation efforts began, the Maasai herds have remained relatively constant in the face of increasing populations, thus lowering the number of livestock units per capita. A shift to more quickly maturing small animals with a market value has taken place.

It is quickly discernible from the statistics offered and from the information in Table 4-5 that over the past thirty years there has been an overall increase in human population and a relatively constant livestock population, thus indicating a reduction in livestock units per person for their support. Because much of the increase in human population is from *in-migration*, the resulting taxing of the subsistence strategy employed by these people has resulted in a reduction in nutrition.[44]

The shift in cattle population as a decreasing resource and the increase of small stock by 36 percent indicates a shift among the Maasai toward faster growing commercial animals rather than the longer-to-develop cattle which were used as a renewable rather than

[43]Rigby, *Cattle, Capitalism, and Class*.

[44]McCabe, Perkin, and Schoffield, "Conservation and Development," pp. 353–366.

Year	Population	Cattle	Small Livestock
1954	10633		
1960		161034	100689
1962		142230	83120
1963		116870	66320
1964		132490	82980
1966	8718	94580	38590
1968		103568	71196
1974	12654	123609	157568
1978	17982	107838	186985
1980	14465	118350	144675
1987	22637	134398	137389

TABLE 4-5 Changes in Population within the NCA, 1954 to 1987

market asset. The smaller livestock have proved advantageous to the Maasai in the face of increased pressure on their grazing lands. Sheep and goats occupy less land and graze on a range of plants and grasses other than those consumed by cattle. They are less susceptible to disease than cattle, who suffer contagion from the large populations of buffalo and wildebeest. Small livestock are open to sale at local markets for ceremonial and dietary purposes. This increases the wealth of the community and the individual without sacrificing the primary source of that wealth—the cattle themselves.

Research shows that an average Maasai family cannot support itself on livestock alone. There are many families in the NCA with less than one-half of the requisite number of livestock necessary to support a pastorally-based subsistence economy. Coupled with this are the facts that the Maasai do not supplement their diet through agriculture and there are few if any income-producing opportunities outside the family. This can only lead to reduced economic welfare. The Maasai are sacrificing long term security for intermediate gain by the composition of their cattle sales as is indicated. By the distribution of cattle stock classifications noted between 1988 and 1989, which indicates that 2% of the herds were heifers, 20% were bulls, 27% were steers (for resale, not reproduction), 47% were cows, and only 3% were young males. Not only are they switching to smaller animals, but the composition of the cattle being sold on the market indicates a movement away from the sale of only steers and bulls to the sale of cows and heifers also—the breeding stock that allows them to maintain their herd sizes.

CONCLUSIONS

Most authors of reports on preservation and development efforts in Africa agree that livestock programs have failed, both in terms of a reasonable return on investment and in terms of improvement to the pastoral peoples affected by those efforts.

One of the best examples of the complexities of carrying out such dual projects of preservation and development of local pastoral populations is the Ngorongoro Conversation Area of northern Tanzania.

Whereas the original intent of the creation of the NCA was to improve the changes of survival for livestock and reduce the negative effects of the intrusion of pastoralists on wildlife and the grasslands, studies have shown this to be an erroneous approach.

One study found that:

1. The current landscape of the area has been created by some 2000 years of interaction between pastoralists and wildlife. Hence, both are essential to the maintenance of that environment in its present form.
2. Fluctuations of livestock numbers have taken place for the past twenty years. No overall increase has been observed in that time, contrary to the economic hopes of those creating the program.
3. The numbers of wildebeest in the area have increased dramatically over the same time period.
4. Disease interactions between wildlife and cattle are to the detriment of the cattle and are on the increase. Increasing numbers of disease-carrying wildebeest and other animals further reduce livestock populations.
5. There is little or no evidence that livestock have contributed in any way to deterioration of the grasslands.
6. There is little or no evidence of erosion, another reason cited for the restriction of grazing by pastoralists.[45]

In retrospect, it seems obvious that the primary results of governmental efforts to foster preservation of the natural resources of these areas and to improve the economic lot of the pastoral populations involved have succeeded in doing the exact opposite. By attempting to artificially restructure the basics of an ecosystem that has successfully flourished for more than two millennia, the "negative" conditions have been exacerbated.

Changes in these pastoral societies, particularly among the Maasai, have occurred. The shift toward a more 'modern' westernized philosophy of both life and economics are slowly shifting these people from their traditional values and beliefs. The traditional view of economics as a division of labor effort centered in a combination of animal husbandry under the control of elders. Cattle raids carried out by the warrior class Moran and foraging and herd care undertaken by the women and children have both been modified to reflect an increasing reliance on market economy trading. Maasai trade began by trading with the Kikuyu for goods in the late nineteenth century, and now includes the participation of Maasai in not only trade of cash crops but the sale of hand-made necklaces, bead work, and other involvement in the tourist trade just to survive. Furthermore, the political structure of the culture has changed from one of loose organization based on communal land holdings with no formal political cohesion to absorption into and subjugation to the political structure of Kenya and Tanzania. In spite of their fierce determination, their simple and successful subsistence strategy cul-

[45]McCabe, Perkin, and Schoffield, "Conservation and Development," pp. 353–366.

ture has been overwhelmed by the larger societies in which they operate, and they are forced to embrace unfamiliar strategies just to survive.[47]

From the information presented, there are four specific caveats that we may offer in terms of attempts to preserve existing ecosystems, if sustainability and ecological welfare is the goal, as stated. This is particularly true of grassland savannahs.

1. Social scientists need to put their analyses in context to be certain that they are based on the local social and economic conditions of the region.
2. Policies should be made with particular attention to the economic and social conditions of a region rather than myth and unnecessary stereotypes.
3. It is critical to limit misunderstandings and foster cooperation and communication between conservationists and the local population.
4. Development among pastoralists should not be limited to the improvement of the livestock-based economy.[48]

This information strongly supports the concept of cultural relativism, an attempt to see behavior and methodology in light of local conditions and cultural belief systems. Imposing an outside paradigm on any indigenous people risks political, social, and economic dislocation and a continuation, if not an intensification, of issues sought to be resolved through modernization and the introduction of new technologies. It is important to keep methods in context with the culture in which they are occurring.

Unfortunately, the deteriorating conditions under which the Maasai have been living do not represent an isolated case. It is rather symptomatic of a failure to put decisions in that aforementioned context that reflects the fiber of the societies affected by those decisions. Progress is not an evil, not even in the case of what has befallen the Maasai. It is only mindless, ill considered, and blind progress that creates such a result. Any insistence on one set of methods over another because of supposed superiority of the methodology is a process of chauvinistically-conditioned presupposition that fails to take into account other alternatives. Perhaps as with all areas of human endeavor, technology would be better used in an atmosphere of 'appropriateness' of application.

BIBLIOGRAPHY

Blauer, Etagale, "Mystique of the Maasai," *The World and I* (March, 1987): pp. 497–513

Cole, Johnnetta B., ed., *Anthropology for the Nineties: Introductory Readings* (New York: The Free Press, 1988).

[47]Shannon Kishel, Emley McAlpin, and Aaron Molloy, "The Maasai Culture and Ecological Adaptation." Research paper, Denison University, 1999. *www.denison.edu/enviro/envs245/papers/Maasi/Maasi2.html*

[48]McCabe, Perkin, and Schoffield, "Conservation and Development," pp. 353–366.

Evangelou, Phylo, *Livestock Development in Kenya's Maasailand: Pastoralists' Transition to a Market Economy* (Boulder: Westview Press, 1984).

Gleason, H. A., *An Introduction to Descriptive Linguistics* (New York: Holt, Rinehart, and Winston, 1961).

Kishel, Shannon, Emley Mcalpin, and Aaron Molloy, "The Maasai Culture and Ecological Adaptations." Research paper, Denison University, 1999. *www.Denison.edu/enviro/envs245/papers/Maasai/Maasai2.html*

McCabe, Terrence J., Scott Perkin, and Claire Schofield, "Can Conservation and Development Be Coupled among Pastoral People? An Examination of the Maasai of the Ngorongoro Conservation Area, Tanzania," *Human Organization* 51, no. 4 (1992): pp. 353–366.

Mukame, Winnie, interviews by Paul Alcorn, January–June, 2001.

Rigby, Peter, *Cattle, Capitalism and Class, Ilparakuyo Maasai Transformations* (Philadelphia: Temple University Press, 1992).

Samovar, Larry A. and Richard E. Porter, *Intercultural Communication: A Reader*, 6th ed. (Belmont: Wadsworth Publishing, 1991).

CHAPTER 5

AN IDEA WHOSE TIME HAS COME

From the previous chapter, we learned that technology is a primary element in shaping the economic structure of the world in which we live. Without technological development, we would not be able to increase output in the face of dwindling supplies of productive resources, nor would we be able to increase efficiency or find new methodology for performing economic functions. In addition, the wide selection of goods and services available to us today to create economic welfare through their consumption would not be available, and our future would be one of stagnation, dropping production, ignorance, and descent into an increasingly primitive social structure.

Thus, it must also be true that society as a whole embraces technology wherever it can find it, rushing to utilize the growing wonders of a technological age, striving ever forward to greater heights of production and consumption. But this is not true. The reluctance of society to embrace new technology goes beyond the simple homeostatic reactions of the population. There is a very real phenomenon, and it is one that follows very natural lines.

To see what that process is, let us first consider the work of an early capitalistic economist and scholar, the Right Reverend Thomas R. Malthus. Malthus was a contemporary of Adam Smith, living and working at about the same time, and his particular interest in economics dealt with the interaction of economic structure and population growth. He is further considered to be the first political economist of the capitalistic age and is credited with being instrumental in having economics labeled the "dismal science." As we shall soon see, this was not without good reason.

THOMAS MALTHUS AND THE MALTHUSIAN PROPOSITION

Malthus presented a paper in 1798 entitled *An Essay on the Principles of Population.* The essay stirred up quite a controversy among economic scholars. Malthus presented his findings on the rise in population compared with the rise in food supplies, concluding that the human race was doomed to misery and extinction through starvation. His argument was as follows.

Consider what takes place when a rich nutrient medium such as agar is introduced into an agar plate for the study of bacterial growth. An extensive though limited (finite) supply of food is made available to airborne spores that can grow once they have landed on the plate. As this happens, it is possible to study the number and size of the bacterial colonies as they develop and experimentally determine in a short time the effect of growing populations in the face of finite resources.

By observation, it can be seen that there is a difference in the rate of growth of the colonies through the life of their development on the agar plate. In the beginning, a small number of colonies appear as small patches in the agar. However, they quickly enlarge, the size and number of patches growing at an increasing rate until a number of flourishing colonies are present. What follows next is a period of relatively constant growth, during which the number of colonies tends to stabilize and the growth rate remains about the same until the agar plate approaches saturation. The third stage is one in which the rate of growth declines and finally comes to a halt, eventually resulting in the decline of colonies and the dying out of bacterial groups in the plate.

The curve inscribed by the growth rate of the colonies in the agar plate is called a *sinusoidal curve* because it has the same shape as the sine curve of trigonometry. It is an S-shaped curve indicating, first, the growth at an increasing rate; second, the growth at a steady rate; third, the growth at a decreasing rate; and, last, absolute negative growth, or decline (see Figure 5-1).

Malthus reasoned that this is analogous to the conditions facing humankind on a planet with limited resources. Whereas our food production is growing, unlike the steady state of the agar plate nutrient, it grows at an arithmetic rate, that is, it rises in a straight line and can be represented

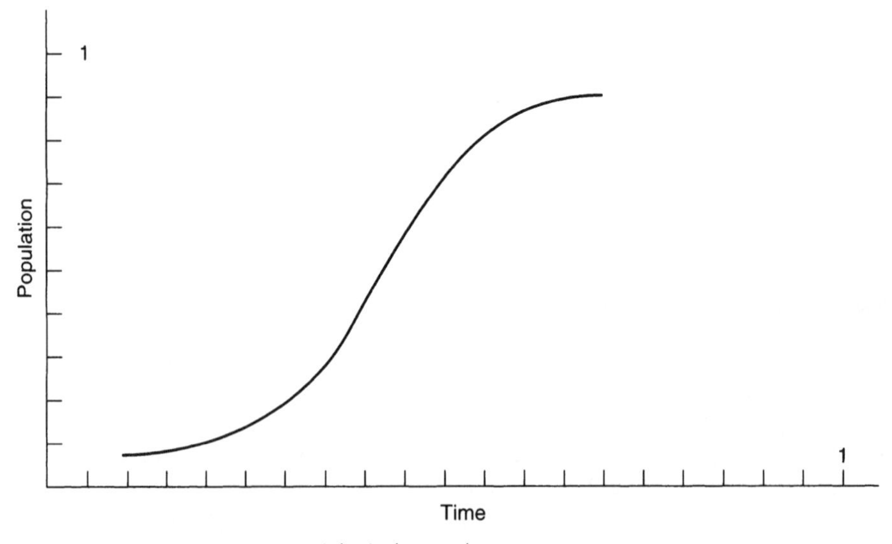

FIGURE 5-1 Biological Growth Curve

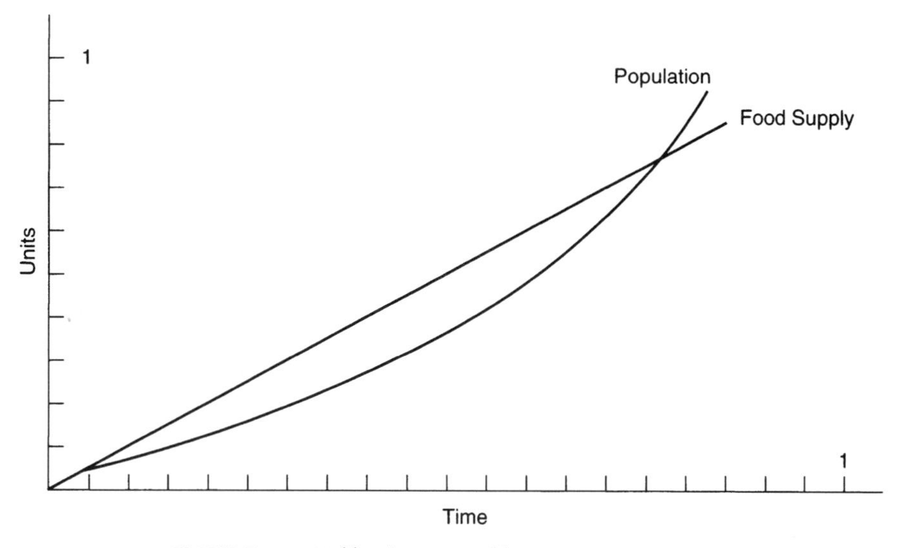

FIGURE 5-2 Malthusian Proposition

mathematically by a linear equation. The population, on the other hand, grows geometrically, doubling and redoubling in ever-shortening time spans, creating a graphic line that curves upward to the right. By comparing the two growth rates, Malthus discovered that the growth in population may soon outstrip the growth in food supplies, and when that takes place, the fate of humankind is the same as it is for the bacteria in the petri dish. In other words, according to Malthus, the human race experiences growth in a sinusoidal curve format through time.

What has happened is that the bacteria were introduced into an environment with an extensive food supply that is benign and offers the opportunity for growth. Being opportunistic in nature, the bacteria grow rapidly, sending out spores to form other colonies, their growth rate thus increasing at an increasing rate. As the petri dish approaches its maximum capacity to support bacteria, the growth rate slows. Spores are not so easily established on the surface of the dish as before, since it is now full of colonies. In addition, as with any biological entity, the bacteria begin to emit waste products that are poisonous to the organism, and this further reduces the amount of available space and agar for new colonies. Growth continues through this period, but at a much stabler rate. Finally, in the third state, the plate becomes crowded and heavily laden with waste products, producing a growth rate that declines and declines at a faster and faster rate, until the organism runs out of food and can no longer support itself in the face of the waste poisons in the petri dish; the bacterial growth ceases and is replaced by a lethal decline in the bacterial colonies present.

According to Malthus, as the human population begins to grow faster than the food supply and to surpass it in growth rate, the necessary result will be famine, pestilence, disease, poisoning of the population, and ultimate extinction (see Figure 5-2). Is it any wonder that economics was labeled the dismal science?

And Malthus was right. There was nothing inherently incorrect in his analysis of the situation. Yet we still thrive, and at population levels far beyond those predicted by Thomas Malthus. Where is the difference?

Malthus was correct in his analysis, as far as it went, but he forgot to consider one important difference between humankind and bacteria or, for that matter, between humankind and any other animal. We evolve and adjust to the environment mainly through external means, and at a very rapid rate. We *innovate,* and Malthus failed to consider this.

Early in the history of the human race, it became troublesome to gather and hunt and thrive in the process. Using gathering and hunting as an economic format requires a great deal of land per person. As the population rose, this became increasingly difficult. Rather than enter a flat period in the growth curve, as Malthus's analysis would have predicted, human beings chose instead to try slash-and-burn farming, clearing land by cutting and burning off the foliage and then planting crops to be harvested. Two things happened as a result of this. One was that humans ceased wandering, at least wherever the new technique was used. The other was that it was so successful that there was actually a surplus of crops, leading to trade and ultimately population growth. Population rose until food supplies could not keep up with the rate of growth. Early farmers then switched from slash and burn, a system that quickly wears out the land, to one of plowing and fertilizing, settling into a life of farming and animal husbandry that was to continue for centuries. The plow was invented when poking holes to plant proved too slow. The cutting blade and steelhead were invented when the soil proved to be too full of roots or too thick for easy plowing. Genetics was developed when it was discovered that what was *really* needed was a method of increasing yield per acre. First the McCormick reaper came along, then the tractor, the combine, chemical fertilizers, hybrid varieties of grains, and on and on and on. In other words, Malthus missed the fact that human beings, unlike bacteria, elephants, or virtually any other animal on the face of the earth, when faced with a deteriorating environment, merely change the environment through the development of technology and go on as before. As Figure 5-3 demonstrates, through this process of innovation, the top of the Malthusian curve is not removed; it is just pushed farther and farther back from the present. This brings us to the principle involved in much of real-world phenomena.

SINUSOIDAL CURVES: THE GENERAL CASE

Malthus had discovered a condition that reflected a special case of a general class of real-world phenomena that describe how things grow through time. This was not limited to the growth of living things. Whether speaking of the

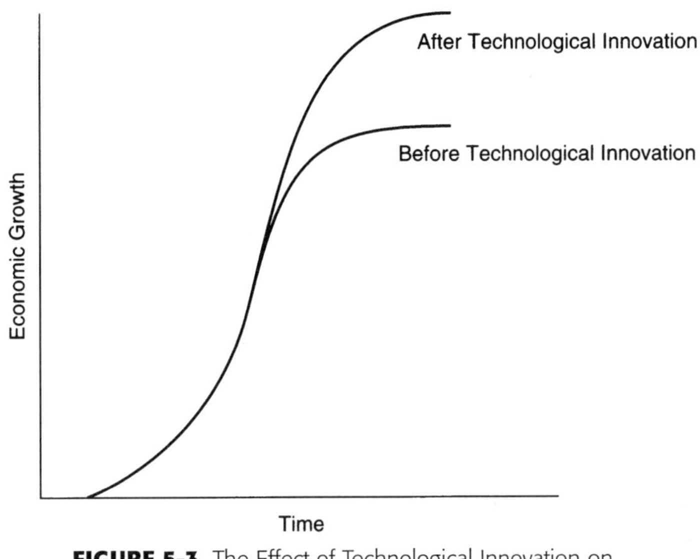

FIGURE 5-3 The Effect of Technological Innovation on the Malthusian Curve

growth of the human population or the change in the number of sunspots over an eleven-year cycle, the shape of the cycle is the same. It approximates a sine curve. Business cycles show the same effect, as do sound waves, ocean waves, electrical currents, the light curves of variable stars, the fluctuations of the Dow Jones industrial average, or the sleep cycles of mammals. In each case, there is a cyclical behavior pattern taking place through time, and that pattern tends to be sinusoidal in nature. Indeed, by definition, if it is cyclical, it must be sinusoidal in nature. The name given to this general description of growth through time is the *biological growth curve*. But what does this general case have to do with technology? As it turns out, the same phenomenon occurs in the technological cycles of modern society.

We already have discussed one such case, at least concerning the lower part of the curve. In the last chapter, the discussion on oligopolies referred to economies of scale as the cause of high market concentration. As industries grow, there is a tendency, particularly where technological innovation is an available means of increasing efficiency, to move toward bigness, thus reducing the number of companies in the market and increasing the market share that each of the surviving companies commands. It is in the economies of scale that we see reflected the lower section of the sinusoidal growth curve.

As technology is used to increase efficiency, the production rate per unit of input tends to rise, and rise faster and faster, thus indicating a geometric growth rate. This is identical to the general shape of the lower lobe of the sinusoidal curve discussed here.

The shape of the upper half of the curve is indicative of what takes place if this movement toward bigness continues. As a company expands and garners increasing levels of efficiency, it approaches a level of output in which the output resulting from additional units of input no longer indicate economies of scale. Beyond a certain point, the returns level off and grow at a constant rate, as is seen through the center of the sinusoidal curve, where the rate of change becomes almost constant. This is the relatively straight section of the curve. Companies still increase their inputs, since the outputs received are still valuable additions to their revenues, yet each unit of input seems to create the same relative output. The returns are constant, and this central area of the curve is said to indicate *constant returns to scale.*

The top portion of the curve represents the limiting case of technology's power to improve efficiency. If a company continues to increase inputs past the area of constant returns to scale, the output received from each additional unit of input *actually starts to decrease,* each unit of input being *less* efficient than the previous unit. Thus we see a characteristic reduction in the angle of increase along the top of the curve. Technically, the phenomenon is known as the *law of diminishing returns.* The law of diminishing returns states that after some point, the increase in output resulting from additional units of input will decline, first relatively and then absolutely. In other words, after a certain point, each additional unit of input is less efficient than the last, and whereas there are gains from employing that last unit of input, the gain will be smaller for each new unit of input employed. This process continues until the result of adding one more unit of input is an absolute decrease in output. There is, therefore, a *maximum level of efficiency of inputs beyond which a system cannot improve, other things being equal.*

This maximum level determines the maximum size of a productive unit. An individual company or an individual plant owned by a company will have some maximum level of output of which it is capable. Beyond that level, further efforts are wasted.

If all three of these factors are joined, coupling the economies of scale with the constant returns to scale experienced through the center of the growth rate and the decreasing returns to scale that exist due to the law of diminishing returns, the result is a sinusoidal curve.

As with Malthus's population problem, the way to push the top of the curve farther away from the present for an industrial enterprise is through the employment of new technology. A *given* technology has a maximum efficiency and a threshold beyond which it is unable to move. However, the introduction of a fundamentally *new* technology can push that limit ever outward. Thus, power methodologies have their limitations, but as we move from foot power to horsepower to steam to internal combustion and electrical generation to fission to fusion and beyond, each new technology increases the available power for our use. The same is true of other technologies as well. There is a limit on the height of buildings that can be produced using wood and brick alone. Adding stone as a material pushes this limit farther but greatly increases the mass displacement of the building without a significant increase in usable space. By switching to steel-reinforced concrete, the mass–space relation is greatly enhanced. By shifting to radically new construction methodology such as the geodesic-dome concept, the ratio is further improved, allowing for monumentally large enclosed spaces at a relatively small outlay of physical resources. In each case, as one technology reached to the edge of its limits, approaching or entering the area of diminishing returns, it became more and more economically feasible to search out and develop new methodologies to overcome the problem.

Other examples are all around us. Compare the DC motors in size and output with the smaller, lighter, and more powerful AC motors developed by Tesla. Compare the load capacity and span capacity of simple bridge construction with that of the box-girder railroad bridge, or that of the huge expansion bridges, such as the Bay Bridge and Golden Gate Bridge in San Francisco and the first great suspension bridge, the Brooklyn Bridge, still a wonder after more than a century of constant use.

So we see that technology represents a method of overcoming the sinusoidal nature of growth curves, representing a way of bypassing the hazards of the law of diminishing returns.

THE EXPLOITATION AND ACCEPTANCE OF TECHNOLOGY

In Chapter 2, we explored the subject of homeostasis, including the tendency of Homo sapiens to reserve judgment on the usefulness of technology as a survival mechanism designed to forestall a headlong charge into unknown new methods of operation that could be disastrous. Continuing this theme, the rate at which a society both accepts technology and exploits it to its fullest can be expressed as sinusoidal curves.

Consider the life of the average product in the marketplace. No matter what the product may be or, for that matter, whether it is an unsuccessful, moderately successful, or highly successful product, it will follow a characteristic life cycle from introduction to discontinuance. This cycle is sinusoidal in nature.

At first introduction, a product is presented with much fanfare, and an advertising campaign is designed to put the name and image of the product before the public as often and as well as possible. This creates an awareness of the product in the minds of potential consumers. During this period, most customers are reticent to try the new product, being locked into their customary modes of purchasing by habit, homeostasis, satisfaction with present forms of the product in question, and a real desire to avoid the time-consuming process of seeking out and testing the new product to determine its suitability. In other words, the new product exists; consumers know it exists. But they do not feel it is sufficiently valuable to spend their money, time, and effort in its procurement and consumption.

There are some people, however, who are ready to try out a new product and are willing to go through the necessary processes to do so. They are called *innovators* in the marketing world. And the name is apt. These are people who are willing to innovate, to experiment, to try new ideas, and to enjoy the process. Their level of fear and homeostatic resistance to new ideas is lower than the norm, and they are more open to new concepts and new products. It should not be surprising that very often they are younger people, who are not so solidly locked into the characteristic cultural norms of their times, having not fully tested themselves in the maturing process.

Innovators will buy the product because it is new, because it is useful (or perceived to be so), or because it is different. This latter is the old "be-the-first-on-your-block" approach to individuality.

Once the innovators have begun buying the product, it will either be found to be useful by these pioneering consumers or found to be worthless. In the latter case, they stop purchasing the product, and it falls by the wayside. If they like it, however, they will continue to purchase it, and the product will survive long enough to establish itself in the marketplace.

Now a second set of consumers discovers the product. These are called the *early majority,* and they represent a large number of people who are willing to try something new, provided it has been accepted by a sector of the population. Rather than rush right out to purchase the good, they will wait a while and see if it is still around next week or next month. They are the "wait-and-see" consumers who are willing to take a chance, if that chance does not appear too great. During this period of acceptance, the sales of the product will grow rapidly, will increase at a fast rate, and the product will become even more widely accepted and seen in society.

At this point, the final holdouts, the *great majority,* will "discover" the product, as it is now seen as useful, has been tested by the more progressive of the population, and has become an accepted part of the culture. Once this part of the consuming population accepts the product, its sales start to level off, and it tends to peak at what is called *maturity.* From here on, it is a matter of finding ways to maintain the high sales levels of the mature product and to keep customers from being lured away from this product to try others. Eventually, the product will decline as new products take its place in the market and consumers move on to other, more interesting ways of creating satisfaction through consumption.

If this process sounds familiar, it is probably because it describes a sinusoidal growth curve. Technology, to be accepted in the marketplace as a means of accomplishing ends, must go through the same acceptance phases as any other product or method that is being introduced. An early majority will always be willing to try it, but unless they decide that the new technology is useful, it will be rejected. If the innovators continue to use it, others will consider it, and the early majority begins to discover it as a means of accomplishing ends as well. Finally, it establishes itself as a benevolent methodology and is generally accepted, being used by much of the society for the good of the society. Innovation is no less a product than any other good and can be viewed as such. The questions to consider are how quickly each of the stages will be spanned and how long it will be before the technology is accepted.

And who is to say that a given technology must be accepted at all? There are many examples of technologies that were perfectly viable and yet were rejected. Their rejection stemmed not from their lack of viability but from the lack of need for what they were capable of doing. Technologies are only used if they are useful. They must wait until their time has come or be passed by.

TECHNOLOGY WHOSE TIME HAS COME

Technology, like any other social element, will only exist as long as it is useful to the ends of the society. Just as the various economic systems disappeared as conditions changed, so will a given technology give way to another when the old is no longer capable of accomplishing the goals of the society in the face of current conditions.

And yet, the time between the development of a technology and its widespread use by a society may vary considerably from instance to instance. In the case of flying, the acceptance of the technology was rapid, being spurred on by the visionary actions of a few innovators who built on the original work of early pioneers such as Claudius Dornier, the Wright brothers, and Glenn Curtiss, and by the added impetus of World War I. The world moved from Kitty Hawk to space in less than sixty years! On the other hand, the steam engine had a very long initial acceptance process. First developed in 1687 by Denis Papin, though not existent as a working model until Thomas Savery's version was finished in 1698, the steam engine was not in common use until after the first model of James Watt's engine, patented in 1769. It was not available for sale until 1775. Why the difference? What creates the delay in one form of technology while another takes nearly a century to even begin to receive acceptance? Two examples should suffice to show the difference.

Consider a technology that is really a method more than a technology, that of Boolean algebra. Anyone who has ever dealt with computers beyond a very superficial level knows well the importance of the Boolean algebraic methods in the way a computer works. The Boolean system is binary, having only two numbers in it, 0 and 1. It is through Boolean algebra that these numbers can be manipulated to represent values as powers of two, and thus allow large numbers to be stored in electronic switches as either on (1) or off (0). Yet the methodology is nothing new.

Boolean algebra was developed in 1854 by an English mathematician, George Boole. It is alleged that part of his motivation in creating the algebra was to win a prize in mathematics being offered for the most original new concept in the field. Boole merely noted that deductive logic was a system that could be symbolically described using the language of mathematics and that if certain rules of manipulation were formulated, these mathematical symbols could be utilized in combination to come to deductively logical conclusions. This is essentially what Boolean algebra does, using 0 and 1 for true or false and utilizing them to build truth tables for the solution of complicated deductively logical problems. This is all very well, but unless you happen to be a deductive logician, it may appear to be of little relative value.

Or is it? In light of what we now know, that is obviously not the case. Claude Shannon, an engineering student, noted in his master's thesis that the Boolean algebraic format could be applied to logical problems using the binary system. This was in 1938, some eighty-four years later! What was at first a mathematical oddity, of use only to those interested in the more esoteric forms of logical construct, has become the basis of an entire revolution in science, technology, and, indeed, culture.

It was not until the time was right for the fruitful application of the method that it gained importance. Thus, when it was first discovered, it was largely ignored because of a *lack of application.* Once the applications arose, the explosive use that it experienced was incredible. It is now in the midst of an ever-increasing growth pattern and is not yet fully utilized by the society.

A second case shows the exact opposite pattern. The first scientific explanation of laser, a device for focusing coherent beams of visible light (*laser,* as you are probably aware, stands for *l*ight *a*mplification by *s*timulated *e*mission of *r*adiation), was published in 1958 by Charles Townes, who shared the 1964 Nobel prize in physics for his work in developing a similar procedure for microwaves, the *maser.* By 1960, two successful working models had been developed, one by Theodore Maiman at Hughes Laboratory and a second toward the end of the year by Ali Javan at Bell Labs. Almost immediately, physicists and electronic engineers were investigating the possibilities represented by this wondrous new device. Because of the special properties of lased light, the applications were staggering. It created a highly accurate, thin, closely focused beam of light of considerable power that could be used in many fields. The laser became known as the "solution in search of a problem." In under fifty years, the laser has been utilized in weapons and weapons guidance, eye and cancer surgery, communications, measurement, fusion research, information retrieval and storage, holography, photomicrography, electroholography, chemical research, nuclear physics, and many more applications.

In this case, the conditions encountered by the Boolean algebraic process were not present with the invention of the laser. The laser came about at a time when there was a crying need for it. Investigators were on the edge of new discoveries in a number of sciences but were unable to continue because of a lack of technology. The laser provided that new technology, and expansion took place in many fields. The time was ripe for laser technology, and as with the Boolean process, we have only just begun to find uses for it.

Yet this "right time and right place" element is not alone in creating the growth curve of a technology. Also to be considered is the matter of awareness. Information expands geometrically if it is made available to a population. One person tells two people, who each tell two others, and so on, the rate of dissemination growing geometrically. And with this comes an expansion of the application of a technology. Early on, when the population becomes aware of the existence of a new technology, it applies that technology to all of the obvious problems for which it is a solution. And as more and more people become aware of its usefulness, that application expands rapidly, as with the laser. Then, as the obvious applications begin to be utilized fully, the less obvious ones are allowed a chance. Since they are less obvious, it takes somewhat longer to realize their potential. This is the steadying growth rate experienced in the sinusoidal curve form after the initial explosive flash of application, as with the laser. Finally, when the technology is mature, it reaches the limit of its application, and the expansion of its use slows, turning the sinusoidal curve at a slower and slower rate, until it peaks. At this point, the technology is fully utilized and has probably become an integral part of the culture.

The steam engine was initially used for a host of static applications, such as pumping water and running mills. Later it was successfully developed into a transportation power source, with the advent of the locomotive and, in the early twentieth century, the steam automobile. In addition, it found application in the production of electrical power, first at the hands of Edison, who opened his steam-generated electrical power plant in New York City at the Pearl Street power station in 1882, and later in municipalities across the nation, where hydroelectric power generation was not feasible. In the production of power for ships it improved through the paddle wheels of the early nineteenth century to a zenith of efficiency with the steam turbine developed by Charles Parsons in 1894 and adopted as the standard propulsion of the British navy shortly thereafter.

The nineteenth century was the age of steam just as the twentieth has been the electrical age (no matter how often we try to convince ourselves that it is the nuclear age, an epithet that is not yet logically applicable). Steam still exists as a motive force in our society, but its useful life as a technology is in the final stages of maturity, having given way to more exotic, modern, and appropriate forms of power production.

Technologies run a course from infancy through maturity in a manner similar to the life of a given paradigm, which slowly replaces earlier, less useful paradigms and is finally replaced by a new one once the solutions the paradigm offers are exhausted. Each of these instances is reflected by a sinusoidal curve. As such, the characteristics of a technology and the way in which it interfaces with the rest of the social system can be understood by determining its position on this continuum.

CONCLUSION

The biological growth curve format, the sinusoidal curve, describes a range of real-world phenomena, particularly phenomena that deal with development or growth over time. It is equally applicable to biological and nonbiological examples. Also, by studying technology in terms of growth, maturation, and decline, it becomes possible to increase our ability to understand what has taken place, what is taking place, and, most importantly perhaps, what will probably take place in the future. The characteristic slow start-up of technology in its developmental infancy can be described as the beginning of the sinusoidal growth curve, with homeostatic communication, and application problems being the main causes. The following rapid adoption and expansion of applications can be seen as the increasing growth rate of an expanding system, followed by the steady growth of a fully internalized technology, being developed to its fullest. As the utilization of the technology reaches its zenith, the high-end diminishing returns will slow growth to the point where the technology is fully utilized. It peaks here and either continues as a useful methodology or declines as new technologies replace it as a means of solving new and extended problems for the culture.

Thus far, we have dealt with the philosophy and the theory of how and why humans create and use technology and why it is so important to our daily lives. We have seen that it is a natural process, that it is innate and inborn to create technology, and that it is natural to resist that technological bent so that we do not move too quickly to accept what may be dangerous to us in the long run. We naturally recognize that such a powerful mechanism may hold within its structure the capacity to destroy as well as create.

In the next section, we will develop both methodologies and approaches to the study of the subject so that we may develop expertise in beginning with raw information and then placing that information appropriately in a framework that allows us to draw logical and demonstrable conclusions. These methods of study, including modeling, game theory, simulations, and, most importantly, a systemic approach, will make the task of studying technological and social changes a much easier, more organized and potentially fruitful venture.

KEY TERMS

Biological Growth Curve
Innovation
Thomas Malthus
Sinusoidal Curve

REVIEW QUESTIONS

1. Who was Thomas Malthus and what did he discover?
2. How does technology affect the Malthusian Proposition?
3. What is the natural relationship between population growth and food supply?

4. What is the difference between the creation and the exploitation of technology?
5. What determines which technological innovations are created at a particular point in time?

ESSAY QUESTIONS

1. How can the Malthusian concept be used to describe and predict other phenomena that we encounter in our society? Give several examples.
2. Expand the Malthusian concept into a statement of general principle regarding the physical world. Explain your statement.
3. If technology gives us the capacity to extend the Malthusian curve through innovation, is there an upper limit on how far this process can be used, or are we ultimately doomed to experience the downward second half of the sinusoidal curve?
4. According to the text, ideas arise at critical times when the sinusoidal curve is undergoing a shift in rate of increase; thus, we avoid the natural downturn to follow. Develop a scenario supported by historical example to describe the expected consequences of a failure of new ideas to develop in a society.

THOUGHT AND PROCESS

1. If you have access to a petri dish or other flat glass dish, try to duplicate the biological growth curve. In the plate, place a solution of nutrient material (pure gelatin works well) or a small piece of moist bread and periodically observe the growth in colonies that develop. For an added dimension to the experiment, prepare two plates. Leave one open to the air to allow a continuous inflow of microorganisms, and place a jar or some other transparent cover over the other to limit further influxes of migrating spores after the first initial colonies appear. In this way, it is possible to study by analogy the effects of immigration on the life cycle of an environment, as in the difference between population growth of North America before and after the migration of Europeans began in the 1500s.

2. The development of applications for technology has been purported in this chapter to follow some type of sinusoidal curve form. To test the truth of this, try a simple brainstorming technique. With four or five others, pick a technology from the following list of hypothetical technologies and speculate on the possible uses to which it could be put. Have one person write down the ideas in brief while an outside party records the *rate* at which the ideas flow through time. (An easy way to do this is to note the time passage periodically by starring the last idea listed each two minutes, say, in ten-second intervals.) Give yourself about fifteen minutes for this exercise to be sure that you explore all possibilities. Also, allow yourself to loosen up enough to verbalize some of the more farfetched applications that you think of. The results of this experiment, when plotted, should approximate a biological growth rate curve.

Hypothetical technologies include the following: (a) an antigravity machine, (b) a teleportation device, (c) a faster-than-light engine, (d) chemical intelligence boosters, (e) a means of tapping the atmosphere for free energy, (f) telepathic communications, (g) a worldwide master computer, and (h) direct interface between the human brain and microchips.

3. With a little research (in some cases, none) it is possible to get some sense of where we stand in relation to our technology. Draw a sinusoidal curve, and place the following list of common technological devices and methods on the curve in terms of their position in their useful life. Disregard whether the individual technology has been in existence for a long or short time. Concentrate on their position on the curve, not on their relation to each other.

Common technological devices and methods include the following: (a) the laser, (b) the steam engine, (c) electricity, (d) automobiles, (e) diesel engines, (f) the computer, (g) the internal combustion engine, (h) genetic engineering, (i) steel, (j) plastics, (k) wood, (l) refrigeration pump technology, (m) airplanes, (n) glass, (o) electronic synthesizers, and (p) television communications.

4. Research the discovery and development of Teflon® as a product. How long did it initially take the innovators to discover the product? How long did it take for the upswing in use by the early majority? Where are we now on the Teflon® use curve? Why?

5. One of the most rapid discovery-adoption-maturity-decline patterns available in the real world is the fad cycle of certain products. Among the products exhibiting this rapid rise and decline are such things as hula hoops; yo-yos (which seem to return every generation or so); pet rocks; and knickknacks with certain themes, such as frogs, cats, mushrooms, rainbows, and cartoon characters. The skateboard is an example of a fad that never quite died. In light of our discussion in this chapter, what other fads can you think of that did not die, and what makes them different from those that did? How does this reflect on the acceptance and maturation cycle of technology?

The phenomenon known as television is the twentieth-century equivalent of the printing press. It offered an entirely new method of communicating information to the general public. Like the printing press and radio before it, television offered a more accessible means of receiving information in the form of entertainment, news briefings, and cultural events. Its primary advantage over the other forms of communication was immediacy, the fact that it happened in real time, and, most importantly, that it was visual.

Television is considered a cold media, requiring little participation on the part of the viewer. It presents information in a variety of forms simultaneously to an audience that may passively absorb the message. Nothing like it had ever existed before, and it continues to dominate much of our lives. How much of culture is the result of the image of ourselves presented by television? How much of our way of thinking has been molded through our observation of "the great cyclopean boob tube"? As you read the following report, note not only the technological history of the television but also the way it has permeated every aspect of modern life and brought the world to a new understanding of who we are, who our neighbors are, and what we believe to be reality.

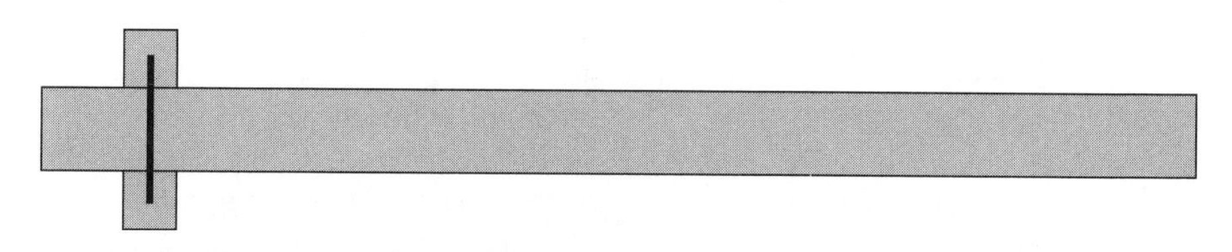

Television and Cultural Change

FRED ABOUNDADER AND JAMES A. HUTCHINSON*

INTRODUCTION

Is television one of the most beneficial inventions ever devised by humankind? Or is it one of the greatest curses? In this report we are going to examine some of the pros and cons of TV and let you, the audience, decide. We have divided our subject into six categories, as pointed out here in our outline.

1. Development of television
2. Influence of television
3. Corruption of television
4. Educational aspects of TV
5. Communication capabilities of TV
6. Future aspects of TV

DEVELOPMENT OF TELEVISION

Modern television, now viewed by millions of people around the world, is a result of over eighty years of research and development. Scientists in Britain, Germany, France, the USSR, and the United States contributed to early experiments in television. However, it was Britain, the United States, and Japan that solved the problems leading to a full television service.

*Fred Aboundader and James A. Hutchinson, "Television and Cultural Change" (Paper presented at De Vry Institute of Technology, Atlanta, Georgia, April 24, 1984).

The dream of extending human vision, as the telephone had extended human voice, began to be realized in 1883. In that year, a German scientist, Paul Nipkon, invented a scanning device that would break down an image into a sequence of tiny pictorial elements. This crude mechanical scanner was used with a photoelectric cell that converted light into electrical impulses.

Experiments with mechanical scanning were pursued in the 1920s. In 1925 Charles Francis Jenkins, an American inventor, used elaborations of the Nipkon disk to broadcast silhouette pictures from his workshop in Washington, D.C. The Scottish inventor John Logie Baird made a public demonstration of television in 1926, but he only produced shadow pictures. However, by 1928 he had broadcast television pictures in color, outdoor scenes, and stereoscopic scenes.

Ernest J. W. Alexanderson began daily TV tests on the experimental station W2XAD in 1928. On September 11, 1928, General Electric (GE) presented the first dramatic production on television. It was "The Queen's Messenger," with the sound carried on the AM radio station WGY. In 1931 the Radio Corporation of America (RCA) made experimental tests over station 2XBS in New York. David L. Sarnoff, president of RCA, predicted that within five years television would become as much a part of our life as radio.

Despite these early successes, mechanical scanning had inherent drawbacks. In particular, it did not provide sharpness of detail. Consequently, further advances depended on the development of electronic scanning.

The idea of electronic scanning dates as far back as experiments by Heinrich Hertz in the 1800s, to the publication of the theory of photoelectric effects in 1905, by Albert Einstein, and to Karl Braun's discovery that he could change the course of electrons in a cathode-ray tube. It was in 1907 that the English scientist Alan Campbell used a cathode-ray tube at the receiving end.

World War I put an end to all but theoretical work, but its aftermath brought to America one of the outstanding scientists in the development of modern television. He was Vladimir K. Zworykin, and in 1923 he had developed a crude, but workable, partly electronic TV system. In 1930 Philo T. Farnsworth developed a new electronic television system that made TV pictures suitable for the home. In 1939, the first television sets were manufactured for the public.

It was in 1935 that David Sarnoff invested $1 million in television program demonstrations. Also in 1935, the British Broadcasting Corporation (BBC) took over the control of British television from experimenting companies.

It was in 1939 that the National Broadcasting Company (NBC), a subsidiary of RCA, began a regular TV service. 1939 was the first time a U.S. president was televised, the first broadcast of a major league baseball game, and the first broadcast of a college football game.

In 1941 frequency modulation (FM) broadcasting was adopted for transmission of sound for television. In 1941 the Federal Communications Commission (FCC) authorized commercial TV beginning on July 1. WNBC, New York, became the first commercial station. CBS presented the first television newscast on December 7, 1941, reporting the events at Pearl Harbor. Consequently, World War II greatly slowed down the development of TV, and what television broadcasting remained was used for purposes of civil defense and air raid warden training, Red Cross instruction, and U.S. bond sales.

By 1948 television was well on its way. There were thirty-six television stations on the air, seventy under construction, and about one million sets in use. However, on September 30, 1948, the FCC declared a freeze on the licenses of any new TV stations in order to study frequency allocations. When the FCC lifted the freeze in April 1952, it allocated more than 500 stations to the VHF band and more than 1400 stations in the UHF band.

Now we have more than 150 million households with television, and the rest is history.

INFLUENCE OF TELEVISION

Indeed, it can safely be said that no other force, in so short a time, has exerted such a powerful influence on so many people. In the last three decades it has had an enormous effect on family relationships, entertainment, education, politics, advertising, news, sports, and other areas of human endeavor. Such cases reflect TV's increasingly pervasive influence on America, both good and bad. In America, where television has become the eyes and ears on the rest of the world, TV is a primary force determining how people work, relax, and behave. Recent studies show that the lives of Americans from their selection of food to their choice of political leaders are deeply affected by TV, and the influence is growing.

Let's take an example here: When one woman's TV set broke down, she said, "It's like somebody in the family just died."[1] Some people admit to being "TV intoxicated," needing a daily fix of it as a drug addict would need drugs. That is an interesting point, because a lot of people are TV addicts and don't even know it. How can you tell? Ask yourself these few questions, and see how you answer them.

1. Do you look forward to the end of the day so that you can watch your favorite TV program?
2. Do you keep the TV on after your favorite program is over and keep watching others?
3. Do you do the above (1) and (2) night after night?
4. Do you keep the set on even when you are not actually watching it?
5. Do you turn the TV on in the morning if you have the opportunity?
6. Are you irritable during an evening if you cannot watch TV?
7. Do you make excuses for watching too much?
8. Do you spend more hours watching TV than all other leisure activities combined?
9. Would you rather watch TV than be with friends or do things with the family?
10. Do you become defensive if accused of watching too much TV?[2]

If you have answered yes to a number of these questions, then this suggests that some degree of TV addiction has already set in. Don't fret. Not only do I have questions, but I

[1]Frederick W. Franz, "What Are People Saying about Television?" *Awake* (April 22, 1982), p. 3.
[2]Franz, "What Are People Saying about Television?" p. 17.

also have some possible solutions. To help control TV watching, some people have put their set in a place where it is inconvenient to spend long hours with it. Some have put the TV in a cabinet or a closet, requiring effort to prepare for viewing. Also, since a bedroom is too conducive to lying down and watching for long periods, many will not have a TV there. Research shows that older women, 55 and over, watch TV the most on a weekly basis. This seems possible because older women generally outlive men, and many may be poor and may not be able to afford other forms of information and entertainment. Also, the elderly and shut-ins have found TV useful, since it helps to combat loneliness.

Another example of powerful influence is the advertisers who use TV as a medium of communication. Their advertising conditions people mentally so that they will buy their products. Advertisers invested some $11 billion in 1980 in order to present their commercial messages before the public.

Another point discovered in doing this report was how TV affects family relationships. Some publications state that families are brought together by television. This may be true from a physical standpoint, but we believe it has actually created a communication gap at home. In essence, families do not get the opportunity to share day-to-day experiences or participate in the normal give and take of family communication due to the competition of prime-time TV viewing. TV makes it harder for some people to relate to others because it is difficult to make the transition from watching TV to real people. This may be because viewing TV requires no effort, whereas dealing with real people requires effort.

Another harmful effect of too much TV viewing is in regard to actual physical health. Many people eat between meals while watching TV, which contributes to excessive weight problems. Television viewing requires no physical participation. Prolonged periods of inactivity are detrimental to the human body; moreover, inactivity can also be detrimental to the human mind.

CORRUPTION OF TELEVISION

Television with all its pluses in communication and information has one big minus—its corruption. I am not talking about the corruption of the television industry, but rather the corruption of American society. This corruption of society is brought about by the frequent use of crime, violence, sex, and abusive language on the home screen. This continual viewing of violence, crime, and sex affects all ages, but none more than children and teens.

Children, the most easily influenced creatures on the earth, view TV on an average of about 4.2 hours daily; this is approximately 33 percent of total daily awareness hours. It is true that most children, ages 6 to 13, spend an average of 3 more hours per day viewing television than studying. This would not be so bad if the television shows they watched were informative and educational. Since 1975, public complaints about violence and sexual items on programs have rapidly increased.

For instance, in Los Angeles in January 1975, Metromedia-owned KTTV dropped a half-hour segment of the *Superman, Batman, Aquaman* cartoon. It was labeled violent for children by the local chapter of the National Association for Better Broadcasting (NABB). Also in 1975, a California woman sued television producers of the film *Born Innocent,* charging the program's graphic rape scene may have stimulated a similar sexual attack on her daughter. It isn't that the public isn't aware of the violence, sex, and abusive language; on the contrary, they (the public) have accepted it as "normal" in their society.

Children cannot discriminate between reality or fiction in television shows. What they view is stored subconsciously in the back of their brains to be recalled later on in life. Studies show that children who have viewed violent TV productions at very young ages are more disruptive and violent than their peers who viewed mostly educational shows, such as *Sesame Street* or *Mr. Rogers' Neighborhood,* or those who viewed very little television.

Every day on cable television's MTV, children as young as five regularly watch women in chains and people being tortured and shot.[3] This was stated in the *Atlanta Journal and Constitution,* April 16, 1984, afternoon edition. What is so ironic is that on December 20, 1982, Jabari Siniama, access director for Cable Atlanta, was quoted as stating, "In no other city in the country will you see anything like it."[4] What he was referring to was a video on New York City cable that showed a rock group chainsawing a female victim to death. Now, less than two years from when he was quoted saying this, Atlanta has its own Atlanta Rock Review, which airs videos often viewed as equally distasteful as the one in question in New York City.

In the May 1982 issue of *Gallup Report,* a monthly magazine that reviews statistics and surveys, the cover article dealt with TV and crime. The report sees a cause-and-effect relationship between TV and real-life violence. As television has increased its broadcasting of crime and violence on public cable, real-life crimes, especially sexually related and violent crimes, have increased. For example, in 1976 Los Angeles police asked NBC to set up a special screening of a *Police Story* episode that some believe may have inspired a killer to slash the throats of three sleeping derelicts on skid row. The plot of the *Police Story* program featured the same kind of crime.

The Federal Communications Commission also noted that nationwide levels of crime and violence are rising; it observed that an inordinate number of high-tension, crime-drama shows are running night after night on prime-time television stations across the country. So why do networks continue to show explicit violence and sex despite the adverse results?

Network officials assert that most programming is determined by extensive survey of the public. "Basically, networks produce the kinds of programs that the audience demonstrates it likes," said Herb Jacobs, president of Telkom Associates. "If the public wants police stories and Westerns, it's pretty hard to dramatize them without violence."[5]

The most obvious change is in the language being used on the air. Only a few years ago words such as ethnic characterizations would not have been broadcast. Evidently, today these slang terms are used commonly in everyday dialogue.

All these changes of more violence, crime, sex, and abusive language on prime-time television reflect changing attitudes in American society. Basically, networks give the public "what the public wants."

[3]Barbara Jaeger, "Violence, Rock Music Counterculture Art Could Pose Problems for Media Culture Kids," *Atlanta Journal and Constitution,* 16 April 1984, p. 1–B.

[4]James Gosh, "TV Video and Children," *Wall Street Journal* (December 1982).

[5]John Anderson, "What Is TV Doing to America?" *U.S. News and World Report* (August 1982), p. 26.

EDUCATIONAL ASPECTS OF TV

One beneficial use of TV is for education; good TV programs can certainly teach children and adults many things. From the beginning, educational television has had a dual purpose: to supply programs for school use as a supplement to the teacher during the day and to supply programs of community interest and cultural enrichment at night. Early studies showed that teaching by television generally was more effective than most classroom procedures; therefore, educational television networks are very promising. As for cultural programming, the need for it increased as commercial TV became less and less a service and more and more a business. Since 1952, the number of channels reserved for educational TV has increased from 242 to 309.

As a matter of fact, some local colleges use TV stations for teaching credited college courses. Dekalb College, for instance, teaches such subjects as sociology, psychology, and computer programming by TV. The students only come in four times a quarter to take examinations. I don't know how effective this system is, but it seems to be freeing the student from actual class time.

Experts say that, when used properly, TV can stimulate reading. It can present ideas that encourage viewers to want more information so they will get reading materials that will add to their knowledge of the subject.

I want to deviate from the subject of education to tie in some other significant point. Not only does good programming teach the public, but bad programming does an equally effective job. For example, convicts have admitted to getting ideas for crimes by watching programs in prison. In one survey, a surprising 90 percent said that they had actually learned ways to improve their criminal techniques through TV viewing. Four out of ten said that they had already tried specific crimes that they first saw on television. One prisoner said, "Television has taught me how to steal cars, how to break into establishments, how to go about robbing people, and even how to roll a drunk. Everybody is picking up on what's on the TV."[6]

COMMUNICATION CAPABILITIES OF TV

Since World War II, television has become the most popular mass communications medium in history—perhaps the first mass medium to reach all segments and groups in a society. Next to sleeping and working, Americans spend the greatest part of their time watching television, and TV viewing clearly is a pervasive social activity in some other countries as well.

TV is an effective medium of communication because it brings information about current events to us faster than do magazines or newspapers. And it does so in a form that is highly interesting to the eye: in motion pictures. We feel as though we are actually present, witnessing what is going on sometimes thousands of miles away.

The transmission of commercial television across an ocean first became possible in April 1965, when Early Bird (Intelsat 1), the first commercial communications satellite, was placed over the Atlantic. International TV broadcasting on a global basis was first established in mid-1969 by a global system of Intelsat 3 satellites.

[6]Franz, "What Are People Saying about Television?" p. 10.

In addition to transmitting television, telephone, telegraph, digital data, and facsimile signals simultaneously, the Intelsat 3 satellite has a multipoint communications capability that is particularly useful for distributing TV programs. What this means is that programs produced in the United States could be broadcast for simultaneous viewing in countries all over the world.

In researching further, TV in the 1960s provided a number of other firsts in the field of communications. For example, TV became a triumph of communications technology, highly interesting, and historical on that memorable day in July 1969. Of course, many will recall that when Neil Armstrong and Edwin Aldrin made the first manned landing on the moon's surface, an estimated 600 million people in some fifty countries on six continents witnessed this event as it was happening.

Also, TV had another first in the area of politics when it was used to air the debates between John F. Kennedy and Richard M. Nixon. These confrontations gave the candidates an opportunity to discuss differences on issues for immediate rebuttal. This gave the public a chance to see and hear the candidates. An interesting comment made by the Mass Communications Research Center regarding the results of the debates was, "Kennedy did not necessarily win them, but Nixon lost them."[7] The debates definitely did much to create John Kennedy's "image."

After Kennedy won the election, he was credited with introducing "live" TV press conferences from the White House.

Some people have grown increasingly troubled by some of the effects of TV on democratic government. In 1980 networks declared Ronald Reagan the projected winner soon after polls in Eastern states closed but before balloting ended in the West. Experts say some prospective voters never went to the polls in the West, believing their votes would make no difference.

To sum this up, we can see that the TV, as a medium of communication, is the most effective means of mass communication known to humankind. It has linked together over 118 countries—all nations in the Western Hemisphere, all nations in Europe, over one-half of Africa, and most of the nations of Asia. These countries all have a privately, publicly, or governmentally owned television service.

FUTURE ASPECTS OF TV

Television is a very important aspect of today's society, and it will be much more of tomorrow's. With technology spreading so rapidly, we might all be wired up to a two-way television system as early as the year 1990.

A two-way television system is where we, the viewers, have a computer tied to our cable system and are able to give the station some kind of feedback. Already in America there are over 90,000 homes with some type of two-way system; however, a large number of these systems just have response buttons and not computers. Two-way systems allow the viewers to respond to opinion polls, advertisements, and shows; they can also allow subscribers to do their shopping and banking from the convenience of their homes.

[7]*Encyclopaedia Britannica: Macropaedia,* 15th ed., s.v. "Television."

For example, an average housewife's morning might consist of going to the grocery store, picking up a few items at the department store, paying the utility bills, and depositing a check at the bank on the way home. All this might take several hours.

However, if she were living in a "wired society," she would only need to walk over to her television set and spend maybe 15 to 20 minutes typing in her morning chores. This leaves her 4.5 hours to do other things.

It may seem now that the telephone system has the market on two-way communication, but it may not be so six years from now. It is already possible to hook the television into the telephone and have vision as well as voice. TV-phones have been around for over two decades, but weren't feasible until now with the invention and application of the microprocessor.

Imagine everybody with a TV-phone: your friends, your family, the hospital, the department stores, or maybe your date for the evening. This would almost eliminate so-called "blind dates." TV-phones will bring people who are far apart very close together. They will see each other's clothes, hairstyles, and facial expressions. As the old saying goes, "A picture is worth a thousand words."

Another aspect of television in the future is the screen size. Television screens will get larger and larger; this is already true of the new projection television.

When you think of a large-screen television you might think of a 21-inch diagonal, but the screen size I am referring to is about a 10-foot diagonal. You might think I'm crazy, but think about it. With the rapid advancements in digital techniques, it may not be too far away.

Also I believe that as technology advances and the screens get bigger, the picture itself will become more realistic. What I mean is that the television will broadcast more three-dimensional (3-D) movies on the air, but the 3-D I am referring to will be more realistic than in the past. Television broadcasts in 3-D combined with television sets capable of receiving 3-D broadcasts will give a much better portrayal of reality.

Overall, however, the most important part television will play in the future is as a two-way communication medium. Once the lines have been installed and the computer-based two-way systems are more common, there will be no end to its limits.

BIBLIOGRAPHY

Anderson, John, "What Is TV Doing to America?" *U.S. News and World Report,* August 1982, pp. 24–29.

Encyclopaedia Britannica: Macropaedia, 15th ed., s.v. "Television."

Franz, Frederick W., "What Are People Saying about Television?" *Awake,* April 22, 1978, pp. 3–20.

Gosh, James, "TV Video and Children," *Wall Street Journal,* December 20, 1982.

Jaeger, Barbara, "Violence, Rock Music Counterculture Art Could Pose Problems for Media Culture Kids," *Atlanta Journal and Constitution,* April 16, 1984, p. 1-B.

McLuhan, Marshall, *Understanding Media.* New York: The New American Library, 1964.

Naisbitt, John, *Megatrends.* New York: Warner, 1984.

Peter, Fred, "Future World of Television," *Science Digest,* September 1980, pp. 16–21.

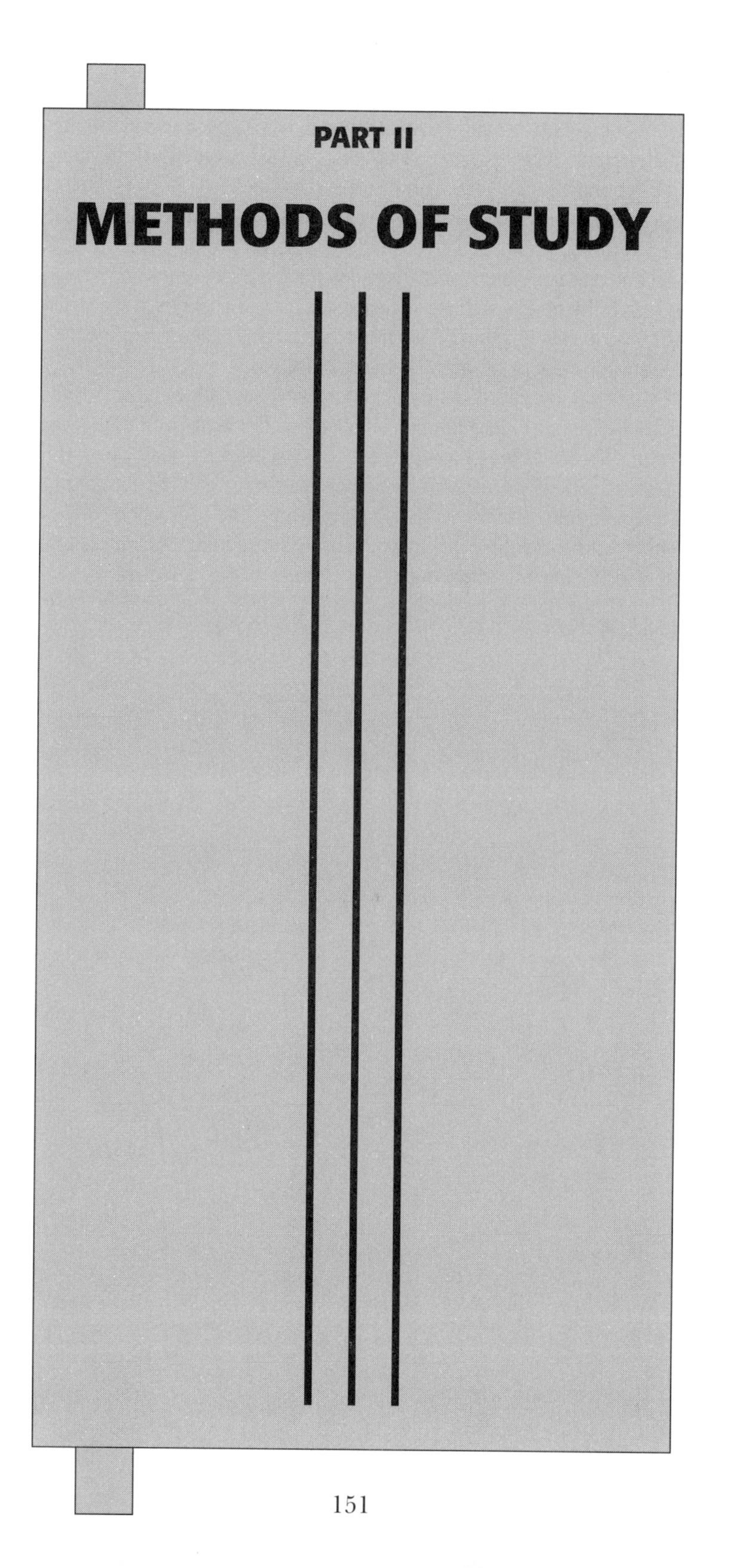

PART II

METHODS OF STUDY

In Part I, we explored the nature of technology and how this natural process is 'built in' to the fabric of being human. We saw that technology is an evolutionary human trait, that it is offset by a natural tendency toward homeostasis (the two survival mechanisms available to humans almost exclusively), and that this combination expresses itself in the creative process to the economic ends of societies. We have seen that economic systems are organized to provide cultures with the tools of survival. We have seen that these tools are largely technological in nature and, further, that we adjust our productivity and level of creativity to match the needs of those societies, reacting to changes in the environment by adapting to those changes through technology, in many cases through a process of adapting the environment rather than changing ourselves.

In Part II, we begin to explore in a more general way how this behavior relates to the general behavior of nature *in toto* through an approach to the study of the universe called dynamic systems theory. As we proceed, the theory will become clearer, and its relevance to our study of technology will become evident. It will show that the behavior of our highly developed technological approach to life is nothing more than a variation on a theme that runs throughout nature and the physical world. By understanding this theme, we can begin to predict how technology will affect our lives and those of our descendants.

CHAPTER 6

THE CAUSE AND EFFECT OF TECHNOLOGY AND SOCIETY

CAUSE AND EFFECT: THE ROOTS OF PREDICTION

If something occurs once, it is an incident. If it occurs twice in the same way and under the same circumstances, it is a coincidence. If it happens again a third time, then we have the beginning of a pattern, a way of proving cause and effect. The comparison of events to determine patterns is something that we have all experienced throughout our lives. If we miss a parking place whenever we arrive at work after 9:30 A.M., we learn that there is a pattern and attempt to arrive prior to 9:30 A.M. If certain foods create discomfort whenever we eat them, we learn to not eat them. If the new boss is grumpy three Mondays in a row, we expect a similar mood on the fourth. And in each case, our expectations about the future are based on a comparison of similar events from the past.

Cause-and-effect relationships are a fundamental element in studying the manner in which various aspects of society affect one another. It can be seen from the preceding chapters that the desires of individuals interacting in the economy caused their behavior so that a given result (effect) was created. Given the desire to maximize profits (the theoretical aim of any economically rational human being), the result was the production and distribution of goods and services in the economy. This is purely a matter of cause and effect.

And the fact that individuals took certain actions in order to create the desired results is an indication of how much people depend on their understanding of cause-and-effect relationships to create the world as they wish it to be. All of us, every day, live in accordance with our understandings of the cause-and-effect relationships that exist in the physical world. As such, we are predisposed to view things in the light of this knowledge.

Imagine what it would be like if the world were constructed in a random manner or, just as importantly, if people behaved as if the world were randomly ordered. What could we depend on? What could we know about the future? What would we expect to happen with our every move? It would be impossible to get up in the morning and go through a normal routine without predisposing ourselves to the belief that the cause-and-effect relationships that we had experienced in the past were going to hold true today as they had before. The sun has risen every day of your life. Without cause-and-effect assumptions, you would not know to expect it to happen again tomorrow morning. In fact, it is this determination of cause-and-effect patterning that allows us to study and manipulate the physical world in the first place.

We make these assumptions about what will and will not happen on the basis of statistics, which brings us back to the statement made at the beginning of this section: One event is an incident, two identical events are a coincidence, and three begin a pattern that could indicate a cause-and-effect relationship.

We are always gathering data and comparing them with other data, counting the times certain events coincide. Based on this analysis, we reach conclusions that are assumptions about the future. The more data we have indicating that two events are connected, the greater our belief that, given one set of causative factors, another set of results will occur. With three events, we have sufficient data for a pattern. With less than three, there is not enough security in our understanding of the situation to allow us to predict with any degree of confidence what will happen next. And we do this whether we are consciously aware of the process or not.

As we record more and more instances of circumstances and events, we increase the odds that our assumptions of cause and effect are correct, and we are more secure in behaving in accordance with those assumptions.

Consider the infant who is constantly going about the process of learning what works in his or her new world. Reaching out and experiencing objects creates a store of information upon which to base expectations. The infant is fascinated with everything. Anything new, anything different that is within reach and within the power of the infant to explore will probably be investigated. Bright colors, interesting objects of new shape or size, the soft cooing of a parent's voice, or the sound of a music box are all eagerly explored by the infant, who is attempting to gather data. And when the same objects are experienced in the same way over and over again, the infant is able to create patterns, to connect characteristics that define objects as different from one another. They can be classified as fun or not fun, enjoyable or not enjoyable, new or previously experienced. They can be classified as something to be sought, such as a ball that makes a delightful tinkling sound when it rolls, or something to be avoided, such as a hot light bulb that causes pain when touched, or an object that the baby is not allowed to touch and, if touched, results in rejection behavior from parents.

Many cause-and-effect relationships are learned in these first years. If you drop a ball, it falls. If you touch a hot light bulb, it burns. If you dribble your food, your patient mother or father merely puts more in your mouth until you learn to swallow it properly.

And we continue to make these associations through life. They are the pigeonholes into which we place our understandings of experiences and form our worldview. As mentioned in the discussion of creativity, this paradigm construction can limit our ability to deal with problems by narrowing our focus. Yet without the cause-and-effect processing, we would be hard pressed to progress beyond a very primitive stage of existence.

RIPPLES IN A POOL

To the social scientist or to the technologist studying the effects of society and technology on each other, the importance of the cause-and-effect relationship lies not only in its basic nature but also in the fact that each effect becomes a cause and each cause is in reality the effect of some other cause. This is an integrated system, in which no single entity can act or react without causing further results (effects). The phenomenon is often referred to as "ripples in a pool," analogous to the spreading wave patterns emanating from the locus of a pebble dropped into a body of water.

When a single event occurs, it causes other events to occur. Each of those events will have consequences, causing, in turn, other events, so that the occurrence of a single event spreads its influence in ever-widening patterns, affecting in some way every element of the system in which it exists. For most events, these effects are small, but for some, they are quite large and, unfortunately, not always immediately apparent. This was the case cited in the example in Chapter 4 of the producer of farm goods who expelled wastes into a nearby stream. The widening ripples of the producer's actions encompassed the entire community within which the plant existed and, if followed far enough, will have some minuscule effect on the economy of the entire world. Primary causes create effects that result in further effects, called *secondary effects,* which further result in more effects, which are called *tertiary effects,* and so forth.

Following this process to its logical conclusion would seem to lead to the assumption that everything is affected by everything, and that may be the case. How then can we possibly study the effects of a single change in technology or the rise of a single sociological factor? How can we reasonably deal with such a massive number of cause-and-effect relationships? For this, we must turn to the study of systems and systems analysis, which reduces huge interactive processes to manageable, organized forms.

SYSTEMS AND SYSTEMS ANALYSIS

The System

A *system* is an aggregate of two or more physical components and a set of disciplines or procedures by means of which they interact. The physical world is itself a system; it contains innumerable systems within it. This broad definition includes any and all groups of physical components that interact through a set of procedures. These procedures are definitions of the cause-and-effect relationships that exist in a system.

An automobile is a system. It is composed of a finite number of highly specialized physical components, each of which has a specific relationship to other components in the vehicle. A piston can only do certain things and only does them in response to certain causes. When there is oxygen, gaseous fuel, and a spark present in the cylinder with which the piston is interacting, the resulting explosion causes the piston to move away from the cylinder head under pressure from the expanding gases. At its fullest recession, the crank shaft inhibits any further retreat in the receding direction, valves open to allow the gases to escape the chamber, and the crankshaft forces the cylinder to again move toward the cylinder head, where the same set of actions takes place again. Physical components—the gaseous fuel, spark plugs, oxygen, cylinder and cylinder head, piston and piston ring, crankshaft and connecting rod—all interact through certain laws (the laws of thermodynamics, the behavior of expanding gases, the mechanical laws of axles and cylinders, and so forth) to behave in the same way over and over again. This is a system.

It is also a subsystem. Each cylinder and cylinder system is only one of four (or six or eight) that all perform their functions in a coordinated manner to create the engine. And a number of other systems also function in this task. Taken together, we have a system known as an engine. In concert with the other components of the vehicle, the engine is part of a system known as a car, which is part of a system known as a transportation system, which is part of the social system, which is part of. . . .

Systems need not be mechanical in nature. They may be animate as well, or they may be a combination of mechanical and animate, as in the case of integrated systems such as a space shuttle, where both people and machine are components of the whole. The Battle of the Bulge was an interaction of two systems. The voyage of Ferdinand Magellan was a system as well, as was the transcontinental railroad and the Standard Oil Trust. Major corporations are systems, as are the human body and the political system of France. These are all systems, each with a given purpose, made up of elements proper to the performance of the functions for which the system was designed, and each acting in accord with other systems in the physical world.

The primary differences between earlier systems and the modern systems that we encounter today are *complexity, size, degree of mechanization, speed of action,* and particularly the degree to which *automatic controls* are necessary to their successful operation.

In early mills, where water was the primary source of power, it was not difficult to construct and operate the system. There was a building and a mechanism—consisting of waterwheel, gears, grinding stone, and bagging room—that was designed to act in concert to accomplish the task of grinding and packaging grain. The system was straightforward and simple in design. The waterpower produced a relatively small amount of usable energy and was fairly constant in its delivery, being governed by the relationship between the movement of the stream and the size and design of the waterwheel itself. The gear mechanism worked in one direction and turned one wheel over one surface as grain was poured into a cavity that ground it to flour as it made its way through the system. Bagging was a matter of waiting for each sack to be filled, then replacing it with an empty bag while the full sack was sewn shut, stacked, or stored prior to delivery. If anything went awry, the mill was shut down, repairs made, problems corrected, or components changed.

Compare this with a textile mill after the advent of the steam engine. There is now considerably more power available, and it can be delivered at a much higher speed, utilizing more complicated delivery systems consisting of belts, shafts, gearing systems, and other means of mechanical translation. Control becomes more difficult, and reaction speed is greatly reduced. Automatic governors become necessary along with valving and pressure gauges to ensure a safe and constant delivery of power. Safety hazards multiply in the more complicated environment, where many machines are connected to the power plant through pulleys and drive shafts. If an error occurs, it must be corrected immediately, and errors are more serious by virtue of the reduced reaction time to changes in the system.

Compare both of the examples above with the operation of a modern supersonic fighter. Reaction time is now so short that an individual pilot is unable to keep up with the changes needed to correct for variations in performance. The machine flies itself. It doesn't have time to wait while a human makes a decision. Failsafes and servos do the job once done by human command. The power being supplied and the rate at which it is supplied are beyond the capacity of human pilots to control. The machine is a complex system of complex machines consisting of complex components produced through the expertise of specialists from many fields of endeavor.

This progressive complexity and speed are present in social systems as well. Compare a feudal system with a modern socialistic–capitalistic society. Under feudalism, there were two classes of citizens: the lords and members of the Church, who were the landowners, and the peasants, who were landless and dependent on feudal masters for sustenance and protection. The main concern of the lords was the maintenance of property by defense and by accumulation of wealth. The main concern of the peasantry was survival in a world where their very existence depended on the protection and good will of their sovereign lord. It was a simple system. There were few changes taking place (the major one being the appearance and spreading use of a more efficient plow), and there was little impetus to change. The lords maintained the status quo for their own benefit, and the peasantry maintained it from lack of choice. It was not until the rise of the middle class, those who were landless yet who possessed skills that could be traded for wealth (making them tradespeople), that we see any changes occurring in the feudal system. When this change occurred, it brought on the Renaissance.

A modern Western society is far more complex. The characteristics and distances among classes are quite indistinct by comparison. The variety of activities available to the people, the variety of social structures through which people interact, the variety of goods available, the richness of the culture, communications, transportation, political points of view, and educational opportunities all far surpass those of the feudal system. With this complexity comes the price of a fuller, more extensively intertwined network of social factors. Rather than a single family structure, there are many types of structures. Rather than a single choice of economic endeavor, the availability is almost endless. Rather than a limited choice of where to live, how to live, and what to buy, the variety is staggering. In such an environment (system), control becomes a problem, requiring more extensive networks of courts and police systems and, in general, a higher degree of structure in human affairs. This is part of the opportunity cost connected with choosing a modern way of life.

Understanding such a system seems like a monumental task, but fortunately it can be eased by using a systems approach to the analysis of what is taking place.

To understand a system, it is necessary to understand the subsystems of which it is made. Understanding these components (subsystems) involves the following:

1. Enumerating and understanding the properties of the component in order to determine the cause-and-effect relationships that make up the next larger system of which the component is a member.
2. Limiting the number and level of components investigated to no more than are necessary to understand the workings of the system of which they are members. Reduction of the system to interacting subsystems should take place only as far as is necessary to give the information about how the components operate that is cogent to our study of the larger system.

This *reductionism,* as it is called, should only be carried as far as is absolutely necessary. In essence, the investigator is attempting to facilitate analysis by reducing a given system to a reasonable number of subsystems or components. Too much reduction results in an unwieldy and useless framework. We must be cautious not to reduce the system so far that we lose sight of our original purpose or to the point that analysis would become so involved that we experience "not seeing the forest for the trees."

If, for example, I wish to investigate the system known as a personal computer, my investigation may lead me in different directions, even within the determination of subsystems, depending on the *purpose* of the analysis. As a programmer, I might choose to consider the computer as made up of a central processor, the primary storage, a logic unit, input–output devices, and so forth. From here I might move to the ways each of these elements figure in the scheme of developing programs. In other words, how does each affect the input of information? What languages are available? What commands place information in certain parts of the main storage? What output devices are available, and how will that affect the way I write the program? Is this an interactive system where I can communicate with the screen as the program runs, or is it passive, where the program is run *in toto*, followed by output without the benefit of communication during the run?

If I were approaching the personal computer from the point of view of an engineer, the analysis might be quite different. In this case, the important issues may be ones of hardware, exclusive of software. What components are needed to create main memory? How many chips? Which ones? What is the mode of interface between computer and peripheral? Are data ports parallel or sequential? What types of ports are involved? What is the computational rate? What type of power supply is required?

The way we subdivide the system depends on the purpose we have in mind. It depends on what we wish to learn about the system's operation. Note that we need take reductionism only as far as necessary to perform the analysis in which we are interested. I do not need to know the design of a chip to know that it will store binary information. I need not know the chemical makeup of plastic to know that a plastic key functions as an input device to input a given symbol.

The main question in describing the components or subsystems of a system is, What is our purpose? In other words, what is it we wish to learn, and *what level of component reduction is needed to describe the way in which the system functions?*

The Analysis

The system is a set of components that interact in a given set of ways called *criteria*. To analyze a system requires understanding the nature and structure of the components and the way that they interact with one another. In fact, the purpose of systems analysis is to observe the functioning of a system and to be able to predict the properties of that system. It must be remembered that the properties of a system are not something innate in the components, but rather they are a function of their interaction. That is, it is the manner in which the components interact that gives the system its properties, not the components themselves. People are components in some systems. As people, they have capabilities and limitations that make them systems themselves, but the way they contribute to the functioning of a system (e.g., a business organization) is the result of how those human capabilities are used *in concert with the other components of the system* to achieve the purpose of the system of which they are a part. The analyst seeks to first find the properties of the system that are predictable and repetitive, and therefore considered likely to remain the same within the system's environment—in other words, to find the cause-and-effect relationships among the system components. Second, the analyst seeks to determine the ways in which the system is dependent on its environment, to find interenvironmental relationships, and to predict possible changes in the system as a result of future changes in the system. Once this is done, the systems analyst will further attempt to predict the performance of the system under conditions not statistically determinable, which is, at best, an act of visualizing future trends and, at worst, an act of faith.

Analysis and purpose

To properly analyze a system, it is necessary to understand the purpose of the system and then to understand the ways in which the system goes about the achievement of its purpose. Basically, all systems have as their central purpose the performance of some function or functions in such a way that they remain in a state of dynamic balance, either through maintenance of the status quo (steady-state systems) or through growth (dynamic-state systems). Starting with this premise, analysis is merely a matter of determining what the end results of the system are *supposed* to be, the methods used by the system to achieve those end results (interaction of components guided by defined limiting parameters), and the nature of the problems that can arise in the process.

A universal model of a system is presented in Figure 6-1. It represents the basic elements necessary for a system to achieve and maintain balance. Without each of these components, the system does not function, nor does it technically exist as a system at all!

The *input* is necessary to supply the system with whatever is needed to perform its appointed tasks and achieve its desired goals. Inputs can range from information and pressure from the exterior environment, as in a computer network, to physical resources, such as raw materials, labor, energy, and capital, as in a textile plant. Whatever the form, if there is no input, there will be no activity, since this is the trigger or impetus that starts and maintains the system.

Process is a matter of the transformations that take place within the system by which the inputs are changed, altered, used, or otherwise utilized to

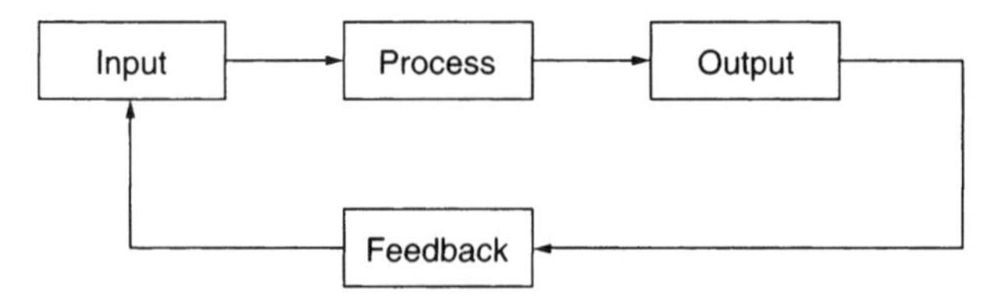

FIGURE 6-1 General Systems Model

achieve the system's purpose. This could range from using data to create financial reports or statistical conclusions to using tungsten, copper, phosphorus, glass, and other inputs to create light bulbs. In whatever manner, the inputs are utilized, and by some process, the system functions, its components interacting to create some desired result.

Output is the purpose of the system. The output is the thing, idea, object, product, or condition that the system is designed to achieve. This represents the *raison d'être,* the reason for existence, of the system. Without the output, there would be no purpose for its being, and the system would simply cease to exist.

Feedback is the control mechanism by which the system determines whether it has achieved its purpose. The environment within which the system is operating will be subject to change over time. In any real-world situation, this is so highly probable that it is often considered a given. The existence of the possibility of change means the necessity within a successful system of a method of correcting activities to adjust for that change. The primary purpose of the feedback element is to supply the system with information so that it can correct itself. Feedback may be either passive or active. It may be part of an automatic system designed to always create certain changes in input and processing behavior as a result of certain criteria, or it may be discretionary, in that it supplies some other decision making element of the system with raw data. The governor on a steam engine automatically puts on the brakes if the speed of shaft revolution passes a certain critical point. In a market system, the feedback on sales made in the market is just a single element in decisions of production, distribution, and advertising. Both are cases of feedback.

Given this basic model of a system, the analyst needs to determine what the system actually does (how it functions). One very efficient means of doing this is through determination of input–output relationships to the process. The analyst seeks to answer questions such as the following concerning these elements, usually beginning with the output and working backward:

1. What type of output can be expected from the system?
2. How does this output reflect the purpose of the system?
3. How fast is this output generated, and what is the feedback time between output and effect of interaction with the surrounding environmental system?
4. What forms of input are required to create the output?
5. How fast can the input be received by the system?
6. How fast can the system change its rate of input reception in reaction to feedback information?
7. How does the system process the input to create the output?
8. In what ways can the process fail?

9. What are the probabilities of failure of any systems component at any time under given circumstances?
10. What are the probable consequences of each type of failure (a) to the system and (b) to the environment?

The answers to these questions are available either through analysis of historical data, through experimentation, or through inspection. The last element, inspection, is really a special case of experimentation. If experimentation is the necessary form of investigation, measurement and analysis using the scientific method should be undertaken.

HISTORICAL APPLICATIONS OF THE SYSTEMS APPROACH

The concept of "systems" is not a new idea. It has been with us throughout history, from the Greek division of matter into fire, water, air, and earth and the corresponding Chinese division of matter into fire, water, air, earth, and wood (all matter considered to be the result of various interactions among these basic elements), through Leibniz, Hegel, Hesse, and Kohler. The concept is evident in the Gestalt school of psychology first developed in Germany, which describes a person's psychological makeup as the result of all things experienced by the person acting in concert to create a given set of response patterns to external stimuli. It is also apparent in the scalar approach of Henri Fayol, who first enunciated the principles of authority and responsibility and who developed the concept of delegation of authority through a well-structured, hierarchical organizational system. It is apparent as well in the concept of bureaucracy, in the taxonomic classification of life on this planet, and in the cybernetics of the twentieth century and the synergetics of Buckminster Fuller. The concept of a system has long been apparent in humanity's attempt to understand itself and its world, yet it was not until the twentieth century that we began dealing with our relationship to the world as a system.

Consider the hierarchy of large corporations. The principle of the hierarchy still overshadows other organizational patterns with its highly structured format. Each job is listed in its proper place (component), designed to carry out certain specific functions (making each component a subsystem in itself), having interaction with all of the other positions in the firm according to highly defined relationships (defining the procedures through which the components perform the process of the system). Inputs in the form of money, market reports, raw materials, and information flow into the organization, where they are used to produce a useful product (process to output), which is then sold on the open market for money, which flows back into the system as feedback. The system operates and controls itself in accordance with its relationship to the environment and to changes in that environment, such as changes in demand for the product the system is producing.

And each element of the system is defined prior to operation down to the last detail. Hierarchical structures such as this are *planned* and *organized* as the company is put together, then altered in response to environmental influences in its search for what works (seeking balance) and in its efforts to grow as a company (implying dynamic growth).

In a similar manner, the military is structured as a system, with each component down to the individual foot soldier defined as to purpose and relationship to every other member of the system. And the more complex the organization, the more need there is for systems organization, as is evidenced by the degree to which military systems codify and specify activities, positions, and so forth.

TECHNOLOGY AND THE SYSTEMIC APPROACH

The systemic approach is a logical one to take in an investigation of technology and society. Technology is a subsystem of what it is to be part of humanity. As already seen, the artifacts that we create are manipulations of physical laws to allow us to structure our society the way we want it to be. In the same way, the social elements of our society are an integral part of what it is to be human, defining the ways that we choose to interact, cooperate, structure groups, and function in dynamic social patterns. This is another subsystem of humanity.

By virtue of this connection, we can study in an orderly fashion the relationships between social structures and technology using various statistical and methodological techniques within an overall framework of systems analysis. The meat of this pie, however, must wait until we have explored a few techniques of analysis in subsequent chapters.

CONCLUSION

We live in a world controlled by natural laws. These laws describe the ways in which individual elements of our physical environment interact with one another. Understanding these relationships is a matter of determining cause-and-effect patterns. Through understanding these patterns of behavior we are able to predict what will happen in the world with some degree of accuracy. This is the way we make assumptions, and on these assumptions, we base our behavior.

The cause-and-effect relationships that we encounter are descriptive of a general overall pattern that involves the entire network of physical existence. Each physical element in the real world is connected to every other element in that they are all part of a single system. We can define a system as a collection of components connected to one another by various interactive procedures. Each system consists of smaller subsystems that, in turn, consist of still smaller subsystems down to the smallest component imaginable, to the world of subatomic particles.

The process of analyzing phenomena by breaking down systems into constituent components is known as the *systems approach.* Breaking down systems into subsystems and following this with the same procedure for each subsystem is known as *reductionism.* It is a method of reducing problems, perceived subsystems, or organizations to a size that makes them manageable and predictable. The purpose of systems analysis is to carry on this reductionism to the point that a system may be analyzed and understood in conjunction with one's purposes. Reductionism should only be carried out insofar as it accomplishes one's purpose, further subdivision being meaningless.

The methods by which one carries out systems analysis are useful in the study of the relationship between technology and social structures, as both technology and social structure are subsystems of the larger system we call humanity—which is, in turn, part of the yet larger world system—and both are therefore subject to reductionism as part of those larger systems. Armed with the methodology of statistical analysis and predictive modeling, it is possible to use this approach.

APPENDIX: EXAMPLE OF SYSTEMS APPROACH IN USE

As an example of how the systems approach may be used to analyze a system and predict outcomes on the basis of that analysis, we offer the following simplified view of a macroeconomic system. For the sake of brevity and to avoid lengthy explanations, the system has been kept to a simplified form.

The purpose of the analysis is to investigate the nature of a macroeconomic system, that is, a national economy, in hopes of developing predictive data on the probable performance of that economy under various combinations of conditions.

Definition of the System and Its Purpose

An economic system is designed to produce and distribute goods and services produced from scarce resources so that those goods and services may be consumed by the final user. The *purpose* of the system is seen as the creation of satisfaction among consumers as a result of consuming goods and services.

Outputs

Goods and services can be defined as final goods and services in the form of produced items of value and valuable services to the public that consumers would wish to consume.

Inputs

Productive resources can be defined as natural resources (land with its original fertility and mineral deposits), labor (the productive efforts of people working with their minds and their muscles to create useful products), and capital (all manufactured productive resources, including machinery and equipment, buildings, and improvements to land).

Process Components

The economy is considered to consist of three sectors and three markets, the former consisting of industry (or the investment sector), consumers, and government, and the latter consisting of the product market, the factor market, and the credit market.

The investment (or industrial) sector

This sector consists of industry that produces goods and services for the general population through the product market in hopes of receiving profits for its efforts and therefore increasing its overall economic welfare. The industrial sector also purchases the means of production through the factor market from consumers who own said means as private property so that they may have the means by which they create the goods and services to be sold.

The consumer sector

This sector consists of individual consumers in the economy who purchase goods and services from industry through the product market for the purpose of receiving satisfaction through the consumption process and therefore increasing their overall economic welfare. In addition, consumers offer their property and labor to the industrial sector through the factor market so that they may receive compensation in the form of rent, interest, or profits and thus have the means by which they can purchase finished goods and services for consumption.

The government sector

This sector consists of all governmental entities, collectively known as the *public sector.* It receives taxes and borrows money from both the industrial sector and the consumer sector and utilizes these funds to provide goods and services to the population as a whole, based on the desires and needs of the members of the society. It purchases goods and services through the product market from the industrial sector and purchases labor and other factors of production from the consumer sector through the factor market, in turn supplying public goods as needed.

The product market

This market consists of all exchange mechanisms through which the goods and services of producers are bought and sold.

The factor market

This market consists of all markets where the means of production (natural resources, labor, and capital) are bought and sold.

The credit market

This market consists of all institutions that function as intermediaries between consumers who save and industry and government that borrow.

Determination of Component Functions and Interactions

As indicated in Figure 6-2, each of the three sectors interacts with the other two sectors through all three of the available markets. This diagrammatic description of the processes inherent in the operation of the macro-

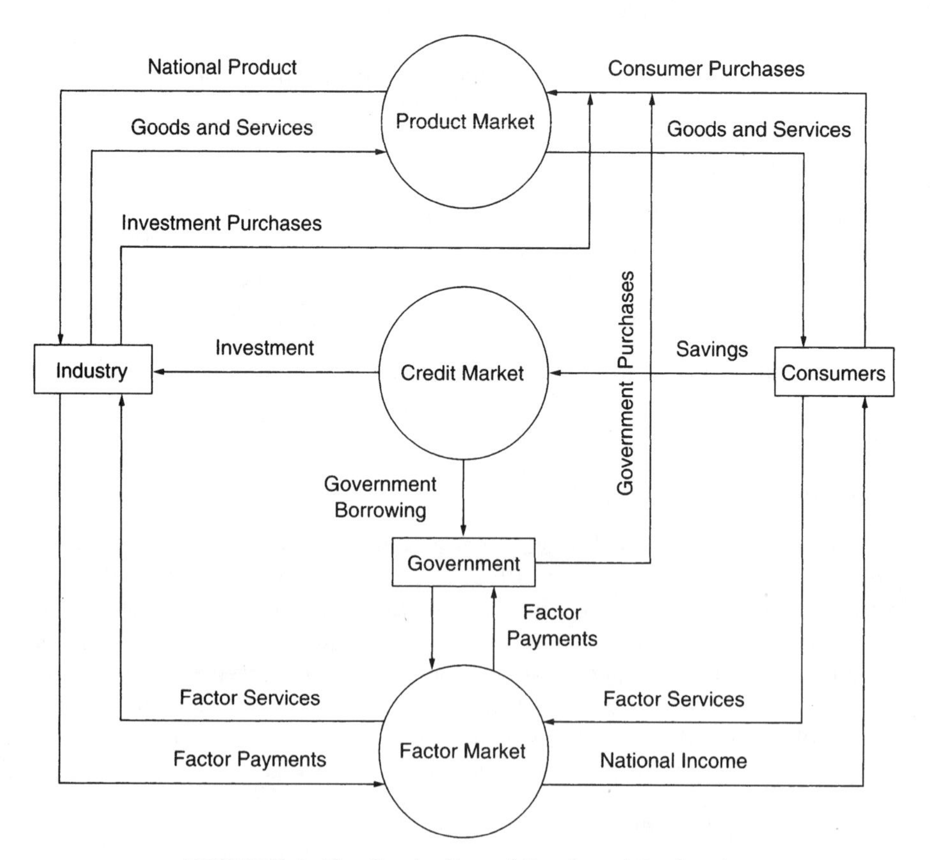

FIGURE 6-2 The Circular Flow of Goods and Services in the U.S. Economy

economic system gives an overview of the types and forms the interactions take. A detailed analysis of each of the interactions yields a more complete model of the system.

Consumer interactions

The determination of consumer behavior is predicated upon two underlying conditions—the availability of income for expenditure (so-called *disposable income*) and the manner in which this available income is spent. To determine the amount of disposable income, the tax rate must be known, yielding the formula $D = (Y - t)$, where D is disposable income, Y is total income, and t represents net taxes, here assumed at an average flat rate to avoid unnecessarily complicating the issue. In addition, it is necessary to know the value of the marginal propensity to consume (MPC), which is a measure of the tendency of the average person in the economy to spend a portion of the next dollar earned. A final necessary value is the amount of autonomous consumption in the society, that is, the minimum level of consumption that must take place for the population to survive. This factor is a discrete number based on historical data concerning prices and the level of consumption necessary for survival. It is usually represented as the constant a. Combining these factors, we find that the *consumption function* is described by the formula $C = a + b \times (Y - t)$.

Investment function

The investment function is the result of two relationships, one concerning the level of expected return on investment (ROI) and the other involving the interest rate. The level of investment within an economy, that is, the amount of money that producers are willing to invest in the production and selling of goods and services, is inversely related to the interest rate (the higher the interest rate, the more costly it is to borrow funds to invest) and directly related to the expected rate of return on investment (since the greater the ROI, the greater the expected profits from the production and sale of the products in question). This function can be denoted by the formula $I = c - dr$, where I indicates the rate of investment, d is a constant of proportionality, c is the ideal level of investment, and r is the prevailing interest rate in the economy.

Government spending

Government spending is defined through the political structure, being indeterminate in the normal sense. The estimate for government spending utilized in economic modeling is based on the official or expected federal, state, and local budgets. In formula, it is expressed as G, where G is government spending as specified by the budgets.

Combining these three sector actions to define the aggregate demand in the economy, we find that $AD = C + I + G$, or $AD = a + c + b \times (Y - t) + G + c - dr$.

A secondary relationship among the sectors of the economy involves their relationship to the money supply, which determines prices and interest rates. The guiding rule here is that the quantity of money demanded in an economy is inversely related to the nominal interest rate and directly proportional to the level of nominal national income. This is called the *theory of proportionality* and is mathematically represented by the formula $Md = e \times Y - f \times r$, where Y represents nominal national income, r represents the prevailing interest rate in the economy, and e and f are constants of proportionality.

One final relationship must be noted—the determination of the equilibrium level of nominal national income, that is, the level of income at which the economy is balanced (in equilibrium). It is denoted by the location of the intersection of the aggregate demand curve, defined above as AD, and the aggregate supply curve, given by the formula $E = Y$. Combining the two equations and solving for Y (in this case, a value found on the x-axis), the equilibrium level of nominal national income is located.

Armed with these relationships, we can therefore define the operation of a simple macroeconomic system as the solution set of the two equations, $Md = e \times Y + f \times r$ and $AD = a + c + b \times (Y - t) + dr + G$, such that the values of r and Y are identical for both formulas. When Y and r are the same for both formulas, the result is a macroeconomic system in equilibrium (i.e., it is balanced).

This brief synopsis of a simplistic solution to the macroeconomic model should serve to indicate the complexity to which systems are capable of moving in a very short time. It is quite beyond the scope of this book to explain in detail how to construct highly sophisticated systems models. However, as indicated in Chapter 7, modeling is a primary source of analysis for quantifiable systems. A more reasonable and less mathematical approach to the analysis of technological social systems is presented in a later chapter.

As part of the methodology available to us to discover the machinations of the universe in which we live, humans develop scenarios of how it might be and then test them to discover whether they truly reflect reality. This process is so prevalent in our lives that for the most part, it tends to go unnoticed unless it is part and parcel with our professional activities. If I am a statistician, I use mathematical models to determine probable outcomes. If I am designing airplanes, I use computer simulations to discover the drawbacks inherent in a given design. If I am a financial analyst, I use market models, charts, and economic assumptions to plot the course of my client's (and my own!) plans for retirement. Yet for the most part, this has become such a common process that we do not give much thought to the roots of that process, why it works so well for us, or, when it does not work well, what the underlying causes of the model's breakdown might be.

In Chapter 7, we look at how the cause and effect of modeling, simulations, and games leads us to our predictions and why this is an integral part of how one successfully goes about the process of discovering the why and how of technological dynamics. The social consequences of technology make it clear that the ability to model technological-societal interaction goes a long way toward an understanding of this relationship. Chapter 7 is a primer in predictive modeling; it presents the main elements and shows when and how they are useful. Though it may seem to diverge from the main theme of the text, it is in itself an example of humankind's creative tendencies at work to give us a better understanding of the environment that we inhabit.

KEY TERMS

Cause and Effect
Complexity
Consumption Function
Economic Markets
Feedback
Gestalt Therapy
Primary Effects
Reductionism
Subsystems
Systems
Systems Approach
Systems Complexity

REVIEW QUESTIONS

1. What is a system?
2. Define systems criteria.
3. Define systems analysis.
4. Define the principle of cause and effect.
5. How are corporate hierarchies related to the systems concept?
6. What is input-output analysis, and what questions would you ask in such a process?

ESSAY QUESTIONS

1. By what criteria would you determine a society to be complex? What benchmark would you use in that determination? Include examples of both complex and simple societies.
2. How is reductionism a result of societal paradigms? How is it a reflection of the development of technology in Western nations?
3. There is a difference between the traditional definition of reductionism and the manner in which the term is used today. How do you account for this difference? What does it mean for the usefulness of reductionism as an approach to understanding reality?
4. Given the proposition that cause-and-effect analysis is effective, why do we not have accurate predictions of the future?

THOUGHT AND PROCESS

1. Anything that is a process can be thought of in terms of a system. It is the fact that the systems approach is so universally applicable that gives it its strength. Choose some activity or process in which you are regularly involved and use the systems approach to analyze it. Some examples might be (a) the process by which a person makes decisions that get him or her up in the morning and to work on time, (b) the process of washing the car, (c) the system known as "Thanksgiving dinner," (d) the process through which one determines what to do on a Saturday night, or (e) the system known as the family.

2. Many systems have specific characteristics that are instrumental in making them what they are. For instance, the only characteristic that maintains a business organization as what it is lies in its purpose, that is, the objectives that it develops to accomplish. Without these objectives, there is no system, no matter how much the physical parts of the former system still exist. With this in mind, analyze the following systems in terms of their distinguishing elements, classifying them by their input, process, output, and feedback elements: (a) the human body, (b) an oak tree, (c) an automobile manufacturer, (d) the solar system, and (e) the play *Hamlet.*

3. If every action taken is the cause of some other occurrence, which, in turn, becomes the cause of other effects, what does this imply for the independence of events in the real world? That is, if we carry this to its logical conclusion, are we saying that all things are really interconnected to one another? Why? Why not?

CHAPTER 7

MODELING, SIMULATIONS, AND GAMING

Charles McAfferty was an unlikely candidate for the kind of success he had experienced. He was a quiet young man, slight, with straight dark hair and deep blue eyes. He was, all in all, quite unobtrusive and passive to the casual observer. All of his life, he had been the favorite of his friends, entertaining, witty in a droll sort of way, and he had gained the reputation for being the least offensive yet kindest of people. But when it came to business, Charles was all business, and that's a fact.

Anyone who knew him well was aware of what to expect when Charles was confronted with the opportunity to make a profit. His entire personality changed. His priorities suddenly shifted. No one was safe if he got in his way in a business deal. He had been in jail six times because of his willingness to take extreme risks for the opportunity to add to his considerable holdings, and he had not always been successful in extricating himself from legal difficulties without a considerable expenditure of time, money, and effort. But to him, it was all just a part of the game. Nothing was more important than his financial empire.

Starting with only a small nest egg, Charles had been able to amass huge holdings and incredible wealth. His shrewd purchases in real estate, his uncanny ability to shift in and out of certain stocks at just the right moment, and his fearless aggressiveness in the face of stiff competition had led him on a road to success that few have ever been able to equal. Charles was truly an amazing person.

He sat at the conference table, his small thin hands folded in his lap, a faint shadow of a smile on his face, waiting for some response from the man sitting opposite him. His eyes gleamed as he thought of the irony of it all. Charles thought of all those people whom he had relieved of their fortunes. All those others, so confident and so sure of themselves and who were now completely wiped out by his chicanery and his greed, meant nothing to him, and never had. But the man across the table from him was something else again. This man would be the supreme victory. Charles was on the verge of his greatest triumph, his most incredible deal, and it was to be at the expense, literally, of his best friend.

Just think of it! Tom Hendrix's fortune was within his grasp. He had been a rival since they were small children, the older brother he had never had, the hero of dreams as long as he could remember. And now he was sitting directly across the conference table, only moments from financial ruin, and it was all Charles McAfferty's doing! Victory would be sweet indeed this day.

Tom, on the other hand, was anything but pleased. He sat silently brooding, his crossed brow reflecting his dismay at his impending financial doom, his folded hands resting lightly on his thin lips, as if to say, "Shhh! Don't bother me while I'm thinking." He shook his head and recalculated his position one more time to be certain. He knew in his heart that he was ruined, but he simply could not admit that Charles had finally destroyed him. It was just too horrible to believe. He had to look one last time for an out, some loophole that would save him from his opponent's avarice. But it was useless. Charles had him, and he knew it.

"Ready to admit that I've got you, Tom?" he asked a bit too casually.

Tom looked up at him. His gaze would have been more than a weaker person could endure, but Charles merely broadened his grin a bit and waited.

"You really think you've got me this time, don't you Charley?"

"I know I have, buddy. Why don't you just admit it and give in. I'm about to destroy you. I'm about to own everything you've spent so long accumulating. I'm going to take the shirt off your back. More than anyone else in this town, I've wanted to beat you, Tom. And now I'm going to have it all."

"Probably," Tom said noncommittally. "Tell me something, Charley. What's so special about me? Why go after me? I've seen you run people out of business before. I've watched you take them for all they're worth and pull them to their knees, but it was never personal with you. For you it was always just mechanical, part of the fun, a way to pass the time, to garner victories, and to add to your fortune. But with me, it's different, isn't it? Oh, you needn't answer, Charley. I know it is. This is personal with us, and we both know it. It's not the money this time, or the holdings, or the power. You've already got more of that than you can possibly use. This time it's just to beat me and for no other reason. Why? What's so special about me?"

Charles released a sinister and somewhat uncharacteristic laugh. He leaned forward and locked Tom's eyes, the steel blue of his own turning even colder as he spoke.

"I could lie to you and just say that winning is the name of the game, but I won't. You're right, of course. It is personal. I've waited a long time for this, Tom. We're terribly different, you and I. You're all caught up in your own self-righteousness. You remind me of a character in a romantic novel, naive and full of goodness, and all that rot. I've watched your empire grow just as mine has. I've watched you lend money to people when they needed it, to forego payments when borrowers couldn't afford to pay. I've seen you give away your wealth to subsidize the folly of those who dared to compete against you. And I've watched them bail you out of a few tight places, too. It was pitiful to watch, Tom, and pitiful and weak of you to allow it.

"Do you remember the last time I was in jail? Did any one of them offer to give me a hand? Did any of them offer to put up my bail or pay my fines until I could get back on my feet? Oh, you did, Tom, but you must have known that I could never accept any help from you. Not my childhood hero. I did it all on my own. I got myself out of my own scrapes, and I went on to watch every one of them fall by the wayside, ruined one by one, wiped out by my hand. They're all gone now, Tom, all but you. And now I've got you. I'm going to really enjoy watching you going under, pal. It's going to be a pleasure to wipe you out the way I did the others. I've won, Tom, and no one can take that from me. Why not just admit it? I've got you and you know it."

Tom shook his head. He took a deep breath and released a long, soft sigh. "If you want my holdings, my dear friend, you'll just have to take them. I'll not give in to you that easily."

"So be it!" Charles hissed. "Let's get on with it then."

Tom sighed again and reached out his hand to the conference table. He picked up the dice and rolled. Any roll but a seven was going to mean disaster. Only a seven could save him from financial ruin. And a seven it was! Tom counted out the seven spaces on the board and landed on Marvin Gardens, the only space on that side of the board that was his.

Charles grabbed the dice and rolled with a vengeance, then advanced his token ten spaces. He groaned when he realized that he had landed on Park Avenue, another of Tom's properties. Piled on one side was a neatly arranged stack of five hotels! He had done it! Tom had wiped him out in a single throw of the dice. His financial empire was toppled, his wealth drained off in a hideous orgy of rental payments! Tom had won the game!

Tom patted Charley's shoulder as the slight young man slumped, head in folded arms, among the scattered playing pieces on the board.

"Don't take it so hard, buddy," Tom said. "After all, you're the one who wanted to change the rules. It could have been just one hotel over there. And it had to end soon anyway. I've got a class at 8:00 A.M., and it's already 2:30 in the morning now. We'll try again tomorrow afternoon at my place. Cheer up, pal. After all, it's only a game."

Gotcha! Or perhaps not. By this time, it should be evident that things are not always exactly what they appear to be. And in a sense, it is that very proposition with which this chapter concerns itself. How does one determine when something is behaving as expected? How does one define what the world looks like? How does one gain experience in formulating an understanding of the real world? How can one learn to predict what may happen and how it will happen under a certain set of circumstances? All these questions can be dealt with in terms of models and their adjuncts, simulation, and gaming.

A small child picks up a toy gun and begins shooting imaginary soldiers. Another child holds a favorite stuffed animal and tries to feed it a piece of apple, frowning in displeasure because Jojo isn't hungry. Scientists peer through small windows or watch closed-circuit television broadcasts to study the passage of smoke streaming across an experimental airfoil in the face of a wind tunnel hurricane. A business executive punches yet another set of numbers into a desk computer and watches as the profit estimates for next fall change drastically. These are examples of different types of models being used to study and to mimic the real world. Their use is a phenomenon that we all experience every day and in many ways.

How can things that are so different all be examples of the same technique? What is the common thread among stuffed animals, toy guns, mathematical formulas, and airfoils? Each is an idea of the real world, designed to be tested for its accuracy. The more accurate a model is, the more useful it is in *describing* and *predicting* what takes place around us. The diversity of models is no more extreme than the diversity of the real-world phenomena one might wish to study. In each case, the model noted is *appropriate* to the type of phenomenon under study and the type of results one may be interested in achieving.

A *model,* then, is a copy or imitation of a physical structure (a thing) or a concept that is designed to demonstrate certain characteristics of that thing or concept in accordance with the purposes of the modeler. In this context, the form the model takes is a matter of how one wishes to use it. An aeronautical engineer would not use a plastic airplane model to study the effects of friction on the surface temperature of a metal alloy airplane wing, as this would be an inappropriate medium for such a test. Nor would she use a series of formulas describing aerodynamic behavior as a means of teaching pilots the "feel" of an aircraft in a steep dive. In the first case, the plastic model is designed to *look* like the real-world object, not to *behave* like it; in the second, the formulas are symbolic representations designed to relay information about performance characteristics (descriptions), not to artificially create physical experiences for pilots. The choice of models is therefore important if we are to find them useful. In this sense, a major factor in picking what models to use is *appropriateness.*

FORMS OF MODELS

The form that a model takes varies depending on need, just as the choice of models varies for appropriateness. Generally, models are said to be an *analog,* an *icon, verbal,* or *mathematical.* This classification embodies the basic types of models that are encountered in everyday life.

Analog Models

An *analog model* is a model that behaves in a way similar to the reality it is designed to represent. A model airplane that actually flies is an example of an analog model. The model behaves similarly to an actual airplane in flight, being subject to and reacting to the natural forces that a full-size airplane would be expected to encounter. Performance characteristics are approximated using this type of model, and the analog model offers an inexpensive and, in this case, safer alternative to producing a prototype at the outset.

Another example of an analog model is the slide rule, a device that translates mathematical relationships into spatial patterns. The movement of the slide along its track and the distance covered are directly analogous to the change in values experienced when one uses logarithmic principles. Adding logarithms is represented on the slide rule by adding lengths, and the results exactly mirror the behavior of the numbers in question. This is the same procedure that is done when clocks are used to measure time, a nonspatial concept, by sweeping out lengths of arc on the clock face, or the use of an orrery, a mechanical model of the solar system in which balls connected to lengths of rod are geared to circle a central hub representing the sun to show how the planets sweep out arcs through time. The movement of the balls around the hub is analogous to the movement of the planets around the sun.

Analog models are extremely useful in investigating and understanding physical phenomena. They often produce large amounts of information in their creation, as the modeler strives to mimic the real world. By going through the thought processes and the activities necessary to produce the model itself, the modeler learns about the mechanisms and behavior of the thing or concept that he or she is attempting to describe. Many airfoil designs, for instance, are tried and rejected before a final successful design is achieved for a new aircraft. It is through the ability of the analog model wings and fuselages to be studied and modified that we are able to learn what will work by correcting what does not work in the model.

The use of analog models in industry should be quickly obvious. New devices are first produced in smaller, less expensive form to allow for refinement of original designs until the producer has a useful, workable "model" that can be manufactured and made available to the public. Yet there is a less obvious place for analog models in the field of technology, particularly as technology applies to the environment and other elements of the social structure. Analog models can predict the effects of erosion on soil. They can be used to telescope time in the study of populations and population control (as we have seen in the case of Malthus). They can be quite useful in studying the effects of architecture on earthquake safety or wind and weather in the inner city. To relegate their use solely to the production of technological innovations would be shortsighted in the extreme.

Iconic Models

Iconic models are somewhat different in that they are designed to look like (resemble) the physical reality that they are describing, but not necessarily to behave in a similar manner. Toy trucks and animals are examples of icons, as are architectural models of houses and office buildings, or the previously cited example of the plastic airplane model. None of these models necessarily behaves in ways indicative of what they are describing, but they do resemble them physically. Paintings, sculptures, and many computer-created illustrations are other examples of icons that are designed *only* to look like the "real thing."

How are they useful to technology? Icons can be utilized to study aspects of real-world phenomena that are not directly related to performance, though every bit as important. Industrial design uses icons to discover the most aesthetically pleasing form for useful products to take. The shape, color, or texture of a product may have little or no effect on how well it performs the task for which it is created, but it may make the difference between success or failure in the marketplace as much as shoddy workmanship or poorly chosen materials can.

Spatial relationships can also be studied using icons, as in the homemaker who uses models of furniture to rearrange a living room many times before deciding on the best combination for comfort and function. Large companies often use iconic representations of machinery and equipment in conjunction with physical models of new plants to best determine how these items should be positioned to maximize efficiency and minimize hazards.

Yet these examples are only a small part of the huge number of uses to which we put icons. The majority of them, and those that have the greatest effect on our society, are in a special category known as *toys*. Through the use of toys, children and, to a larger extent than you might imagine, adults practice the skills that they will need later for survival in our society. Iconic toys in combination with a fertile imagination allow children to develop and walk through numerous scenarios of possible and expected futures without risk to life and limb or fear of suffering trauma. With this in mind, it should be realized that the actions and interests of adults are at a minimum partially determined by the icons that they dealt with as children. The child who plays computer games today is already intimately familiar with and comfortable with computers as an adult. The child who plays with toy trucks and cars is more aware of the hazards of these modes of transportation, their capabilities, and their possible uses. Children practice through play, increasing their ability to deal with the adult world once they find themselves faced with the responsibilities and opportunities of that world.

Verbal Models

Verbal models are descriptive in nature. They are designed to convert thoughts and concepts into language to establish relationships and restrictions of the real-world system and to organize them. It is in this establishment of relationships and restrictions and in their organization that verbal models excel.

In order to create a verbal model, it is necessary to *conceptualize* what is being modeled. A process must be undergone through which the modeler structures her understanding of the phenomenon being described and organizes the information about the phenomenon into some logical pattern. Verbalization forces inspection. Describing something, as for instance in a simple definition, means that the person must decide, first, what is included in the definition, second, what must be specifically excluded from the definition, and third, what sort of logical structure to put the definition into. By way of example, notice how this process is illustrated by entries in any dictionary, then read the instructions provided with your favorite game or the instructions for assembling the lawnmower you recently purchased and are still trying to figure out, or, for that matter, take a second look at the table of contents of this book. It is nothing more than a verbal representation of an idea about how the world is put together, how it functions, and how to study that functioning.

Verbal models have one inherent weakness, and that is interpretation. They have a tendency to become ungainly due to the necessity of exactitude of description to produce accuracy of description. The old saying about a picture being worth a thousand words has been around for a very long time, and its survival stems from it being true. Words have a tremendous power to describe, and to describe accurately to the smallest detail, but if the interpretation of the words used varies, the meaning suffers. Simple concepts take few words to describe. Complicated concepts require paragraphs, chapters, and indeed volumes to adequately explain them. When are they useful? When we need to include, exclude, and organize our ideas about real-world phenomena and to offer knowledge in a way that can be easily understood by a large number of people. Verbal models greatly enhance the dissemination of knowledge and, along with it, understanding.

Mathematical Models

The fourth form of models is the mathematical form. *Mathematical models* are manipulative, symbolic representations of reality designed to describe the *relationships* among certain factors of a thing or concept, to establish restrictions (limits) on the thing or concept, and to use behavioral characteristics to *predict* with some degree of accuracy the manner in which behavior will change under given sets of circumstances.

Mathematical models are usually in the form of a formula or set of formulas describing and predicting behavior. Much of the information we have about real-world behavior is in the form of formulas. The physical sciences, such as astronomy, physics, and chemistry, rely heavily on this predictive and descriptive form of modeling to interpret what is observed. In these sciences particularly, the behavior that can be quantized (i.e., reduced to numbers, ratios, and so forth) is the behavior that can be investigated, studied, and used. Technology depends heavily on mathematical models, as do the social sciences, which utilize the concepts of statistics to take nondeterministic phenomena and generalize them into manipulative form. Mathematical models are among the most successful at predicting macrocultural and microcultural behavior. Examples are found in nearly every discipline, from the models of market behavior encountered in economics to the laws of electrical behavior in physics.

THE MODELING PROCESS

Fortunately, there is a simple and general process by which modeling can be done. It is equally applicable to all types of models and involves determining what type of model to utilize and how to go about it. The process is presented below in brief.

The Modeling Process

1. Gather information about the concept or physical structure to be modeled.
2. Based on this information, reach conclusions about the nature, characteristics, and behavior of the reality to be modeled.
3. Determine an appropriate form for the model, the degree of detail required, what elements are important in understanding the nature of the reality, and, of those elements, which should be included in the model itself.
4. Build the model.
5. Compare the model with reality to determine the degree to which the model actually approximates the reality.
6. Adjust the model as necessary to achieve the desired "fit."

Note the distinct similarity between this modeling process and what is known as the scientific process used in the physical sciences. This is not too surprising, since most scientific theories and laws are models of reality. All that has been done here is to present a more generalized format for the application of the technique to a less strictured range of problems. For clarity, we discuss the steps in somewhat more detail below.

1. *Gather information about the concept or physical structure to be modeled.* A model can be rendered useless if this step is not done properly. The ignorance of information can change the entire complexion of a problem and greatly influence the direction that an investigator takes in creating her or his final concept of reality. It is imperative that all possible information pertaining to the aspects of the reality under scrutiny be gathered for study. A single fact can totally change the meaning of research by either its inclusion or exclusion. How much research is enough? When is it too much? This is a question that the investigator has to answer. With experience, it becomes easier to make this determination. A certain understanding develops over time to tell the investigator when the law of diminishing returns will set in, yielding less information per unit of effort than is acceptable. In general, research suggests research, and the details present themselves in the course of investigation.

2. *Based on this information, reach conclusions about the nature, characteristics, and behavior of the reality to be modeled.* What did you find in your research? What facts presented themselves? What relationships did you observe? Just thinking about the subject will yield a wealth of possibilities. It is in the combining of the information gathered through research and its ordering into logical cause-and-effect relationships that produces the first mental picture of the reality we are attempting to model. The first efforts tend to be macro in structure, dealing with sweeping, overall statements about the way the reality is structured, the form it takes, and the way it behaves under different sets of conditions. Observations and the observations of others often suggest possible pictures of the reality that can be tested, incorporated into our present concept of what is, and rearranged or modified to fit, yielding a much better understanding of what we are dealing with.

As an example, consider the case of an infant exploring the world of the nursery floor. Toys lay round about, bright and inviting, awaiting inspection. The infant focuses on a ball and reaches for it. The toddler notices its shape, its color, the fact that it shines and reflects light, that it is smooth to the touch, that its weight is not too great to pick up, and that it can be grasped with both hands and held up, whereas one hand will not do the job. These are all observations that could be considered firsthand research (or perhaps hands-on investigation?).

The infant reaches certain conclusions: So far, the object is not dangerous, that is, it does not elicit feelings of discomfort and pain. It is pleasurable—it feels nice and the bright red color looks nice. It is enjoyable to learn as much as possible about it. At some point, however, the infant tires of it, being attracted by some other toy, and lets the ball drop to the floor. At this point, the child observes it bounce once or twice, then roll toward the corner of the room by the closet.

The child now has a picture of a real-world object, the red ball, and an idea of the nature and behavior of the ball that can be used later when encountering similar objects. The memory model of the red ball's behavior can help the child to react in appropriate (successful) ways with other spherical objects. Needless to say, the infant will find out that other spheres do not necessarily behave the same way as the red plastic ball, as, for instance, in the case of a bowling ball (should the child try picking one up with two hands) or a tennis ball (more bounce, fuzzy surface, smaller size, less weight). But the child has a model to guide her in supposing what to expect, and it is one that can be modified with experience to include the differences encountered with other similar objects.

In the same way, when we build a model and gather information about its behavior and characteristics, we formulate a mental picture of expected results from encountering the real-world phenomenon that the model represents. The difference between what the child does with the ball and what an adult does with, say, the concept of acid rain, is one of content, not one of procedure. The *process* is essentially the same.

3. *Determine an appropriate form for the model, the degree of detail required, what elements are important in understanding the nature of the reality, and, of those elements, which should be included in the model itself.* Through steps one and two, we have obtained an understanding of the nature of the reality in question. Using this mental picture, it is now possible to consider what should go into the model that we are going to formally build. The elements to consider involve what is important to the purposes of the investigator, what restrictions exist on the possible nature of the model, and how these affect the final form that the model will take.

By way of example, let us return to the orrery mentioned earlier. As indicated, an orrery is a physical representation of the solar system that demonstrates the movements of the planets on their yearly journey around the sun. The positions of the planets are indicated by small spheres on long rods attached to a central hub and geared to change position in the same ratio as the various planets change position through time. The moon sphere is geared to circle the earth sphere every 28 revolutions of the earth sphere in its axis. The earth sphere is geared to circle the central sun sphere once every 365.4 (approximately) revolutions of the earth sphere about its own axis, and so on. Certain characteristics of the physical system known as the solar system have been faithfully mimicked in the orrery model, while other obvious and important ones have been ignored. The spheres are not placed at a ratio of distance that is physically analogous to the distances the real planets demonstrate. The composition of material of each sphere is far from commensurate with that of the real world. The central sphere of the model does not consist of mostly hydrogen gas with helium and a trace of the other known elements mixed in. The earth model does not have a liquid iron–nickel core (as some "models" predict the real earth does) surrounded by a rocky lithosphere and mantle. There are no orbiting rings of rock and ice about Saturn or Uranus. The modeler has ignored all of these elements of the physical reality under investigation.

Yet the *spatial* relationships of the planets are maintained to as high a degree of accuracy as the modeler can obtain. And this is because of (a) the characteristics that are important to the investigator, (b) the possibilities, given the structure of reality, and (c) what appears to be the most logical form for the model to take in order to copy those aspects of the real phenomenon of interest. It is not possible to accurately measure the distances, even on a reduced scale, among the planetary bodies. The distances are too huge and the model would be useless. It is equally impossible on a small scale to construct a physical model incorporating all of the true mineral and chemical combinations existing in the real planets. And why would we want to? The purpose of the model is to show the *movements* of the various planets about the sun, and that aspect of the reality is successfully and clearly demonstrated by the orrery.

Had the modeler wished to discuss distances, he or she would have used different forms of models. Mathematical relationships could have been formulated and arranged into tables or charts and graphs (symbolic icons). If the modeler had wanted to demonstrate the chemical and physical phenomena illustrated by the real-world system, he could have formulated the laws of gravitation as Newton did, or dropped balls from the leaning tower of Pisa (as, alas, no one of importance has ever done), or demonstrated the activity of the sun quite handily by detonating a fusion bomb! Copying nature in models is a matter of choosing the model appropriate to the purpose of the model. *What* you are attempting to illustrate will dictate the *form* that your model takes.

4. *Build the model.* If the previous steps have been done well, the actual construction of the model should be a simple task indeed. Armed with an understanding of the nature of the reality, what characteristics and behaviors are important to the investigator, and the form that the model should take to most perfectly demonstrate those important aspects, construction of the model itself is no more than physically doing what has been planned.

In building the model, many overlooked problems will present themselves. No matter how detailed an investigation of a subject, the modeler is likely to forget some aspects of the problem, resulting in stumbling blocks, unforeseen changes in approach, and, most importantly, new understanding about what is being represented. The building process is really a combination of building, discovering, reassessing previous steps, and restructuring of the model itself. Piece by piece, the parts of the model fall together and are adjusted to fit in with the other parts until some cohesive creation exists that the modeler believes to be a representation of the phenomenon being investigated.

Consider the iconic modeler who constructs a detailed ship model. She may first search out designs, specifications, and photographs of the subject ship. She may even seek out other models of the same subject, noting as well as possible the techniques of fabrication and detail to assess what can be done and how reasonable it is to expect a given type of model in the end. Proceeding from this, the modeler reaches conclusions about the final design, how detailed the model should be, how large it should be, what materials to use for the various elements in the model (Should small bronze cannon be cast, or should I use painted lead or wood? How about sails? Are normal cloth weaves too heavy to create a realistic effect?), and generally how far to go with the process.

This is all well and good, but as soon as the modeler begins constructing the model, she discovers new information not considered previously. What if the authentic woods are too grainy to be easily worked in the smaller scale? What if thread is not available in the proper sizes and turns? What should be done about catlines too small in the model to be woven and tied accurately? The modeler constantly adjusts the concept of the model to changes in the way the problem is understood in order to put the various elements together in a useful way.

An equally valid example would be the structuring of an economic model to deal with "simple" market responses to changes in price. On first inspection, it would appear that an observation of market behavior could create a useful, though crude, model, and that is partially true. The laws of supply and demand are well documented and have been borne out over years of observation. Yet the market process is far more complicated than that, and if the modeler is interested in any degree of accuracy, the list of considerations with which she or he must deal grows rapidly. What about advertising? Are we going to incorporate a model for the effects of advertising into our market model? What about quality differentials among manufacturers? Are we to assume that every producer's product is equally valuable and equally well made? If this is a distortion of reality, how important is it to correct that distortion? Will it affect the market model at some later, critical point?

These examples illustrate the value of the model construction to the modeler and to the understanding of any reality. It is in the process of the model construction that the modeler can gain a true sense of the nature of the object of study and is able to satisfy the initial motives in investigating the phenomenon to start with.

5. *Compare the model with reality to determine the degree to which the model actually approximates the reality.* Does the model we have created accurately mimic the real world? Have the relationships and factors that we have sought to demonstrate with the model been demonstrated? How like reality is the model? Is it very accurate? Too accurate? Not sufficiently accurate? Inspection of the model and comparison with the "real thing" allows us to fine tune and make last-minute adjustments to bring the model into line with our concept of what it should do, be, and indicate. The comparison can be a simple inspection, as with an icon such as the ship model, or can involve one or more trial runs to test a dynamic model's response to motive forces, as with a glider or, for that matter, the economic model. The ship need merely *look* like the real thing. The market model must *behave* properly in reaction to market changes and *be predictive* of change that might occur in the future. And that means testing the mechanization as well as the format. A well-formatted model that does not behave as it should is useless for purposes of predicting behavior. Anyone who has spent the slightest amount of time building computer programs is well aware of *that* reality, to be sure.

6. *Adjust the model as necessary to achieve the desired "fit."* Final adjustment in accordance with the comparative inspection of the fifth step is the last action in building a model. It represents the finishing touches that a modeler puts on the model, since all that could be done has already been done by the modeler to achieve accuracy and purpose in the construction of the model. It is the last step, but in a changing, dynamic world, where the interrelationships among factors of society and technology are constantly rearranging themselves, it is a never-ending step, involving the modeler in a constant reassessment of the model, its usefulness, and its ability to describe the real world. In the future, the model may change dramatically, or it may remain stable, or, what is most likely, it may simply be replaced with a better, more accurate, more useful model. Models are never static if they have accuracy, because the world that they are designed to describe is dynamic, ever changing to offer new challenges and new information to be gained from its investigation.

MATHEMATICAL SIMPLICITY

Before leaving the general subject of models to delve into two specific forms of modeling useful to the investigator of social and technological structures, a brief explanation should be made of the general class of models known as *mathematical models.*

Mathematical models are considered to be a type of symbolic model. *Symbolic models* deal with ideas and abstract approaches rather than physical or mental constructs. They are descriptive and predictive as are other useful models, and the particular importance of the mathematical model form of the symbolic category is that it can succinctly and briefly describe extremely complicated real-world relationships in very little space. Unlike the verbal type of model, also a symbolic form, the mathematical model is not easily misunderstood. Mathematical relationships are very tightly defined, and their interaction cannot easily be mistaken. Any given formula will always describe the same relationship among factors, the only difference being one of magnitude, as represented by the variable quantities that can be plugged in. The formula $Y = 3X + 7$ will always result in an upward-sloping curve when plotted on a Cartesian coordinate system. It can do nothing else, no matter what real (positive) number quantities are entered for the independent variable, X. There can be no confusion. Every graph of this function will be identical in plot to every other, by definition. This is not always true of verbal models. Mathematics is exact and is therefore valuable as a diagnostic tool, since we can be certain of unvarying answers to formulas, if the information we seek can be quantized. A reduction of a picture to mathematics, as with a computer scan of a lunar landscape or the Mona Lisa, ensures absolute identical reproductions if translated into the original medium. Verbal models such as descriptions, on the other hand, are vague, inexact, and, by their nature, subject to the interpretation of the observer. Try asking four people to describe a building, and compare that with the results of asking four scientists to describe mathematically the results of dropping a ten-gram weight from a height of sixteen meters in a vacuum, measuring the height from a reference point of sea level. The difference in the models and their exactness becomes quite apparent.

Mathematical models are classified in accordance with a series of dichotomies that define their characteristics. They are said to be either *quantitative* or *qualitative, probabilistic* or *deterministic, general* or *custom constructed, descriptive* or *optimum.* Briefly, the dichotomies define the mathematical models as follows.

A model is either quantitative or qualitative depending on the ability of the user to accurately define information numerically. In the case of *quantitative models,* the measurements used to determine the values of dependent and independent variables in the formulas are accurate and discrete. I can count the number of seconds it takes a ball to drop sixteen meters with a very high degree of accuracy. The result of an experiment to do so would yield a specific, discrete value for elapsed time. This would be a quantifiable measurement and fit nicely into a quantitative model. As an example, consider the formula, $P = TR - TC$. This formula describes the profit received from the sale of goods as the difference between total revenues (TR) and total cost (TC). All that is necessary to find a definite value for profit is to know the total amount of money received for the sale of goods and the total amount of money paid out in producing those goods. The difference is profit, and that had best be a definite, discrete number. (If it is not, the IRS will be most happy to assist the businessperson in finding out why.)

Qualitative models are not exact. A *qualitative model* is a method of encoding inexact concepts numerically. This is done through the use of a branch of mathematics dealing with continuous rather than discrete functions known as *statistics*. Qualitative mathematical models indicate the *general* results of behavior rather than the exact results. The difference lies only in the ability to accurately measure the phenomenon under study. An excellent example of a qualitative model is the use of a survey to determine public opinion on nonmathematical issues. By asking a group of people (sometimes referred to as a "statistically significant population," meaning a large enough sample of opinions to represent a general attitude among the general population) to rank in order of importance a series of ten characteristics that they would like to have in a spouse, it would be possible to determine to some degree the types of traits that people look for in potential mates. The results would not be equally true for all participants, and there is no way of physically measuring the importance of one trait over another, yet the attitudes and expected behavior of a population can still be predicted as a result of the survey. Qualitative models are growing in importance, particularly with the advent of the computer and advanced statistical methods that allow investigators to process huge amounts of data and test models for accuracy. This is a growing area of modeling that holds great promise for the future.

In a similar manner, mathematical models can be considered either deterministic or probabilistic. *Deterministic models* are formulas that always result in the same, precise answers, whereas *probabilistic models* express tendencies. The probabilistic models use such concepts as the mathematical mean and standard deviation to describe accurately what the *tendencies* of a system are, or to develop probabilistic formulas based on repetitions of trials to calculate "the odds" of something happening under a given set of circumstances. An example would be determining what percent of the population is viewing the first half hour of the 7 o'clock news over the NBC network affiliate in San Bernardino, California. We could conceivably go around to every household on a given night and ask, but the likelihood of success and the expense of achieving that success would probably far outweigh the benefit we would receive from the information. And that would only tell us how many people watched the broadcast at that time over that channel on that one night. What if we wanted to know the figure for the entire year of 1997? Accuracy in this case would be expensive and impossible from a practical standpoint. Does the modeler simply give up and decide it can't be done? Not if she or he works for a marketing firm and wants to continue to work there. The modeler *estimates* with as high a degree of accuracy as possible and then indicates how much confidence she or he has in the estimate. Chances are the exact number is not even a useful thing to know. The trend or tendency of people to behave in a given way is more usable and more important.

In the case of the trial-and-error approach to probabilistic models, consider the local weather report. What is presented as a predictive forecast is actually a report of the *odds* of certain weather patterns developing in the future based on an analysis of what has happened in the past when conditions were similar or identical to current conditions. No weather forecaster is rash enough to tell his or her audience what *will* occur tomorrow, only what is *expected* to occur.

With a deterministic model, this problem of probabilities does not exist. *Deterministic models* are so accurate that they indicate exactly what to expect in a given situation. Deterministic models predict measurable phenomena, such as force of impact, speed, or the number of electron volts created. The results of these models are predetermined and exact rather than relative.

One caveat is in order concerning deterministic models. Technically, they are also estimates of true quantities, albeit highly accurate ones. The exactness of the model (the degree to which it is deterministic) is relative in nature, being only more or less accurate than some other system of measurement. A foot is exactly twelve inches, and that is a deterministic quantity, yet the measurement of the length of a foot or even an inch is something that becomes more and more exact with improvements in the technology of measurement. At present, the measurement can be made with laser beam technology to a high degree of accuracy useful only for a very limited portion of the population. For most of us, a ruler or yardstick does just fine, inexact as they are.

The dichotomy between ready-made and custom models indicates whether the model is a normal mathematical relationship that is generally accepted for a wide number of applications or a model specifically designed to deal with a singular phenomenon faced by the modeler. Many custom models contain ready-made models as part of their structure. By way of example, the formula $V^P = V^t(1 + r)^t$ is a standard model form that allows anyone to calculate the present value of money received in the future after allowing for the time differential involved. It is a very useful model that is used in business, banking, finance, and other money-oriented disciplines. It is a ready-made model of the idea of present value. In contrast, a company may wish to create a custom model of the inventory system used by the company as part of an attempt to computerize routine operations. Such a model is based on the inventory system and model needs of the one company. It would be *customized* for that company's individual situation, though it may contain a large number of standard functional models used in inventory control.

In the case of descriptive models versus optimizing models, we deal with a difference in purpose. *Descriptive models* have as their purpose to do exactly what it sounds like they should do, describe. They are mathematical representations of real-world phenomena and nothing more. The *optimizing model* is distinguished from the descriptive model in that it seeks to find the optimum combination of actions to create a desired result. The optimizing model compares combinations of variable inputs to find the maximum, minimum, or most acceptable type of result for the purposes of the modeler. Determining the most efficient speed for an automobile through modeling is an example of this. There is a trade-off between speed and fuel consumption for most automobiles, and this conflicts with what we would like to have, that is, minimum fuel expense and maximum speed. The two are simply not possible together. However, by calculating how fast fuel consumption changes with a change in speed, it is possible to find the *optimum* or most efficient acceptable combination. How much reduction in speed am I willing to trade for an increase in fuel economy? It is this sort of optimizing analysis that leads to decisions on speed limits. Other models of this type can be used to determine the size of a manufacturing plant, the design for city plans, the size and shape of office buildings, the use of materials, the horsepower requirements of an automobile, and a wealth of other expense-benefit–type considerations.

GAMING AND SIMULATION

At first inspection, the terms *gaming* and *simulation* appear to have identical meanings. Yet as with any closely associated terms, the fact that there are two distinctly different words available to describe the same primary process indicates a difference, a shade of meaning, that separates the two. In essence, simulation can be considered to be a very specialized form of the game. It is because of the specific characteristics of the simulation that it is particularly useful in studying the interactive systemic nature of society and technology.

Games are considered to be any recreation or sport incorporating specific rules that require the participant to compete in some way, either against other players or against herself or himself in attempting to achieve some specified goal. Note that in order for an exercise to be a game, there are rules to be followed, a goal or winning condition to strive for, and competition against one's self, as in comparing past scores with present scores, or competition against others. Competing against the game itself, as in solitaire or computer games, can be considered competition against one's self. Even team games whose object is to optimize scores against the game itself can be considered self-competition. The level of skill one has or one's team has is an indication of improvement in playing the game.

Note further that a game is considered to be a recreation or, more exactly, a re-creation, that is, something that re-creates a condition or set of circumstances reflective of a real-world phenomenon. Therefore, even formal games are obvious models.

Games can be loosely classified as games of chance (poker, roulette), games of skill (competitive sports, chess), or a combination of both (bridge, Monopoly™, war games, and so forth). They are also often classified according to the type of equipment and competition involved in their play (card games, contact sports, board games, computer games, and so forth). All of this may be interesting to the inveterate gamer but somewhat academic for our purposes, with the exception of a particular type of skill-chance–oriented game—the simulation.

A *simulation* is a model that copies the behavior of some aspect or aspects of reality. Simulations attempt to describe and test behavior patterns of interactive systems and then to be predictive and descriptive of changes in the system that may result from changes in one or more specific elements (subsystems) within that system. The creator–player is interested in re-creating real-world conditions on some limited basis for the purpose of aping (mimicking) the behavior of the real-world system and then inputting changes to gain skill in dealing with the results of environmental change. Certain characteristics of simulations make them specific to their form:

1. Simulations deal with real-world systems with a level of detail ranging from the simple to the quite sophisticated.
2. Simulations are interested in the interactions of subsystems that exist in the overall system that is being studied.
3. Simulations are designed to mimic real-world behavior in accordance with not only known relationships but also the innate uncertainties of the real world by including reasonable uncertainties in the form of chance variances in output as a result of given inputs.
4. Simulations are system specific, dealing with individual problems or scenarios to allow the participant to gain experience artificially in interacting with real-world conditions.
5. Simulations have the capacity to predict the results of many different combinations of conditions in rapid succession, allowing modelers the opportunity to use the strengths of trial-and-error decision making without many of the shortcomings.
6. Simulations are both quantitative and qualitative, offering probable outcomes to given actions.

Not all of these characteristics are obvious in all simulations, but they are nonetheless present.

Examples of the use of simulations are present everywhere. Board games such as Monopoly™ and Risk™ are obvious examples of scenario simulations, dealing with real estate markets and world military domination, respectively. Computer arcade games are often simulations, offering the participants the opportunity to test their skills against electronic foes in everything from space warfare to guiding supertankers through the coastal straits of Alaska. Flight trainers are highly sophisticated simulations utilizing the analog model to give pilots the feel of an aircraft and allow them to work their way through all of the possible hazards that they might encounter without risking life and limb. In business, simulations are utilized to sharpen management and marketing skills, to outthink the competition, to study the effects of proposed strategies, and to test products. Engineers use computer and analog simulations to design and test equipment, machinery, and hardware. The military uses simulations to practice the art of war, often quite realistically. There are even computer simulations designed to simulate the operations of other computers!

SIMULATION DESIGN AND CONSTRUCTION

Designing a useful simulation is essentially the same as designing a model. The differences lie in the purpose of the simulation. Briefly, the process is as follows:

1. *Define the problem.* This may seem self-evident, but it can cause a great deal of consternation in later steps of the process if not properly done in the beginning. People often find themselves hampered by not completely understanding what it is that they are trying to do. A poor definition of the problem results in a severely limited understanding of how to solve it. The modeler may end up simulating the wrong system or not developing true relationships. The way a problem is defined dictates what solutions will and will not be sought.
2. *Conceptualize the system.* This is nothing more than a restatement of the mental model building used in initially conceiving a model's characteristics.
3. *Create a model representation.* Build a primary "rough draft" of the simulation as conceptualized to find the hidden factors that you have not yet considered.
4. *Observe the model's behavior.* Find out if the initial representation behaves as was initially expected.
5. *Evaluate the model's behavior.* Do you need to alter your conception of the problem? Do you need to adjust the factors involved in the

simulation? Are different elements more important than those you initially took into account? Are the results consistent with your initial purpose?

6. *Adjust the model as necessary.* As Alcorn's corollary to Murphy's Law states, any computer program that runs the first time is either worthless or you've missed something. The same is true of simulations.
7. *Use the simulation.* Test it to discover how closely it reflects reality. Some of the most useful simulations have been the ones that did *not* mimic real-world events and yielded information about contributing factors of major importance that were *not* included in their construction. Just as a well-formed experiment can be as valuable if it fails as if it succeeds, so the simulation can be useful in *not* initially working. Constant adjustment of criteria will, in all likelihood, be necessary in the case of complicated computer simulations, if they are expected to ape the activities of a dynamic, real-world system.

CAUSE AND EFFECT: ONE, TWO, THREE

Interactive real-world systems demonstrate cause-and-effect relationships among their subsystems, as do the sub-subsystems within each subsystem. From the overall to the smallest subset of a system, cause-and-effect relationships link elements together in behavioral relationships. In simulations, these cause-and-effect relationships can be used to organize activities cohesively. It is a prime method of construction to approach systemic simulation from this viewpoint.

In creating a simulation model, the patterns of behavioral relationships indicated by cause-and-effect interactions fall into a natural hierarchy. Since cause-and-effect relationships are seldom isolated, each reaction (effect) resulting from some initial causative action is in itself a causative action leading to some other reaction. The occurrence of rain, as an example, could be the cause of plant growth, the plant growth being a primary reaction to the increase in available water. However, the plant growth triggers other activity, such as an increase in parasitic insects in a farmer's field that only occurs because of the presence of the plants. This is a *secondary effect* of the rainfall. And as a result of the increase in the insect population, the population of insect-eating birds may rise; this is a third-level or *tertiary effect.* This multiplicative expansion of cause and effect from primary to secondary to tertiary illustrates the *chaining effect* in real-world systems.

All systems are related to all other systems. They are all part of the same process, all interconnected, and all quite interactive. Because of this, the modeler must decide which elements to include in the simulation and which to leave out, based on the aims of the model. Likewise, there are times when important secondary or tertiary effects may exist as a result of a given action, yet they may pass unnoticed because of the limits of construction of the model. Simulations will differ from reality in accordance with this fact.

It is additionally important to consider the concept of *feedback loops* in the construction of simulation models. These loops describe the interactive nature of certain give-and-take interactions of systemic subsets among themselves. Loops feed on themselves, so to speak, by each element contributing to the functioning of the other in some way. As an illustration, consider the relationship that exists between education and wealth. It could be argued that the occurrence of wealth and the level of education in a population tend to be mutually supportive. The greater the wealth of a culture, the greater its capacity to devote time and effort (and money) to educating that population. Similarly, the greater the level of education, the more productive and therefore the more wealthy the population as a whole can be expected to be. It would seem that these two factors are indeed mutually supportive. This is what is known as a *positive loop.*

An equally valid case could be argued for the relationship between farming and soil erosion. As farming increases in an ecology restricted by finite farmland, the intensity with which that farming takes place rises, resulting in overuse of the soil and eventual erosion. The erosion results in poorer crops, which necessarily means more intensive farming to keep up with the demands of the population for foodstuffs, resulting in more erosion. More farming produces more erosion, and more erosion produces more intensive farming. This is a *negative loop,* a situation in which each subsystem negatively affects the other subsystem, thus resulting in a net downward spiral of conditions.

Both positive and negative feedback loops can be found in real-world situations. The job for the modeler is to determine which loops are most important, how they affect the overall system being modeled, and how they should be taken into account in building the model. The modeler must also keep in mind that each subsystem is in itself a collection of loops, resulting in secondary and tertiary effects to take into consideration.

To illustrate the simulation technique, the following deals with the development of a simple model to explain some relationships that exist in a real-world ecosystem. For our illustration, we will modify the farming example. As a primary cycle, we will use the relationship between farming and population. With increases in farming, the availability of food initially rises, resulting in more food and better health for the population. A healthier, better-fed population is economically successful, and as with all successful economic systems, it grows as a result of the success, and its population rises. The rise in population results in the need for more food, which means more farming. Here we may consider the primary cycle to be positive and self-supporting in that farming creates more population and more population creates more farming. The loop is represented diagrammatically in Figure 7-1. The arrows in the figure indicate the direction of the cause-and-effect cycle to remind us of what is triggering what. The use of the plus sign in the center indicates the nature of the loop, in this case, positive.

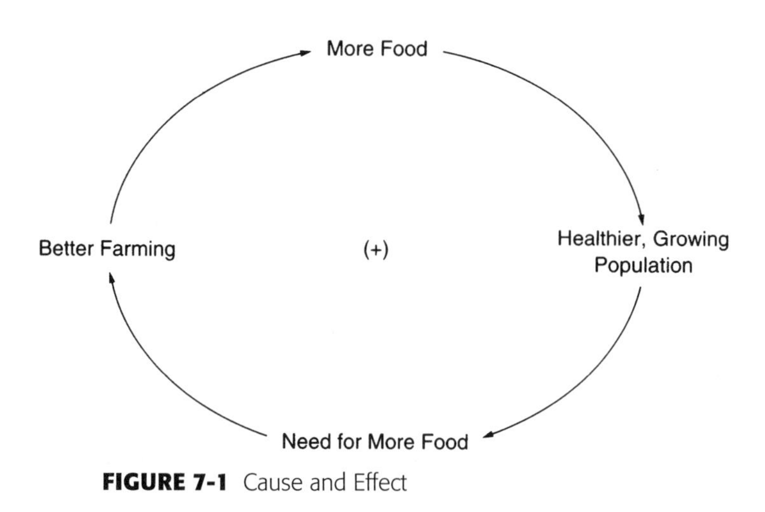

FIGURE 7-1 Cause and Effect

If we stopped at this point with our model, we would find a constant increase in farming and in population through time. And yet we know that this is not a very true picture. Too many secondary and tertiary effects have been left out of the analysis. There are too many other relationships to be dealt with before we may obtain even the simplest degree of accuracy. For further clarity, we should closely examine the observable subsystem elements. One of these we have already shown in illustrating the negative effects of farming on the erosion of soil. Figure 7-2 illustrates this erosion–farming mechanism.

The loop in Figure 7-2 is negative, indicating the dilatory effects of soil erosion on increasing farm output. Erosion acts as a governor, slowing down the farming process as more and more effort is put into intensive farmland use. These secondary effects usually present themselves in the real world and are easily taken into account. They also help to explain other secondary effects that may arise in connection with the primary cycle. As an example, we have already stated that technology is a method by which humans control nature through the production of artificial constructs to counteract negative pressures in their lives. If people are cold, they learn to build fires. If they are hunted by beasts, they learn to defend themselves and become the hunter. It is equally true that if farming becomes difficult, technology is developed to improve it. The reduction in farm output due to erosion may well be counteracted by an increase in farm technology, creating the opportunity to produce more crops on less land or, alternately, to reduce erosion. Taking this aspect into account yields still another loop by which erosion and farming needs create technology. This is represented in Figure 7-2.

The discovery and inclusion of secondary effects in a simulation model pose little difficulty. Tertiary effects and possibly quaternary effects and

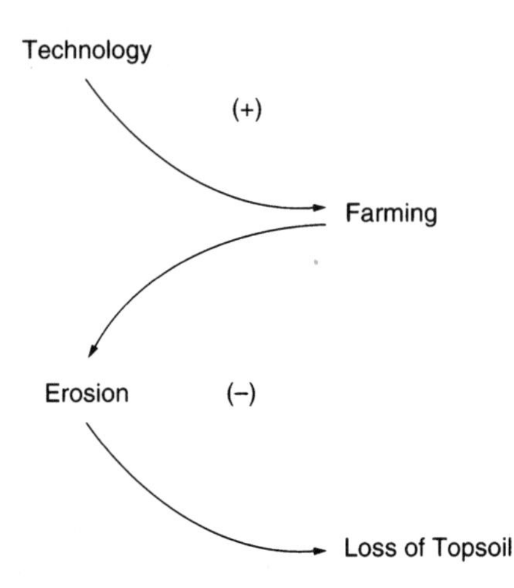

FIGURE 7-2 Technology and Erosion in the Farming Cycle Model

beyond can be more of a problem. It is often the tertiary or quaternary effects that prove to be the most important in analyzing a real-world system not necessarily because of their impact per se but rather because of not being considered in the analysis. It is difficult to react to a factor you have not considered.

A case in point would be the result of soil erosion and the loss of topsoil on other elements of the ecosystem. If erosion takes place and topsoil washes away, that soil must go somewhere, often into rivers and lakes that silt up as a result of the increase in dissolved solids in the water. The lakes become muddy and clogged, and rivers become unnavigable. Fish populations decrease as oxygen supplies in the water decrease, and aquatic vegetation thins. Local industries are hard pressed to dispose of waste once easily handled by the ambient water systems. Stagnation may result. Mosquito populations and possibly disease rates increase as a result of the process. These effects may be the most serious and least understood of all the consequences of the initial increase in population and farming. It is this type of effect, the hidden, not immediately yet potentially dangerous consequence of some system change, that it is most necessary to understand. A complete model of the simplistic system appears in Figure 7-3.

How does the simulation technique help us in our study of technology and society? It is exactly for this type of complex, multilevel system that the simulation technique was devised. It is a method of using the systems technique of defining and subdefining a real-world structure into an understandable network of cause-and-effect behavior and influence so that we can both comprehend the nature of the problems of technology and society and predict the effects of changes before they take place. A further refinement of the process is to quantify as many of these relationships as possible to develop the predictions. The exercises at the end of the chapter offer an opportunity for you to gain firsthand experience in using simulation to enhance your own understanding of systems and the manner in which they operate.

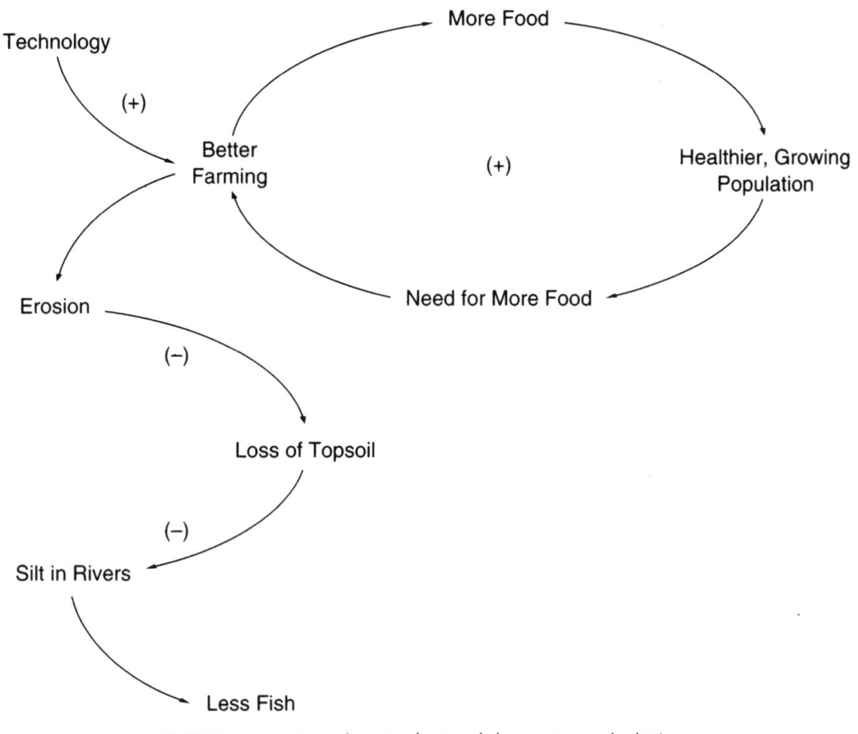

FIGURE 7-3 Farming Cycle Model—An Extended View

CONCLUSION

In pursuit of understanding the physical world and the various ways in which its components interact, humanity has developed the facility to create artificial constructs that behave in similar ways to observed patterns in the physical world. These constructs are known as *models.* They take several forms, depending on the nature of the phenomenon being studied and the characteristics that the individual investigator desires to understand. That is, the model's form is determined by the limitations of the environment and the nature of the individual's questions about the real-world phenomenon in question.

Models can be usefully classified as *analog,* a type that behaves in some way similar to the reality that it is designed to represent; *iconic,* which is designed to resemble a physical reality though not necessarily to behave in an analogous manner; *verbal,* which is a model designed to convert thoughts or concepts into language, establish relationships and restrictions of real-world systems, and organize them in an understandable form; and *mathematical,* a symbolic manipulative representation of reality designed to describe relationships among certain factors of the reality that it is designed to represent.

Of special interest to the investigator of technosociological relationships are simulations and games. These forms of models may involve any or all of the above-mentioned types of models. Games are chiefly useful in creating experience artificially by providing a competitive environment in which an investigator may mimic interaction with a particular environment based on assumed rules or parameters used to describe the environment. Simulations provide a construct for describing and predicting behavior in real-world situations within the confines of specific parameters. By the use of gaming techniques and simulations, an individual is able to develop expertise in dealing with complex systems and in predicting outcomes before a commitment to some line of action is finalized. In addition, the very construction of games and simulations forces the individual investigator to discover the nature of the relationships of the real-world phenomena in question.

Models in general and games and simulations in particular offer organization, structure, cause-and-effect relational concentration, and predictability so important in the study of technosociological systems.

In this chapter, we explored the nature of games, models, and simulations as a tool in the study of the relationships that exist between society and technology. Yet without principles that reflect reality that can be inserted into the models, there is no useful benefit from constructing them in the first place. For that reason, we continue with our exploration of the nature of reality, beginning in the next chapter with a discussion of the universal laws of systems behavior, which, it will be explained, define and predict in general terms how every element of nature behaves. Chapter 8 is a short chapter, yet it contains the essence of our study and should not be considered lightly. Brevity does not always indicate superficiality or inconsequentiality. If the reader is able to master the concepts in this one brief presentation, all else falls into place. By discovering these laws and thinking about our own experiences in life, it is possible to better understand what is happening around us and thus, as with any new knowledge, increase our ability to survive and thrive. As you read Chapter 8, consider how you would apply these universal laws in a model of some technology of societal structure if you were to develop one on your own.

KEY TERMS

Analog Model
Descriptive Model
Deterministic Model
Games
Gaming
Iconic Model
Modeling
Modeling Process
Probabilistic Model
Qualitative Model
Quantitative Model
Spatial Relationships
Symbolic Model
Verbal Model

REVIEW QUESTIONS

1. What is a model?
2. Name four types of models and define each, giving examples.
3. What are the steps in the modeling process?
4. How is the modeling process related to the scientific method?
5. What is the difference between a qualitative model and a quantitative model?
6. What is a simulation?
7. Define game and explain how it is related to the concept of simulations.
8. Give three examples of each of the following: (a) simulation, (b) game, (c) qualitative model, (d) quantitative model, and (e) cause-and-effect relationship.

ESSAY QUESTIONS

1. How is the use of models indicative of the manner in which humans organize information? How is it a result of our organizational processes? That is, which leads to which?
2. If we were limited to only non-mathematical approaches in our modeling, how would it affect our ability to produce simulations? How would we overcome the resulting difficulties?
3. Discuss the importance of recreational game playing and modeling in developing an understanding of our world.
4. Discuss the relationship between the scientific method and the processes of model creation, simulation, and game play. How essential is this connection to the successful development of predictive models of the world?

THOUGHT AND PROCESS

1. In light of the information provided in this chapter, return to the questions at the conclusion of Chapter 6 and try your hand one more time at the process of using the systems analysis technique.

2. The world is filled with models. We encounter scores of them each day. Look around and find an example of each type of model discussed in the text among the common experiences of your daily life. Why do you suppose the particular form of model chosen by the modeler was preferred in each case over the others? What do you believe her or his purpose was as indicated by the form chosen?

3. Design a board game based on one of the following themes. The thematic titles should be self-explanatory. Let your creative instincts run free. Here are the themes: (a) a date with a childhood sweetheart, (b) the Battle of Atlanta, (c) a stock market tycoon, (d) a space travel adventure, (e) college registration survival, (f) invasion of the mechanical man, (g) a zombie's feast, and (h) gridiron glory.

4. Using the modeling process outlined in the chapter, develop a model for some real-world phenomenon in which you are interested. Go through the process and document your choices of form, method, and so forth.

5. Build a diagrammatic simulation model of the various cause-and-effect relationships that are included in some real-world phenomenon in which you are personally interested. Carry your analysis to at least the third level of effects (i.e., tertiary-effect loops).

CHAPTER 8

SYSTEMS BEHAVIOR: THE UNIVERSAL LAWS

As we discussed in Chapter 6, the systems approach is useful for organizing information and for describing the relationships that exist among elements of a process or other aggregation. However, its usefulness goes far beyond its descriptive abilities. According to systems theory, the behavior of all systems is essentially the same. If this is true, and if we can determine what that behavior is, we may predict the results of any external or internal event that takes place in reference to any system. That is, if all systems have the same behavioral characteristics, then they should be *determinant.* We explore this cause-and-effect determinacy and its limitations in Chapter 9. In this chapter, we discuss the nature of systems behavior.

THE THREE LAWS OF SYSTEMS BEHAVIOR

There are three universal laws of systems behavior. These laws describe the general manner in which systems behave; how this behavior manifests itself depends on the nature of the system in question. After stating and understanding the laws, applying them is merely a matter of looking at the relationships among the subsystems to determine what will take place as the system operates.

Each of the three laws is immutable, as is indicated by the use of the term *law,* and is true of all systems; thus, they are referred to as *universal.* There are no instances in the physical world where these laws are not at work. That is quite a statement to make. There are few universals in this world, yet as we proceed, the truth of this statement will become evident.

Briefly stated, the three universal laws of systems are as follows:

1. All systems are synergistic.
2. Systems are absolutely reciprocal.
3. All systems seek to achieve and maintain balance.

All Systems Are Synergistic

By *synergy,* we mean that the whole is greater than the sum of the parts. Something extra exists in the system when these parts are assembled and associated that does not exist until that connection is made. Examples abound in the physical world. R. Buckminster Fuller noted the physical synergistic relationship as it exists in three-dimensional structures. Beginning with triangles, he noted that they consist of three separate lines that when positioned so that each touches the other two, create a stable, self-supporting, three-sided structure. Anyone can easily demonstrate this by gluing or taping three toothpicks or soda straws together to form a triangle. The object distributes the weight to give the structure stability. If one tries this with four or five struts, the system can be easily collapsed with minor pressure exerted on a given side or corner. Three triangles aligned and attached edge to edge form a *tetrahedron,* a pyramid with three sides. Yet the name of the object implies four sides, *tetra* coming from the Greek for four. And indeed, if you lift the pyramid and look at the bottom, you will find a fourth triangle formed by the intersection of the other three. In other words, from three triangles we gain a fourth, simply by putting them together in a system. This is a simple example of synergy. It appears to be almost a parlor trick, more semantic illusion than real. Yet it is the basis for an entire process of both geometry and architecture that Mr. Fuller developed, which culminated with his famous geodesic dome. (A geodesic dome is a spherical structure made of interlocking triangles that each exert pressure and create support for those contiguous to it, resulting in a light, efficient framework of amazing strength—an application of the synergistic nature of the system created.)

Virtually everything that has physical existence can be seen as a synergistic system. Business organizations consist of large numbers of subsystems, including equipment, buildings, people, paper, energy, transportation, and communication systems. Collectively, these elements (subsystems) form a cohesive whole with the capacity to do work and efficiently produce products, services, or other results. This is as true of companies as it is of nations, tribes, organisms, or mechanical objects. In each case, the elements have combined or have been combined because of an inherent advantage in being part of a system. In other words, it is more efficient to act in unison than to act alone.

Consider a group of ancient troglodytes, or cave dwellers. Though living in a community, each person is acting alone. Each goes out every day to hunt, to gather wood for the fire, to fetch water, and to pick berries and other edibles. When they return, they must cook their kill and tan the hides for clothing; when possible, they must construct clay pots and other vessels for storing their goods.

One day, a particularly bright gentleman by the name of Og has an idea. He realizes that he enjoys making projectile points, and that he is very good at it; he notes that his neighbor, Glog, who lives in the next cave, enjoys playing with mud and creating clay pots. Og has neither the desire nor the skill to create good pots, which seriously reduces his capacity to survive. He approaches Glog and suggests that he, Og, produce enough projectile points for both of them and that Glog produce enough pots for both. Glog, delighted with the idea of not having to mess with projectile points, a process he truly despises, readily agrees. To their mutual surprise, this arrangement works out beyond their wildest dreams. They soon discover that they both have more projectile points and pots than they had before and that the quality of both goods is higher! This is not surprising, since each of them is now doing what they most enjoy, meaning they are probably very good at the task. They are avoiding both the time wasted moving from one task to the next and the excessive time spent creating ineffective technology they are not good at producing. They have discovered the synergy inherent in dividing up the work and concentrating on what they do best. They quickly realize that they can extend this process, and soon another neighbor, Slog, joins them, doing what he most enjoys most—hunting. The system grows, and as it does, so does its synergy. In essence, the whole structure is greater than the sum of its parts in that the cooperation of the elements creates a higher degree of efficiency, resulting in more for everyone, and of higher quality as well.

Nature is notorious for efficiency. Though it will try anything, it tends to move toward those constructs that achieve the greatest result with minimum input of time, energy, and resources. This is nothing more than a statement of systemic synergy. Trees are not columnar, consisting of countless parallel tubes, by chance, but rather because it is the most efficient and most synergistic form. The order of the solar system creates a high level of stability and wastes less energy in random impacts than an unstructured system would. Governments exist as a method of organizing human society not because of some fluke of nature but because it reduces waste and allows for a more highly successful species. These are all examples of systems expressing their synergy. If systemic structures did not have advantages over random structures, there would be no reason for them to exist. By virtue of their synergy, they are what nature tends toward, and through time, systems refine and redefine their internal structure to increase that synergy. Without synergy, a system simply does not exist.

Systems Are Absolutely Reciprocal

The second universal law refers to the give and take of a system, from the smallest relationship to the entire physical universe (the ultimate known system). *Reciprocity* in its simplest form means that for a system, you get back what you put in. Again, we find that a number of generally known principles reflect this idea. What goes up, must come down. For every action, there is an equal and opposite reaction. What goes around, comes around. Yin–Yang. Karma. There is no free lunch. All are enunciations of the reciprocity principle.

In economic systems, there is an input of land, labor, and capital into a productive process that results in finished goods that are exchanged in the marketplace for money, which is then returned to the individuals involved in the productive process so that they may purchase other produced goods. This is a gross oversimplification, to be sure, yet it is a clear example of the principle of reciprocity. Exchange takes place, in which both buyer and seller have exchanged something of relatively low value for something of relatively high value, and both are better off. The process is reciprocal (and, please note, synergistic). Restated, the price of a good is exactly equal to all of the income received by everyone connected with the construction of that good, in the form of wages, rents, and profits. It is give and take.

Reciprocity is expressed in natural, physical processes as well as in cultural interactions. A gallon of gasoline oxidizes rapidly when mixed with oxygen and exposed to flame, and either burns or explodes. Given the quantities of gasoline and oxygen involved in the reaction, anyone with the necessary formulas may predict exactly how much energy will be released in the process. The potential energy created when the gasoline was produced is exactly equal to the kinetic energy released when the gasoline is burned. An even simpler example is that of a large iron ball raised to a height. As it rises, potential energy is created by moving it from the earth, thus separating it from the source of gravity that is constantly exerting a downward pressure on its mass. Upon dropping the ball, the kinetic energy and force with which it hits the earth equal that created potential. What you put in is what you get out.

Note that this principle is reciprocal not only in quantity of reaction but also in result. Suppose two individuals in a class decide to cheat on a test. They design a methodology for sneaking answers into class, create a secret document for that purpose, and take it with them. Let us further suppose that they are successful in their attempt and falsely are given a high grade on the test. Is this reciprocal? Absolutely. Others also make a good grade on the test by studying and mastering the material. Yet at the end of the process, the benefits received from the process are quite different. The cheaters have good grades to show for their efforts. Those who studied have not only good grades but also knowledge of the subject. Quite often, in fact, it requires as much energy and effort to design successful strategies for cheating as it does to simply study the material.

There appears, however, to be a fundamental problem with our "universal laws" of systems theory. The first law states that the whole is greater than the sum of the parts, and the second states that you get back what you put in. How can something be both greater than and at the same time equal to its components? We seem to have encountered a *paradox,* defined here as any two mutually exclusive events that appear to occur simultaneously. Fortunately, in the real world, no true paradoxes exist; what actually happens is that we lack information about the nature of the events in question. In other words, we simply do not sufficiently understand the processes taking place to see how both events can simultaneously occur.

To return to the example of the gasoline, we find that both synergy and reciprocity can exist simultaneously. If a large quantity of gasoline is placed in the center of a field, exposed to oxygen and then to flame, we have an explosion. In the process, we create a shock wave in the air that can be felt if we are close enough, and we dislodge an amount of the surrounding terrain and send it flying in a cloud of dust, perhaps producing a small crater. The energy released is reciprocal with the potential energy contained in the gasoline. If an identical quantity of gasoline is confined in a fuel tank and attached to an engine, it will send a rocket skyward with the same total release of energy as with the explosion in the field. Both are examples of reciprocity. In the first case, the energy is released randomly. In the second, we have the additional creation of useful work because the explosion is contained and channeled to propel the rocket. This extra result is the synergy of the system, and it is the *focus* of energy in the system that produces the synergy. By directing the energy, we are able to choose the result we want. What systems do is focus the energy that is poured into them. (Note that our story of Og's and Glog's projectile points and clay pots is equally an illustration of focus. After cooperating and dividing the labor, the individuals involved begin to focus their energy on what they do well, with a minimum of downtime for "switching gears" from one task to another and attempting tasks for which they are ill equipped.) Because of this focusing characteristic of systemic structure, the paradox disappears. The principle applies universally. With fuel, as with all else, we have a choice as to how we will expend our energies. We can, as in the first case, simply blow up a field and in the process blow a lot of smoke, make a lot of noise, and perhaps make a small dent in our environment. Or, with the same amount of energy, we can focus it, and send a person to the moon!

All Systems Seek to Achieve and Maintain Balance

The third universal law of systems behavior deals with the purpose of a system. It was stated in the beginning that a system consists of elements that cooperate to achieve some goal. As it happens, the goal of every system is identical, though the manifestation may be different. That goal is to achieve and maintain balance. By *balance,* we mean that a system has achieved its goal and continues to maintain that goal through time. An organization sets goals of profit, product, and/or service to the community. It then proceeds to reach those goals and intends to continue achieving them. As it does so, it is in balance. Whenever external events threaten the goal, the system adjusts to reattain its goal and return to balance. This is true of the human body, a political system, the ecosystem in which we function, the solar system, or the internal mechanisms of your automobile. In each case, defined criteria are set up for determining what the system's goal is and then for reaching it.

The orbital structure of the solar system consists of opposing forces, one of velocity and one of gravity, that balance each other to create the path a planet or satellite inscribes around its parent object. When some outside force, such as a passing comet or stray asteroid, alters the elements of that orbit, the satellite adjusts and forms a new orbit that brings it back into balance. When your automobile is operating, it has a host of systems (steering, brakes, electrical, fuel, cooling, heating, etc.) that function in concert to achieve its goal, to get passengers and driver from point A to point B with a reasonable degree of speed and comfort. If something changes, such as the driver applying pressure to the accelerator, the system quickly adjusts to go back to balance and run smoothly. The changes that occur in an ecosystem also are the result of internal adjustments by the systemic elements to pressures exerted by events. If a field is left to its own devices, it becomes meadow, then brushland, then pine forest, and finally hardwood forest as it seeks balance. The number and types of insects and animals change and adjust to the new structure of their world. At some level, even temperature and moisture gradients change in accordance with the alterations to terrain, flora, and fauna in a constant dance of creating and re-creating balance. Note that this balance takes place in the face of constant change. Is this not also a paradox? How can we have both stability and change simultaneously?

Consider the human body. It is a system consisting of subsystems, including the circulatory, endocrine, digestive, and nervous systems, as well as others. How does it maintain balance? What if it receives external stimuli? Imagine you are asleep in bed. It is 6:00 A.M. and you currently take classes in the morning and work in the afternoon and evening. While sleeping, you experience shallow breathing and slow respiration, since your limited movements require little energy. Your body temperature is relatively low because of slow metabolism, your brain is churning out alpha waves as you drift in a dreamy state known as REM sleep, and you are physically still. You are in balance.

In the next instant, the alarm suddenly begins to buzz loudly on the bedside table. Immediately you react to this external stimulus in an effort to reattain your prior state of balance. You may (a) simply reach out and slam the button that turns the alarm off, returning to your original state, (b) lay quietly, aware of the offensive object but secure in the knowledge that if you do nothing long enough, your spouse will grumble once, growl twice, and reach over and turn it off, allowing you another ten or fifteen minutes of snooze time, (c) be smart enough to have placed the machine from hell across the room so that you have to get up out of bed to shut it up, or (d) if you're like many of us, suddenly find yourself bolt upright in bed, trying to figure out what just happened. In any case, you react. Now everything is busy. To begin with, you are startled. Your body has activated an automatic program designed to secure your survival, a program that has been around in the human race for aeons; it's called *fight or flight.* You decide almost instantaneously whether you are going to fight or run away from this threat. But whichever decision you choose, you need energy. You find yourself with an adrenaline rush of momentous proportions. Your breathing is rapid as it attempts to send oxygen to your muscles to fight or flee. Your body temperature is suddenly less than what you need, and you may begin to shake (the process is called *shivering thermogenesis,* a fancy name for warming the body with rapid muscular contractions). Your alpha waves have all but disappeared, replaced with the beta waves of waking. You are in a much different state than only a moment ago; you are in balance for emergencies.

Fortunately, this usually doesn't last long, and slowly, as you come to your senses, you realize it's just time to get up. Your body begins to calm itself, though not to the extent of going back to sleep, and you rise, shower, grab some breakfast, and head off to school. About halfway there, you finally come to complete consciousness. Sound familiar? Once at school, you are now wide awake, with the proper levels of awareness and bodily function, and you enter a classroom, where you are expected to sit quietly for fifty minutes or more while remaining mentally alert. Needless to say, your body, which has just adjusted to being active, revolts and, as it readjusts, works off excess energy with fidgets, conversation, and other forms of extraneous behavior. Finally, you arrive at yet another level of balance, where your body is relatively calm and your mind relatively alert, unless, of course, you simply revert to your earlier state and fall asleep in class! Of course, you have achieved this acceptable level of balance no sooner than the bell sounds for the next class, and you are off again, seeking to readjust physically and mentally.

Our whole lives are lived this way. We are constantly reacting to new stimuli that force us to reattain our state of balance. What constitutes balance continually changes, but the need for that balance is constant. It is dynamic balance, or balance in motion.

CONCLUSION

All systems seek to attain their purpose by acting in accordance with the three universal laws discussed in this chapter. For a human being, the purpose is, at its most basic, to survive, thrive, and reproduce. All else is just variations on a theme, in which we find many different combinations of behavior that allow us to adjust to circumstances to achieve this goal. Some combinations of behavior are more successful than others, but the successful ones available are so numerous that we can follow a virtually limitless variety of paths to achieve the goal. Thus some people choose to be in business, others to be artisans, still others to be soldiers or circus performers or thieves or drug dealers (or any combination of these and other roles). In each case, the underlying purpose of the behavior is to achieve the goals of being a human being: to survive, thrive, and reproduce. We watch it happening all the time.

We see all of nature doing it, constantly involved in the process of dynamic balance, even when what appears to be detrimental patterns of behavior occur. We may, for instance, not like the way nature reacts to our imposition of human endeavor on the world, but from the point of view of the whole ecosystem, it is simply reacting and adjusting to another set of stimuli in order to keep the whole system in balance.

Think of a juggler standing on a board that is shifting back and forth over a cylinder. With his hands he is juggling six balls, keeping them in constant motion by throwing them from one hand to the other. On his elbows are rings that rotate in opposite directions simultaneously. And on his chin is a long pole, atop which is a saucer, then a cup, then a saucer, then a cup, then a saucer and a cup, with a single spoon in the third cup. Everything in this system is in constant motion in a fashion that appears to be almost chaotic, yet the spoon never falls. The goal of that system is to keep the spoon in the air. The individual motions, each countered reciprocally by a motion in the opposite direction, collectively create and maintain the goal. This is dynamic balance, balance in motion, and it is the basis of all systems behavior. Synergy serves to more efficiently achieve the balance. Reciprocity balances every action with an opposite one, and the system performs its function through time.

We can therefore be certain of the behavior of any system that we study. Behavior may be varied and specific to the nature of that system, but it will never defy any of the three universal laws. If it tries, as we shall see, the system either fails or ceases to exist.

In the next chapter, we consider the "fly in the ointment"—the apparent error in the systemic theory, that of random events. Thus far, it would seem that we live in a clockwork world, where every event is predictable and determinate. In Chapter 9, we explore the apparent absurdity of this idea, since many events are totally random and without either determination or predictability . . . or are they? If the systemic approach is to prove useful, it must not have obvious fallacies that disprove its tenets. Chapter 9 clears away much of this confusion by adding a single element, one that has been explored for the past thirty years and continues to unfold in a dance that brings much of the confusion of the world into focus. This is the theory of chaos.

KEY TERMS

Balance
Laws of Systems Behavior
Reciprocity
Synergy
Systems

REVIEW QUESTIONS

1. What are the three universal laws of systems behavior?
2. What happens to a system that loses its synergy? What does this tell us about the reason for systems existing?

3. Give an example of reciprocity in each of the following types of systems: (a) a political party, (b) a corporation, (c) a culture, (d) an M-1 tank, (e) a flower garden, and (f) a marriage.
4. Systemically, how do we define balance?
5. What takes place in a system that is not in balance? Give three examples.

ESSAY QUESTIONS

1. Discuss alternative schema (other than the systemic one) to explain the machinations of the real world and develop a defense for them as equally valid or more valid ways of viewing reality.
2. Discuss the existence of order in a non-systemic world. That is, assume that there is no reciprocity, no synergy, and no inherent movement toward balance. What would the resulting universe look like?
3. Discuss the systemic laws in terms of technological development using examples, either historical or current, that reflect the manner in which these laws operate.
4. Discuss the systemic laws, first as a natural observance of natural phenomena and second as a paradigmatic predisposition toward a given point of view developed by societal indoctrination. Show the weaknesses of each point of view.

THOUGHT AND PROCESS

1. To explore the synergy inherent in systems, enlist the help of two friends. First, each of you (a) fold a piece of paper into an envelope, (b) make a paper boat or hat, and (c) pull ten matches from a pile and place them parallel in a bowl. Have each person repeat this process five times. Next, each of you perform just one of the tasks for fifteen repetitions. Now compare the quality of the resulting products and the time taken to complete each process. The difference represents the synergy inherent in a cooperative systemic approach to work.

2. For each of the following systems, determine where the synergy, reciprocity, and balance exists. What is each system's goal? Here are the systems: (a) the solar system, (b) the political system of Great Britain, (c) your family, (d) the class in which you are now enrolled, and (e) a drill press.

3. Create a hypothetical system to carry out each of the following tasks. Note for each system the subsystems involved and the defined goal, and show how the interactions in the system are reciprocal. Here are the tasks: (a) raising water from a well, (b) automatically closing a door two minutes after it opens, (c) watering a garden while the owners are away on vacation, and (d) taking roll in a class without calling out each name.

CHAPTER 9

DIVERSITY, RANDOMNESS, AND SYSTEMIC INTEGRITY

In the previous two chapters, we explored the structure and behavior of systems, noting that everything either is a system itself or is part of one and that the behavior of these systems is predictable. It would appear that the world is a determinate place where, given the tools of analysis, predictability is absolute. Yet we all know from personal experience that predicting what will happen next is all but impossible. Is the theory worthless, then, yielding only superficial advantages when analyzing observed behavior? Once again we are faced with a paradox that we must explain if we are to believe that systemics really is a reflection of reality. We live in a diverse and complex society that operates in a still more complex and diverse world, where the number of possibilities is practically infinite and it appears that anything can happen. Something must be missing if systems theory is indeed true. As it happens, that something is what is generally termed *chaos.* The body of hypotheses researchers have built up to analyze and explain the effects of chaos is known as *chaos theory.* Actually, *chaos* as used here is a misnomer, since the word describes a total lack of order, by definition, a state of utter disorder without structure of any kind, the exact opposite of systemic structure. The behavior we discuss here is not "utter disorder"; rather it is order operating in a mathematically unpredictable fashion.

SYSTEMS THEORY AND CHAOS

Chaos theory is a relatively recent body of knowledge, having developed since the early 1960s, when a coherent look at nondeterministic behavior became valuable. The theory was first enunciated in regard to the weather by MIT meteorologist Edward Lorenz.

> In 1960, Lorenz was using his computer to solve a number of nonlinear equations modeling the earth's atmosphere. Repeating one forecast [he rounded] off the figures in the equations to three decimal places in place of the six he had used in the previous run. The new result wasn't an approximation of his previous forecast, it was a totally *different* forecast. The small, three-

> decimal-place discrepancy had been grossly magnified by the iterative process inherent in solving the equations. That the results were so far apart means that complex nonlinear dynamical systems such as the weather must be so incredibly sensitive that the smallest detail can affect them.[1]

This sensitivity became known as the *butterfly effect,* wherein the beating of the wings of a butterfly in Hawaii can affect the weather in San Francisco six months later. Unless we are willing to measure the wing velocity of every butterfly in the world (along with every other flying creature), our predictions about the weather can only be approximate.

> At first it may seem unfair, or at least an exaggeration, to call a weather system chaotic just because we can't predict it. If our ability to predict is faulty isn't that because we lack all the necessary detail or we don't have the right equation? The answer is no. What Lorenz had seen was that because of the iterated nature of nonlinear equations (which represent the interconnected nature of dynamical systems), no amount of additional detail will help perfect the prediction. . . . A striking property of iterative equations is their extreme sensitivity to initial conditions. If X_1 in [a] number-doubling equation is changed very slightly, then the [new] sequence will soon diverge from the original. It was this property that was discovered by Lorenz in his weather calculations. In the nineteenth century scientists had always assumed that a small error in initial data would either be averaged out, or would, at most, produce a small effect. But where iterations are concerned, small errors can be rapidly amplified.[2]

It was soon seen that the same principle applies in all scientific endeavors, both in the "hard" sciences and in the social, or so-called "soft," sciences.

This approach is most useful when studying highly complex systems. The greater the degree of complexity, the greater the number of variables and, therefore, the higher the probability of small, seemingly insignificant factors having major import on events and processes. It is also significant that this represents a change in paradigm, leading investigators to change their investigative methodology and enter new realms of thought. Since the beginnings of the Industrial Revolution, empirical sciences, such as chemistry and physics, have shunned association with the "soft" sciences because these disciplines are by nature unpredictable. In physics, a single formula

[1]John Briggs and F. David Peat, *Turbulent Mirror: An Illustrated Guide to Chaos Theory and the Science of Wholeness* (New York: Harper & Row, 1989), pp. 68–69.

[2]Briggs and Peat, *Turbulent Mirror,* pp. 69, 70–71.

can describe relationships that are determinate to the point of being laws of behavior. In sociology, biology, or anthropology, relationships becomes less predictable and therefore more difficult to reproduce due to the incredible number of interrelated factors. Thus they are considered less exact and inherently of a "less scientific" nature. Yet if we remove the bias toward the need for exact answers, we find that we may gain a great deal of useful information about the behavior of the physical world.

Quantum physics definitely gains information without demanding exactness, beginning with the Heisenberg uncertainty principle (you can determine the energy of a particle or its position, but not both simultaneously) and operating on the basis of probability rather than certainty of events. But to expand this to a general description of all natural behavior is quite a leap. Yet it appears from chaos theory that indeterminacy, rather than predictability, describes the general case.

Actually, physical phenomena are not chaotic, but rather they are examples of what can be termed *determinate randomness.* That is, any phenomenon that has multiple outcomes dependent on random events can still be determined mathematically, given a set of actual events. That makes it determinate. However, if we begin some time into the operation of the phenomenon, we cannot trace our observed measurements back to the original source event, because there are multiple paths that will get us there (a result of the aforementioned "extreme sensitivity to initial conditions"). This means that where we are is the result of a series of events, some of which are random, that proceed in a logical order of cause and effect from one event to the next. However, it may be impossible to trace the cause and effects back because of the large number of random variables involved.

Tracing Cause and Effect

By way of illustration, there is a phenomenon that almost everyone has consciously encountered in their thinking processes. You make a statement or have a thought that seems random, or at least a non sequitur to the context in which it occurs. For example, you are sitting on a bus, and you notice a woman in a red dress sitting across from you. Within a few minutes, you find yourself smiling at the thought of an event that took place in the sixth grade when one of your schoolmates fell in a trash can. On first inspection, this event seems totally unconnected to the woman in the red dress. You think to yourself, "Where did that come from?" In thinking about it, you trace your thoughts back, remembering that the red dress, not the woman, reminded you of a dress your mother had. In fact, you imagined your mother wearing it one night to a PTA meeting when you were a child. That triggered memories of the school you attended, bringing you to thinking of the year you transferred to a new school, resulting in the loss of a close friend. That friend used to play baseball with you, and you continued to play at the new school, where a classmate played the same position as your friend had. You then suddenly remembered the arrangement of desks in the sixth-grade classroom and who sat where. You can remember everyone's name except that of the person who occupied the desk in the front left row of the room. All you can remember is her face and the fact that another student once fell in a trash can and ended up on the front of her desk, which proceeded to tip, spilling both onto the floor in a slapstick series of rolls and bumps that sent the entire class into laughter. Thus, a woman in a red dress elicits a totally unconnected memory from early childhood.

Imagine now if you were a friend of this person on the bus. She smiles and chuckles, and you say, "What's so funny?" Her answer, the memory of the trash can event, seems totally random, and you as a companion would probably be incapable of determining where it came from.

Much of human experience is like that trail. We have the experience, then seek for an explanation of how it occurred, particularly if it is one we do not wish to repeat, but for all our effort, the connections do not seem clear. It does not mean that they are not there, or that they are random, but merely that the system that created them is too complex to really trace. Thus what appears to be chaotic is not, and what appears to be orderly can at any moment become chaotic.

Numerous examples of deterministic randomness exist in nature. Almost all unexpected events that occur when well-documented evidence predicts that they should not are cases of this. An airliner crashes, and it is an unexpected event. The plane has been designed to survive every possible contingency, yet inexplicably it crashes. An investigation discovers a series of unlikely events culminating in the crash. The probability of each individual event occurring is minuscule, and the combination is even more unlikely, yet in one instance, on one flight, they do occur and the plane crashes. We call it a million-to-one chance. We would be more correct in calling it determinately random.

Before the development of chaos theory, events of this sort, where multiple outcomes are possible, were merely referred to as indeterminate. With the theory's evolution, particularly since the development of a mathematics designed to describe determinate randomness, these events are seen as natural, as capable of analysis, and as valuable in explaining the natural world and real-world events, providing clues about how to operate within the system.

Patterns in Randomness

Mathematically, a series of equations developed by B. Mandelbrot that demonstrated the order in random systems has led to a new understanding of nature. Beginning with a curiosity about what were referred to as *strange attractors*, an apparent tendency of totally random phenomena to center or be attracted to only a small range of results even though any result has an equal theoretical probability of occurring, he began to notice patterns where there should be none. Plotting formulas designed to create totally random answers in multiple dimensions, he found himself creating patterns rather than random plots. Bifurcations arose, where the random plots suddenly diverged into separate groupings, going in slightly different directions and leaving areas of the graph with no points at all. Repetitive patterns emerged as complicated figures that repeated themselves whenever the specific formulas were iterated, and many of the patterns had a familiarity about them that mirrored nature. What most fascinated Mandelbrot was that the patterns repeated themselves. By focusing on a small graphical representation resembling a paisley pattern, or a bulbous turtle, he discovered that the same pattern repeated over and over again, alternating with levels of data that appeared totally random and without form. There was apparently order in the random behavior and randomness in the ordered behavior. Each was a part of the other. These formulas and their graphs became known as the Mandelbrot set and Mandelbrot plots, respectively. Another set of formulas, developed by G. Julia, further explored this phenomenon, and in a short time, many researchers in many fields were utilizing the principles of what was called *fractal geometry* to investigate phenomena in their fields.

CHAOS THEORY AND DYNAMIC BALANCE

The beauty of chaos theory is that it appears to be universally applicable to all natural phenomena. It can be used to explain how the human mind works, how ecosystems shift and adjust to changes in stimuli, and why some societies choose one form of government and other societies another. It is used on the edge of physics and in chemistry, biology, and psychology. It is a paradigmatic shift in how we view our world and its operation.

It is not the purpose of this book, however, to explore the intricacies of chaos theory. Indeed, any chaoticist worth the title would probably groan at the oversimplification here presented. Yet it is necessary to lay the foundations of chaos theory to understand our own subject, that is, the nature of technology and its creation, relevance to, and influence on our culture, our society, and our world. Chaos theory holds the key to explain how a deterministic, systemic approach can have truth in a world of unpredictability. Earlier we explored the concept of dynamic balance. The balance of systems is seen to take place in a dynamic setting, where conditions constantly change and the system reacts to those changes to reattain or maintain the balanced state. It is the randomness of those changes and stimuli that is the meat of chaos theory. Though we may not be able to predict the *exact* nature of a systemic response—there are simply too many variables for that—we can predict with accuracy the direction and general import of what will occur. I cannot predict what the stock market will do tomorrow, but I can note that since reciprocity is absolute, behavior in the market now will be offset by behavior of a reciprocal nature in the future. I cannot tell exactly what changes will take place in a nation or in a community next year or ten years from now, but I can predict the nature of those changes based on what is taking place today. What you put in is what you get out. To use a biblical allusion, what you sow is what you reap. This is true of studying for or cheating on a test, of working hard or hardly working, of seeking political solutions or military solutions to international problems, and, for that matter, of the outcome of relationships within families, nonfamily groups, or whole societies.

As M. M. Waldrop points out, "A lot of nature is not linear, including most of what is interesting in the world. . . . Everything is connected, and often with incredible sensitivity. Tiny perturbations won't always remain tiny. Under the right circumstances, the slightest uncertainty can grow until the system's future becomes utterly unpredictable—or in a word, chaotic."[3]

This sounds as if we can never know what's coming next. Actually, if we use a broader interpretation of events, we find a great deal of simplicity inherent in diverse systems. This simplicity comes from the universal behavioral laws. The trick is to understand that the broad patterns (strange attractors) of events can be identified and predicted, although the specific events cannot.

For example, in a growing industrialized economy with a population that is rising because of greater wealth, better nutrition, and better health, an improvement in the transportation system can be seen as inevitable. This is particularly true because, as we have seen, systems such as nations or cultures will change in the face of decreasing synergy (diminishing returns) or cease to exist. Thus we have the development of a transportation net based on automobiles, individual freedom of movement, and easy expansion into areas not heavily settled before. How this comes about is, however, a matter of random events, a series of small probabilities reinforced by unrelated phenomena and thus increased. In the case of transportation, we find that the popularity of bicycles leads to a demand on the part of the public for better roads for them to ride on. When the automobile appears, its success is partially because a good road system is already in place. The very engines that are used in automobiles were a matter of chance. Why was gasoline internal combustion used rather than steam, a more efficient and more completely developed system? The truth is that the gasoline engine won out due to a set of unconnected events, including favorable early publicity, the development of the spark plug, and an outbreak of hoof-and-mouth disease in this country early in the twentieth century that led to the demise of horse troughs, the steam-powered automobile's main source of water for making steam.[4]

Further developments in the transportation net led to the movement to the suburbs, a process facilitated by the development in this country of military roads patterned on the autobahn (the highly successful military road system Hitler developed in Germany), which in turn led to the development of multiple warhead missiles, as the Russians interpreted this decentralization in the United States as a move to spread out the population in case of

[3]M. Mitchell Waldrop, *Complexity: The Emerging Science of the Edge of Order and Chaos* (New York: Simon & Schuster, 1992), p. 66.

[4]Waldrop, *Complexity,* p. 41.

nuclear attack. And so it continues. Events lead to other events in an apparently unpredictable pattern. Yet the overall outcome can be more closely predicted. No one could predict, and still cannot with accuracy, the exact nature of transportation developments in this country, but that the developments have taken place and will continue to do so is relatively well established. It is a natural consequence of the process of the system dynamically adjusting and changing to grow and maintain balance.

PREDICTING RANDOMNESS

For those studying the social impact of technology, the inclusion of randomness helps to explain observed phenomena. In the chapter on Malthus, we noted that one of the factors altering circumstances and resulting in cultural growth is the tendency of human beings to change the rules of the game when things get tough. As we reach the diminishing returns of the sinusoidal growth curve, we employ technology to create a change in the curve, thus putting off the "evil day" when diminishing returns will cause the system to collapse. It is in that interim section of the curve, the period of constant growth, that chaos and indeterminacy are most important. It is in the diversity present in the complex system we call our culture that the change takes place. Long before the system actually starts to bog down in its own inefficiency due to using up all of its synergy, internal elements begin to work for a change in structure that results in new synergy through the use of new methodology, the development of new subsystems, the dismantling of old subsystems, and the changing of the collective worldview. Which alternatives we choose depend on which of the legions of possibilities find a slight edge in the battle for recognition. Random events combine to reinforce certain paths and reduce the probabilities of others. It is for a short period of time chaotic, the culture apparently going in many different directions at once as it seeks to settle into a new workable pattern of thought and behavior. Eventually, some technologies and concepts win out over others, and the new pattern settles into a period of growth that will be maintained until the next nexus of crisis. The trick as an individual or organization operating in this environment is to know when to jump, when to change strategy in the face of coming events.

Practically, futurists do just this. They look for the trends in determining what will probably happen next. They realize that they will always be wrong in details, but if they are correct in general, they will be prepared for the future and have a better chance of landing on their feet when the change occurs. A restaurateur doesn't sell out or change when business begins to drop off. If she waits that long, she will be too late. The trick is to sell out or change while business is still growing but on its way into the slowdown. Then she will be out of the business when it fails. A restaurant with strong sales and growing customer approval is far more valuable than one experiencing losses. Similarly, manufacturers of consumer goods—particularly fad items, which rise to popularity very quickly, have a relatively short overall life, and die out quickly—have ceased manufacturing a product long before it begins to fail. By the time a fad item hits its peak, the smart producers are already in the production of the next new item rising to take its place. All of these decisions take place during a period of chaos, when many possibilities vie for preference and one finally takes ascendancy.

CONCLUSION

The indeterminacy of events in complex systems arises due to the large number of equally possible variables that exist within that system. In a random fashion, one or more of these variables can become important to a process and change the outcome from the expected. Because of this, complex systems are constantly in the process of changing, thus forcing the system itself to adjust to new and unexpected stimuli, resulting in the need to maintain as well as establish balance. When more than one equilibrium state is possible in a given system, it is the random events that will determine which will actually take place.

However, this does not render systems unpredictable or deny the overall order inherent in systemic processes. Because of behavioral law, we know that (a) systems will always be synergistic, since nonsynergistic systems are inherently inefficient and unnecessary and will eventually fail, (b) they will be reciprocal, and thus we can predict what will come about based on what has come before, and (c) they will seek dynamic balance, since that is the ultimate goal of any system, no matter how it is defined. What form that balance will take is the indeterminate part, because of the multiplicity of diverse possibilities for equilibrium states. By inclusion of random factors in a systemic whole, the capacity to adjust is assured, as in nature, where diversity naturally leads to the capacity to change in the face of changing ecosystemic stimuli. Complex systems are most efficient where there is a balance within the system between order and chaos, between the constancy of systems design and the diversity of internal elements that allow for efficient and appropriate response to changing internal and external conditions. This diversity and randomness is a necessary and integral part of systems at all levels.

Chaos, we have seen, is imbedded in every ordered system, and, conversely, systemic order lies at the heart of every chaotic system. Each is a different state of the same phenomenon, and which of the two we experience is dependent on the level and moment of time in which we are experiencing the process. Yet if we can become aware of the moment at which the change from chaos to order or order to chaos occurs, we are better able to react to those changes and perhaps avoid a movement toward chaos by adjusting more vigorously within the ordered system itself. The study of where these "shifts" occur, the final step in understanding the systemic process, is known in mathematics as catastrophe theory. In Chapter 10, we take a nontechnical tour of the phenomena connected with the point of shift between order and chaos and discover how that manifests itself in social, political, economic, and physical terms.

KEY TERMS

Butterfly Effect
Chaos Theory
Deterministic Randomness
Diversity
Dynamic Balance

REVIEW QUESTIONS

1. Define chaos theory.
2. What is the relationship between chaos theory and systems theory?
3. What is a dynamic system?
4. How is complexity related to dynamical systems structures?
5. Define dynamic balance.
6. What is determinate randomness?

ESSAY QUESTIONS

1. Assuming the truth of the proposition that systemics and the chaos theory represent the yin and yang of the universe's organization, discuss how each is a necessary balancing element for the other. What does this say about the concept of systemic balance?
2. Discuss the positive contribution of chaos to the development of new technologies. Include current examples.
3. It is argued that technology creates regimentation, standardization, and mediocrity, even as it reduces uncertainty and insecurity. How does the chaos theory act to combat these tendencies in a society by supporting individuality and freedom?

THOUGHT AND PROCESS

1. To get a sense of the nature of random behavior in systems, turn on a faucet just enough to create a drip. Notice its regularity. Slowly increase the flow of water and notice the increasing rate of regular dripping. Continue this process until the drips become irregular. Can you locate the exact moment that this occurs? It is here that the chaos enters the system.

2. Think of a time when chaos has entered a major system of the following types: (a) a machine, (b) a political system, (c) a living organism, and (d) a meteorological system. In each case, what do you think caused the destruction of the system's order?

3. Think of a personal experience when a well-planned activity or project ended up completely different than you expected. What minor random details may have served to "upset the applecart"?

4. If you are mathematically inclined, try the following experiment. Build a simple computer program to generate a number of iterations of some first- or second-degree equation such that the answer for each iteration becomes the independent variable of the next. Run it for two hundred iterations using a nonzero positive number to one decimal place as your initial value for the independent variable and find the answer. Using this answer rounded to only two decimal points, add a third random digit and run the program again for two hundred iterations. Now go back to your original beginning value and run it for four hundred iterations. Compare your answers. The difference is the result of a single random digit being changed in the middle of the four hundred iterations, an example of the influence of minor random variables on system behavior.

CHAPTER 10

CATASTROPHE THEORY: THE PLAGUE OF TECHNOLOGICAL COMPLEXITY

Liu Chin was a very unlikely candidate for the role of world leader. He was a slight man, no more than five feet nine or ten, with a thin frame and long slender limbs. His skin was sallow, mostly from years of laboring inside huge complexes, and it sometimes gave the impression of jaundice. His thin wispy hair was unruly, but then, it had always been so. Yet it was none the less true that as the director of the New World Health and Research Council, he was indeed one of the most powerful men on the planet.

As usual, he was scowling, considering some deep conundrum with which to deal. He sat quietly behind his very traditional desk, which was devoid of electronic devices or other technological aids. He was looking through a tent of fingers formed by his upraised hands to a large cube on a pedestal some six feet away. The cube was featureless, smoky gray in color and about thirty inches on a side. In a moment, he knew, it would come to life and present for the entire world the holographic image of his superior and chief mentor, Horace Grovenor, director general of the World Council.

This was Liu Chin's moment of triumph. In a few moments, he would be lauded for all his years of work and praised as the savior of humankind. He was at his peak, at the very zenith of his career, and would probably become the most famous man in the world. None of this was what he was presently pondering as he stared at the still blank cube. At the moment, he was more vitally interested with the ongoing efforts taking place some twenty-seven stories below his office. He was wondering what Pzworski, Garrett, and Clarke had found. He sighed. After all these years of research and devotion to science, he knew better than to be impatient. He would wait, he would wonder, and he would worry. He would know when there was something to know, and nothing could hurry that along. Yet, he worried, as he always did. "Nature of the beast," he said to himself aloud.

A beep and faint crackling sound drew his attention to the cube. At first, it glowed an electric pale blue, then shifted seamlessly through the spectrum to violet before coming to life. It wasn't supposed to do that, he thought. He must have a technician look at it when he got the chance.

The smoky gray surface of the cube had given way to a sharp, clear image of the council chamber of the World Order Center. The hall was filled with the usual array of council members, assistants and assistants to the assistants, media representatives, and the requisite entourage of ordinary citizens who, by law, attended council sessions to assure that they were directly represented at such proceedings. It looked no different than any other media event put on by the council, but Liu Chin knew better. This was to be a different show. Horace Grovenor was standing at the speaker's podium, directly before the council. He looked very confident and very much in command.

"Fellow citizens," he began, and the room became silent. "I am delighted to address you all today on the advent of a new era, an era that portends a bright and glorious future for the human race."

Liu Chin snorted involuntarily. He didn't like Grovenor and never had. The man had been useful in furthering Chin's own research, but they both understood that below the caring comments, the enthusiasm from the council chairman, and all the charm, the two of them were simply involved in a mutually agreed upon political process by which each could achieve his own goals. Grovenor was smiling back at him from the cube, as if speaking directly to him as an individual. It was one of the council head's more effective political characteristics.

"As you know," he was saying, "we have come through a terrible time in this century, from the famines of the '20s and the wars that followed in the '30s and '40s, through the years of devastation from biological and chemical weapons released and the pestilence of an environment in revolt. We have emerged from nearly fifty years of chaos, bloodied, weary, diminished in numbers, but none the less victorious over our infirmities."

Chin smiled. In spite of his distaste for the man, he had to admire his charismatic charm. His speeches were electrifying. It wasn't the words; they were often contrived if not mundane. It was his delivery. He could have been reading the Latin text of some thirteenth-century medical treatise, and he still would have captured his twenty-first-century audience.

"The human race has been reduced by nearly three-quarters in number from where it was in the past century. We have all lost friends and family, have suffered the indignities of privation, and have witnessed the destruction of our individual cultures and traditional ways of life. We have seen what can happen in a world gone mad, where the pursuits of governments and industry were bent on maximizing their own rewards and increasing their power at the expense of all else. We have seen an ecosystem brought to its knees, strained to the very edge of its capacity to function. We have seen our oceans begin to die, our atmosphere depleted, world temperatures rise and totally restructure our agriculture, our geography, and our lives. And we have watched helplessly as our own ignorance, greed, and depravity have compromised all that we have said we worked for.

"Yet today, we emerge triumphant. In the last twenty years, we have also seen the remarkable resilience of nature as waters have become once more potable, forests and plains have begun to replenish themselves, and nature has once more begun to be put right. The reduction in population pressure and increasing isolation of peoples one from the other have been as much a salvation of the race as they have been a misery for those who have lost so much. Like the phoenix, we have begun to rise out of the ashes of our own self-ignited conflagration into a new era of unlimited human potential!"

The council chamber exploded with thunderous applause as even the media representatives found themselves caught up in the fever of the moment. Grovenor raised his arms and quieted the crowd, calming them after a perfectly timed period of adulation.

"We have learned bitter lessons, but we have learned them well. No longer will we unthinkingly rape the environment. No longer will we pretend that our individual actions have no effect on the lives of others. No longer will we set aside human values for progress for its own sake. In a new spirit of consideration and mutual need, we have learned to work together rather than at cross purposes and to move forward in a rational, orderly fashion, deciding future courses on the basis of how it will affect not only our own lives but also those of future generations to come. We are a species reborn, now worthy of our position as dominant on the planet, willing stewards of this world and of the other worlds in our solar system. As we move outward to the stars, we do so carefully and with forethought, not charging ahead unthinkingly, but carefully. All this we have achieved. Yet there has been one obstacle to it all."

"Ah," thought Liu Chin. "Now that arrogant bastard delivers the news. In a few minutes, this man will change the lives of nearly a billion people. Now comes the other shoe."

Grovenor paused dramatically, waiting for everyone listening, worldwide, to be at just the right level of anticipation for maximum effect. Liu Chin had to admire his timing as well as his gall.

"After all we've accomplished," he began solemnly, "we have still not conquered disease, the ravages of age, and the remnants of man's attempts to kill his fellow man through biological manipulations. Genetic anomalies still plague us, as do rare and unchecked exotic illnesses. Our children suffer from diminished immunity, from brittle bones and fragile internal organs. Our morbidity rate is still so high that in spite of all our progress, our population continues to decline.

"But that is no longer the case. I have asked you to join us here today to announce that a new initiative is about to begin that will change all of this. Dr. Liu Chin, director of the New World Health and Research Council, has announced that the Panacea Serum is a reality! After many years of diligent work in the fields of biochemistry and medicine, the Research Council has developed a remarkable drug, part microorganism and part chemical compound. It will, upon injection into the bloodstream, systematically go about the process of restructuring every cell in the body, which will be brought back to a condition of perfect function! This is a truly astounding event in our history. Not only will it prevent future disease and genetic anomaly, it will actually repair damage already present. Every person on the planet can be endowed with perfect health!"

Again there was thunderous applause, but Grovenor continued through the din.

"We have taken into account all possible objections to the project. We have designed it meticulously, maximizing individual freedoms while providing for the common good of the body politic. The Panacea inoculations will prolong life only to an age of one hundred years, after which it will allow for normal aging processes. It has no negative side effects that we can determine, and it is completely free of toxicity for humans or any other living organism. It has been specifically engineered to support life, not to limit it, and we are currently making it available to all of our citizenry. Even as I speak, inoculations have begun in selected major municipalities, beginning with the ill and those suffering from genetic anomalies. Sufficient stocks of the serum are available to take care of all and have already been distributed for rapid deployment. My friends, in a very few weeks, months at the most, all human beings on the planet will find themselves perfectly healthy and free of disease, able to look forward, barring accident, to a life span of well over one hundred years!"

This time there was no stemming the tide of enthusiastic response. Applause, cheers, and howls filled the hall as those present felt the oppressive gloom of the past century melt away in one great moment. Grovenor grinned broadly, basking in his new glory, no doubt, thought Chin, aware that he had assured his own political supremacy for as long as he desired. Liu Chin would reap the rewards of being the man who discovered the Panacea Serum, but Horace Grovenor would be the one to rule the world as it was he who had the vision and created the opportunity for the development to take place. Liu Chin sighed, aware that this was probably the last private moment he would ever have and acutely aware of how undeserving he was of the praise that was about to be his . . . he hoped. There was still the matter of the final tests that Pzworski, Garrett, and Clarke were completing for him now.

There were always the nagging questions. The beauty and misery of research is that there were always more questions and more problems to be solved. Everyone rushed, even after all the lessons of the past, to use new knowledge for practical benefit, yet there were always more questions to be asked and answered. If only people would understand that the joy lies in the discovery, not in the application. That was true, wasn't it? Wasn't that why he entered the field in the first place? Wasn't that what really brought him joy? It was the understanding of processes buried in nature's enigmatic back pocket, waiting to be discovered, that mattered to him, not what others did with that knowledge once obtained. A mystery solved was no longer a mystery and therefore of no interest to him. He always needed the next mystery.

The cube was still alive with applause and cheers, people milling about the chamber, smiling and congratulating themselves on their own genius. As Chin reached to turn off the offensive display, the doors to his office opened and Garrett entered. Liu Chin looked at his face, knowing what he was about to say. All that remained to know was the seriousness of the situation.

"Well?" Liu Chin said as calmly as possible.

Garrett swallowed, his throat visibly bobbing beneath the salt and pepper beard. "Well, it's true," he said simply. "There's no doubt."

"How bad is it?"

Garrett's shoulders slumped. He was obviously fatigued beyond endurance. "As bad as it gets. We get a one hundred percent response. I've checked all of the test animals we've used and, where feasible, repeated the experiments of the last two months. Rats, rabbits, lemurs, monkeys, and human volunteers all seem to exhibit the same genome group. Of course, with humans we cannot demonstrate the effect directly, since it does not happen until the third generation, but with all of the other animals, it has already occurred. The effect is one hundred percent . . . thus far."

Garrett swallowed again, hard, as Liu Chin showed no sign of emotion. The director would hear the echo of those words in that Southern drawl of Garrett's throughout the rest of his life. Internally, Liu Chin was seeing a very different scenario for his relatively short-lived fame and his future. Garrett started three times before he could get the words out. "Dr. Chin, we've got to stop the inoculations."

"It's too late," said the director calmly. "They've already begun, and by the time we were able to convince anyone that we were right, it would all be over. Remember how much political clout the council has invested in this project. They'll not take our word for it. They'll want more tests, independent corroboration, and then they'll discuss it to death before they make a move. We may be witnessing the end of human dominance on the planet."

"Worse than that, I'm afraid. Dr Chin, we're looking at extinction. Don't you understand? They're all sterile! Three generations from now, every person who is the offspring of a recipient of the Panacea inoculation will be sterile! What happens then?"

Dr. Chin hesitated for only a moment. It was the same question he'd been asking for weeks. "I don't know, Dr. Garrett, but we'd better get busy finding an answer. We've a new mystery, and tomorrow we begin again."

That's a fairly gruesome scenario—sterility of the human race, the end of our species on the planet. Or was there hope at the last? If tomorrow they "begin again," will they find a solution? On first inspection, this short tale may seem like just another horror/sci-fi/fantasy story to tickle the spine of the reader or perhaps induce a bit of reflection on the possibilities. In any event, it does appear gruesome. Even if it couldn't happen, it is food for thought.

And yet this is a scenario that reflects the daily conditions of our lives. Admittedly, there has been no great world war, plague, and pestilence destroying five-sixths of the world's population, and the environment has not yet become completely polluted, but the types of problems suggested in the story are indicative of the kinds of issues we face in this world every day. We tend to think of the world as ordered, structured, and predictable. The sun comes up the same time every morning as it "always" has. Water still flows downhill, and it still rains to replenish the land. Governments argue and rattle their sabers, occasionally sending out armies to butcher each other to solve problems of territoriality, economic dominance, or political and philosophical control. Children still come down with the flu, as do adults, and the doctor or hospital is usually there to deal with it. The news still broadcasts the same array of misery and sensationalism and pathos that they have been serving up for years. It's a very predictable world. Is that right?

It is hoped that anyone who has read this far in this text knows better than to believe that to be true. In our discussions of the systemic nature of the universe in which we live and particularly the systemic nature of our complex human cultural process, it has been pointed out again and again how subject we are to chaos and the whims of oddities, large and small, in our lives. At virtually any moment in time, the unexpected and unanticipated can pop up in our lives to create heretofore unknown conditions and trigger new directions in our lives. It happens all the time, in a myriad of small, seemingly insignificant ways.

The large changes we notice. The sweeping events catch our attention because of their scope and their obvious impact. The small ones do not. Interestingly enough, however, there is a constant array of effectively catastrophic events and changes happening around us every day, and many have no effect on our lives at all, thus going unnoticed.

In fact, the vast majority of random factors introduced into our ordered, systemic lives go unnoticed. There's a poem about a nail missing from the shoe of a horse that leads to the downfall of an entire nation. One small nail causes the shoe to be thrown, laming the horse and causing the rider to not deliver a message that causes a battle to be lost that results in the loss of an entire war and ultimately the entire country. All this because of a small piece of iron not being where it is expected to be. Truly the devil is in the details. We see it often enough, where there are totally unexpected consequences that are caused by innocuous events or circumstances that no one has bothered to check. In a complex world, there is simply too much detail to look for to do it effectively and simultaneously with efficiency. What are the odds? Which variables have the greatest possibility of affecting the outcomes of our actions? These are the questions we ask, and must ask, unless we are to experience paralysis by analysis. It is, in fact, not the details of which one is aware that will cause the greatest problems in our plans and in our lives, but rather the details about which we do not know—the unexpected events and circumstances and their outcomes—that will cause us the greatest harm. Much of this is due to the multiplicative nature of the effect of details. Recall the chapter on cause and effect and how each step in the sequence of events leads on to the next. Recall also from the chapter on chaos that the magnitude of that effect can grow exponentially as the deviation is continued through each step. By way of example of how a detail can escape notice, I present a single case of a minor detail that no one considered.

A management team was hired to do a systems analysis for a major aeronautical manufacturing firm that produces, among other things, military and civilian helicopters. The purpose of the study was to review the entire production process by which this popular helicopter was created and look for methods of improving efficiency, safety, and economy. The team reviewed all of the PERT charts, went over the CPM data and other planning functions, and then proceeded to examine each department and phase of production individually. So far we are talking about a fairly standard approach to analyzing a production process.

They checked everything from procurement to warehousing of parts to assembly and quality control. The review went well, and much to the credit of the firm in question, there was little found to be available in way of improvement. In other words, this was a very efficient operation.

There was, however, one glitch that arose toward the end of the process of analysis. One of the final phases of production of these helicopters involved placing the prop (the large propeller that actually provides lift and velocity for the vehicle) on the helicopter just before it left the factory. The operation involved the physical arrival of the prop at the assembly point, the placement of the prop over the hub, a sizeable and critical bolt, the securing of the prop to the hub with a large nut, and finally locking the nut with a cotter pin. The nut is called the "Jesus Nut," because it is the one critical element that must be in place in order to keep the aircraft in the air. In reviewing the process, it was all neatly laid out in an orderly fashion with appropriate safeguards. Yet upon questioning the engineers and managers, the team discovered that no one was assigned the task of securing and locking the Jesus Nut! They were not able to discover anyone who knew whose job it was to place that nut on the prop and lock it in. It was the single most critical element in the actual production process, the final step before the helicopter was rolled out of the plant and set for delivery.

Finally, finding no other alternative, the team proceeded to the assembly area to watch the final step; they observed one of the assembly crew secure the nut and lock it. Upon questioning, the man revealed that he did it "because I've always done it." No one told him to do it; no one made sure he did it; he just knew that before a unit left the line, he would carry out the final step of locking the Jesus Nut. Obviously, the procedure worked, since there had never been a failure in this part of the process nor had there ever been an incident of a Jesus Nut not being properly locked. But the problem was that no one had been aware of this detail. When asked who did the job when he was sick or on vacation, the worker replied, "I've never been sick, and I don't take vacations."

According to the old saw about not fixing something if it's not broken, it would seem that the procedure employed was a reasonable and successful one for building helicopters. Yet what would have happened if this particular worker did get sick? What if he retired or died? He was, after all, an older worker. What would have happened then? Would someone else have secured the Jesus Nut? Would every helicopter on the assembly line the day this critical worker was out be rolled off without a Jesus Nut or without one properly secured? If so, the helicopters would have either never gotten off the ground or had a very rough landing!

There was no tragedy in this example. It was a potential disaster in the making that never presented a problem. It is likely that the system would have been corrected when the need arose, since others would have probably realized the need for the nut to be secured. Probably, probably, probably . . . or there could have been a disaster. The question is how many such glitches exist in our complex individual and collective lives? How many minor possibilities are there that could cause a chain reaction in our society? This is the issue of the current discussion. In this chapter, we deal with two related issues. The first is the principle of unexpected consequences. The second is that of catastrophe theory, a branch of chaos theory that deals with severe reactions to systemic stress. Each will be explored in terms of its relevance to how we go about the process of creating and living in a technologically developed world.

THE PRINCIPLE OF UNINTENDED CONSEQUENCES

Simply stated, the *principle of unintended consequences* maintains that every process and every system contains consequences that were not originally intended when the process or system was developed. This is also true of every purposive change that we instigate within any of the many nested systems within which we exist. Thus, regardless of changes to a system that take place by design, there will be consequences in addition to those expected. Complexity makes it difficult to determine outcome; it is as simple as that. The variation may be quite minor, or it could be major. It may be beneficial and, therefore, serendipitous, or it may be detrimental and even catastrophic. In any case and to whatever degree these changes occur, there will be differences between intended, expected results and actual results. We continue to live in a probabilistic world.

This should come as no surprise to anyone who has ever attempted to create a change in his or her environment. Whenever we decide to do things differently, we encounter unexpected consequences. It could be something as simple as taking a new route to work or changing the hour in which we eat lunch. New discoveries and new "problems" almost inevitably pop up, and some are not as evident as others. With a new route, for instance, we may discover new sights along the way that we have not seen before. In the time since last we came this way, we may find that a service station or convenience store has been built on the corner or that it is now a one-way street. We may discover that the new route is rough on our automobile because of less well-paved roads and that sometime in the future we develop rattles or have a blowout because of it. All of these events are unanticipated and unexpected, though perfectly understandable once they occur.

Actually, this is no more than an extension of the phenomenon discussed in the second chapter concerning homeostasis. We resist change because there are such things as unexpected consequences. Therefore, we prepare for the unknown to occur. We test for it, plan for it, and make sure, for the most part, that we do not venture too far into the realms of the unknown before we are satisfied that these factors do not exist or that they are discovered. All this has been discussed before. If so, why bring it up again? It is because we simply cannot catch all of the unintended circumstances, and that creates potentially serious problems that may not be realized until it is too late to correct them. It is a basic tenet of complexity that one simply cannot catch all of the glitches that will occur or can occur. The system itself continually evolves, and what was perfectly acceptable yesterday can cause major problems today. Equally, what was untenable earlier may be quite acceptable now. We can further recognize that the safeguards we build into a system can become the source of even more seriously harmful conditions at another point in time.

Earlier we discussed secondary, tertiary, and quaternary effects and that because of the law of cause and effect, these unexpected effects cause the problems. However, these are not the effects we are addressing here. This is not a matter of the original body of effects that are inherent within the development and implementation of a piece of technology or a technological process. The effects we are discussing here are the ones that develop because of the changes that take place to that system as it adjusts to changes in other systems with which it is linked. For example, the automobile, in the late nineteenth and early twentieth century, was a means of protecting us from the growing pollution caused by horse dung and the accompanying flies. Yet we are faced with far worse levels of pollution of all sorts as a result of automobile technology, not only directly, from emissions of automobile exhausts, but also indirectly, from the pollution that results from the changes in the way our society operates because of the existence of the automobile. The automobile is partially responsible for the pollution caused by industry attempting to fulfill the desires of a suburban and urban population, an entertainment oriented society, a society that insists on speed of operation, and on and on. Just as technological successes are evolutionary and networked, so are the problems.

To be fully appreciated, the law of unintended consequences must not be viewed as a "point-source" problem, where an event results in consequences that are not expected. It must be seen as a "class-reaction" problem, where every action sets in motion an entire spectrum of events, all related yet none specifically following from some one source. It is a chain reaction, not a linear link-step set of events. It does not proceed; it spreads. It does not penetrate; it permeates and engulfs, and we see it happening all the time. The problem is not that it occurs; that is natural and a very important part of the evolution of our society and our species. It is when it is detrimental in ways that can cause catastrophic outcomes that we are in the greatest danger. This brings us to the second topic of the chapter, that of *catastrophe theory*.

CATASTROPHE THEORY

Originally, catastrophe theory began as a mathematical treatment of continuous action producing a discontinuous result. It was developed by the French mathematician René Thom. Thom was attempting to describe the behavior of systems where gradually changing forces lead to rapid changes. These rapid or abrupt changes are known as catastrophes. To put it in laypersons' terms, think of the straw that breaks the camel's back. It is a system where slow changes finally create, at some critical point, untenable structure dynamics, and the entire system adjusts rapidly and often violently. By definition, a catastrophe is an occasion in which a small, gradual change in the environment of an object (hence a system) brings about a sudden and abrupt change. Remember that a system is continually in the process of attempting to create and maintain a state of balance so that it can operate with a minimum of energy expenditure and a maximum of efficiency. In this process, small changes may not require any substantive adjustment in a system until their cumulative effects become so great that the system adjusts all at once.

Examples range from the sudden toppling of bookshelves to the game where stacked blocks are removed one at a time until finally the structure becomes so weak that it collapses. In each case, the stresses built up by small changes (adding one more book or removing one more block from the structure) finally become too much, and the system collapses. One of the chief features of this theory, as it turns out, is the assumption of stresses being imposed by forces outside the system itself while the system is seen as initially in a state of balance, or stasis. It is only with the introduction of the added elements that the system becomes unstable in small increments, until it "fails" or jumps to a new state of equilibrium. A moment of thought will produce numerous examples of such phenomena in the real world.

Consider the dripping of a faucet. Water is operating under two principles as it flows: The first is cohesion, the tendency of water particles to stay together rather than scatter into smaller droplets, and the other is adhesion, the tendency of water to adhere to surfaces in a way that resists flow or gravitational surfaces. As long as the flow of water is small, the drops simply get larger and larger (cohesion) and hang on the edge of whatever object they are flowing over (such as the end of a faucet). Eventually, these slow changes in the droplet of water accumulate to the point where it can no longer counteract the force of gravity, and a cohesive drop of water falls away, no longer adhering to the lower surface of the faucet. Put simply, it drips. Interestingly enough, if we observe the droplet of water as if falls, we will see that it increases its speed (a la Newtonian gravitational laws of motion) and that when it hits, its higher velocity and resultant higher kinetic energy may actually be enough to overcome the adhesion principle, and the droplet may shatter into smaller droplets as well. In each of these two actions, a relatively steady state within a system (equilibrium state) has been subjected to minute stresses that collectively result in an eventual jump to another state of equilibrium.

This is all very well and good, but what does it have to do with my daily life? If a faucet drips, whatever the cause of the behavior, I merely turn it off. True enough, but just as all systems are subject to the universal laws of systemic behavior and the context of chaos, they are also all subject to the process of catastrophe theory. What is true of a dripping faucet is also true of economic structures, mechanical devices, political systems, ecosystems, and virtually everything else in the known universe. Thus, it is important to understand the concept and how it affects our lives.

As used here, a *catastrophe* is a sudden and widespread change that results in one or more major upheaval in the order of a system. It is by necessity a probabilistically rare event, since closely repeated catastrophes would result in either the demise of a system or in serious enough alterations to that system that it would evolve into what is effectively a new and different system, either in part or in entirety. In other words, systems cannot tolerate too many catastrophes in a short period of time and still maintain their structure. Of course, what a short period of time is or how many catastrophes are too many varies from system to system. Some systems are by their nature more tolerant of stress than others. A pyramid, for instance, constructed of thin sheets of paper held together by wheat paste would not tolerate strong winds as well as a solid wooden block carved into the shape of a pyramid.

Each system contains its own set of characteristics and therefore different tolerances, though they may have superficially the same basic physical shape and size.

In geology, a similar term, *catastrophism*, is used to describe the theory that major periodic geological events occur to change the entire process of development on our planet, as opposed to steady, slow evolution. These events, because they create rapid change, are called catastrophes. An ice age, which is geologically a very rapid event, or a meteor strike would be examples of such stresses. A related and similar variation on the theme is the naturalistic concept of *punctuated evolution*, by which evolution is a combination of slow change through mutation, adaptation, and survival of the fittest, punctuated by periods of catastrophic change, forcing major shifts in the process itself. In both of these cases, the state of our ecosystem is the result of violent, rapid changes in circumstances and the resultant reaction of systems within the ecosystem to adjust to new circumstances.

This is a probabilistic concept, a matter of beating the odds. Since the events are rare, the probability of them occurring is very low. But because the changes are so violent, rapid, and widespread, the consequences of such an event can be horrendous. A mundane example of taking catastrophe into account in our daily lives is the purchase of an umbrella insurance policy, which takes over if the normal insurance policies are unable to cover contingencies. The odds are very low that one's conventional insurance will not be able to take care of problems that may arise, but if it were to happen, the consequences would be so dire that we protect against it just the same. It is the same for fire protection. No one expects their house to catch fire and burn to the ground, yet we still insure against it, since such a loss would be considered truly "catastrophic."

On the surface, it may seem as though we are talking about two different types of catastrophes here; one type involves a rapid change brought about by single massive events that assault a system, such as a meteor impact, and the other involves a cumulative effect from many small increments of stress that assault a system over a long period of time. The truth is that they are both the same process. Often, all we see are the cumulative effects rather than the small ones as they happen. If a house burns to the ground because of faulty wiring, we see the failure of the wiring as the cause. This is a single event. Yet the wiring may have failed due to a long period of changing temperatures, changing current flux, being physically rubbed by a surrounding structure, or being gnawed by rats or squirrels. But we do not observe these small steps, just the big ones. Tornadoes begin as small changes in atmospheric conditions, as do hurricanes. A ladder tips over because of that last bit of lateral strain put on it by someone reaching to one side or because of one last increment of frictional slip that finally causes it to skitter away from a wall. They are still cumulative stresses.

Consider the postal worker who suddenly decides to go on a killing spree after years of small irritations. Consider the housewife who shoots her husband after picking up socks from under the bed just one time too many. Consider the nation that goes into open revolt after years of miniscule losses of freedom and increased oppression by a central government. All of these are examples of catastrophe theory.

This is also the key to understanding the importance of catastrophe theory in our technology and how it can affect society. The odds of a catastrophe are very low, but the consequences, should one occur, are so serious that we are forced to take into account the remote possibility of such an event in our planning. Thus, though the probability of accidentally triggering a nuclear war is rather small (I hope!), the consequences of such an event would be so extreme that we safeguard against it with system after system, employing redundancy and nested fail-safe mechanisms to ensure that it does not occur.

If we combine catastrophe theory with the principle of unintended consequences, we find that eventually we can expect systemic behavior to produce a rapid, violent change. In other words, given a high level of complexity, we can assume a high *catastrophic probability* as opposed to merely a catastrophic possibility. If the system in question is technological in nature, it follows that a catastrophe will eventually occur by virtue of our tendency to depend on technology and to increase the complexity of our technological systems. Indeed, it is with increases in complexity that the probability of catastrophic events rises. Eventually, we reach a level of complexity in which we can assume that there is no longer merely the *possibility* of such an event but rather the inevitability of such an event within a given period. This is true even if we have no way of determining what the nature of that event might be. The very complexity of the system dictates a catastrophic event will occur, though what it is may be a total mystery. Complexity leads to chaos, and increased complexity leads to increased chaos. The catastrophic event brought about as an unintentional consequence of our purposive attempts to control our system is merely the trigger that brings about that chaos.

Our job is to build our systems so that they can withstand these events; the study of that process is known as *contingency planning*. Even in our own lives we make contingency plans. We have savings accounts for unexpected expenses or opportunities. We carry insurance, as mentioned above. We fill our vehicles with fuel early enough to ensure that we do not run out when we have an emergency. This is no different than the contingency planning of corporations, governments, military organizations, or engineers who create technological devices. The problem is that we'll never be able to second-guess all of the contingencies or, if we do, understand the importance of the stresses bombarding our lives. *Catastrophes are inevitable!*

Early in the development of computers, there was very little memory available to programmers and users. Because of this, efficient use of what memory there was dictated that programming be done with as much economy of code as possible. Among other techniques used, memory was reduced by programmers through the adoption of a truncated date convention. Only the last two digits of the year were used, the *19* preceding these digits being understood. In this way, valuable storage space could be reserved, particularly in the case of database management, for important information. As we all know now, that one simple detail in programming convention has created serious concerns for all of us some fifty years after the fact.

When this numbering convention started, it could be assumed for all practical purposes that any date the computer dealt with as far as chronological determination was concerned would be a date in the twentieth century, and that the first two digits did not need to be specified. Computers keep track of time; they must in order to function. They therefore all keep track of that time, with the assumption that they are operating in the twentieth century. This worked well for nearly half a century. Yet as we approached the twenty-first century, as most of you already know, there was a problem. We found that dates recorded as current dates after midnight January 1, 2000, would be assumed by the computer to be 1900 rather than 2000. This results in not only misinformation but also faulty logic within programs and older operating systems. How, for instance, is a computer to actually compute the interest due on a loan taken out in 1995 when it believes the current date is 1901? It cannot logically handle the input and will thus simply freeze, spinning its wheels so to speak, trying to figure out what to do.

Actually, this was not the issue causing the bulk of the problem. If all that was required was to change the dating protocol of a few programs, it would be relatively easy to solve the problem. The real issue stems from the fact that we have embraced the computer and the computer chip in virtually every area of our lives. It is not a few programs that need to be "fixed." There are millions, if not billions of programs that use dates in their functioning, both in software and in internal operating systems. No one is quite sure how much of the computer operations out there are date sensitive or what the resultant anomalies might be if they are not corrected. Further compounding the problem are the computer chips whose programs require keeping track of dates and whose programming is literally hard wired and unalterable. How many such chips are there? The number is over 50 billion. How many have date sensitive programs that would fail? In many cases, there is no way to tell until the failure occurs.

This has not always been a problem, nor would it have been even if the programmers in the 1940s and 1950s had thought ahead to the year 2000. Originally, the computer was seen as a rapid calculator, and that is how it was used. They were expensive, massive, and difficult to use. It was only under certain circumstances, where a large volume of information was being processed, that they were believed to be of any practical use at all. Under such circumstances, this one little detail was an easily corrected problem that could be taken care of before the turn of the century. But that was before the development of personal computers and the widespread use of microchips in machinery and equipment. By the end of the twentieth century, the functioning of computers had control of virtually every aspect of our lives. One cannot function in daily life without encountering computers at every turn. If money is required, it is either obtained from an automatic teller machine (ATM) linked to the bank's central computing center to determine if you have money in your account or from a teller, who merely instigates the same process. There are credit cards for payment, bank cards, cash cards, debit cards, and checks (for the moment), all of which are linked to massive computer networks. If one goes to buy food, the checkout process is one of scanning prices, not punching keys (except, of course, when the computer program has not kept up with codes and changes). If one drives to work, it is through streets with computer operated traffic control systems. We drive computer controlled automobiles and listen to microchip operated radios and talk on computer controlled communications devices while our computerized radar detector tells us the location of the computer controlled police speed traps. The very distribution of electricity and water is controlled by computers and computer programs. Our government functions, our defense systems, our weather services, and our entertainment from television to movies to amusement parks are all computer controlled. Fifty years ago, nobody imagined this problem.

By late 1998, it was predicted that there were only two types of computer programmers in the United States, those who were working on the so-called Y2K problem and those who were not. As of that time, there were an estimated 550,000 programmers working on correcting problems in COBOL alone. And in spite of all the efforts to that date to deal with the issues, it was predicted that there still would be many instances of computer failure with the coming of the year 2000.

COBOL was introduced in the early 1960s, and since then, the amount of code that has been written, rewritten, expanded, and modified is so great that we literally are not sure how much there is. Nearly forty years after its introduction, many of these programs are so massive and so indecipherable that they must be investigated line by line. Much of the original code is not even available anymore. Many of the original programmers have died and cannot help with the problem. This is only a single aspect of the problem.

The Y2K problem turned out to be a far less serious one than doomsday prophets predicted, yet there were unforeseen consequences, and the head-in-the-sand critics were also incorrect in their assumptions. There was at least one case of a penal system whose database failed, eliminating all records of who was held in the city jail and for what reason. In another case, an entire governmental division received paychecks of the same amount—less than $2. Perhaps the results were not catastrophic to society, but if you were a school bus driver who received one of these paychecks, you would certainly view it as personally disruptive, if not disastrous. But this is not the main purpose of our discussion here. Y2K is merely symptomatic of a general principle—it is that general principle that is being explored in this volume.

Y2K is merely the example, a symptom of the general condition under which we operate in a technologically complex society.

The truth is that our culture is a very complex one and is becoming ever more complex. The advent of a problem of the potential magnitude of the Y2K event was inevitable, and the existence of other potential examples in the future are equally inevitable. We can expect them. Considering the number of determining factors that exist in the fabric of our sociopolitico-economic system, the building stress from a wide range of determinates can only serve to eventually create conditions ripe for catastrophic change in our society. Fortunately, we also live in a very flexible system that is eminently capable of adjustment and that has developed the homeostatic tendency to search for and recognize a potential catastrophe before it happens.

Yet all systems change, and they do it either slowly or rapidly. We have experienced both types of change before and will again. The question is how well prepared we will be to meet those changes and those challenges when they arise. We were aware of the Y2K problem and diligently moved to minimize the effect of this one small detail that has become so important to our systemic survival through years of inclusion in an ever-widening computer-based technological infrastructure. A normal distribution of probabilities dictates that both the doomsday prophets and those in denial are wrong in their estimates of catastrophe and no effect, respectively. The truth lies somewhere in the middle. The severity is a matter of facts and events. But what about the next such "glitch"? What about the next minor detail that no one bothers to correct or consider? What is being done about the ones that have already been put in place? The responsibilities of technology require us to continue vigilance and increase the care with which we restructure our world. Each change is part of the continual change and evolution of our species. It is up to us as technologists and as those who implement, embrace, and utilize these new ways of life to see to it that they do not hold within them the seeds of our own destruction.

CONCLUSION

The systems in which we live are complex and dynamical. Because of that, they are constantly under stress from internal and external forces creating shifts and adjustments so that they may stay in a state of equilibrium. Two consequences of this process are (1) that there are often unintended consequences to the actions taken by systems and systemic efforts due to the extreme complexity of the system itself and (2) that small stresses caused by the pressures exerted from within and without the system can build until there is a sudden wholesale change in the structure/behavior of the system. The first we call the *principle of unintended consequences*. It can often act both serendipitously to our benefit and disastrously to our detriment. The second we call *catastrophe theory,* and this refers to the sudden shifts brought on by an ongoing assault by minor pressures and changes. In practical terms, the modern technologically complex sociopolitico-economic system must guard against such eventualities both by anticipating them and by maintaining sufficient flexibility to be able to react to them as they occur. A prime example of how such a minor change can have manifold effects on a complex system is the Y2K issue of the late twentieth and early twenty-first centuries.

Chapter 10 completed an explanatory process begun in Chapter 1 in order to present both the theory that explains the relationship between technology and society and to present tools and methods for studying the relationship as it exists. The purpose has been to prepare the student for the role as investigator, armed with an understanding of what is going on and ready to actually apply this knowledge to current and future issues. In Chapter 11, the final step in this process is presented. A set of three systemic models is offered that will allow the student to investigate any socio-technological issue rationally and thoroughly. The models are similar yet different enough to give the researcher a variety of approaches to the investigation of the subject. I offer them here as a starting point in developing systemic models. They are of necessity suggestive rather than definitive. By following one of these schemata and remembering the universal laws, it is possible to predict and examine in detail the consequences of our technological development. We will never be completely correct in our predictions; the system is far too complex. However, with care and this structure, we can go a long way toward understanding what is occurring and the nature of what will follow.

The unfolding message of this book has brought us to this single point. We have developed a theory of technology and behavior that offers the opportunity to explore logically and intelligently how any given technology can affect a society or societies. We have the structure now, and it is systemic. By systematically breaking down a structure into its constituent parts and observing the behavior of those parts as they interact through the universal laws of systems behavior, we can develop an idea of why past events have taken place and what future events may occur. With the further addition of chaos theory and catastrophe theory, we can begin to explain and predict how systems will diverge from a given path and why. We can better see the dangers inherent in any system and how complexity at all levels of development tend to contribute to these dangers. We are in a better position to plan and implement technology in a positive manner. We have what we need to begin developing models of our world.

KEY TERMS

Catastrophe Theory
Catastrophic Probability
Catastrophism
Principle of Unintended Consequences
Punctuated Evolution
Y2K

REVIEW QUESTIONS

1. What is *catastrophe theory* and how does it apply to technology and society?
2. What is the *principle of unintended consequences*?
3. What is *Y2K* and how does it engender the principle of catastrophe theory?
4. What is *punctuated evolution*?

ESSAY QUESTIONS

1. Discuss catastrophe theory in terms of the positive benefits and the inevitability of the development of human society through technology.
2. Discuss the probability of catastrophic events developing due to new technology in the first half of the twenty-first century. Give details of the technologies involved and the possible consequences.
3. Discuss examples of the punctuated evolution in the development of modern technologies and their relationship to catastrophe theory.
4. By its very nature, catastrophe theory is seen to describe events that are considered negative. Discuss how this is purely a point of view—that what may be a catastrophe for some, might be a benefit to others. What does this say about our objectivity in viewing consequences?

THOUGHT AND PROCESS

1. In Ireland in the nineteenth century, the predominance and dependence of mono-cropping potatoes led to massive starvation and one of the largest migrations to that time, all because of a blight that wiped out the crop. Much as we have become dependent on computers for our economy, the Irish had become dependent on the potato for their agriculture and to feed their families. The entire system of villages and tenant farming was centered in this one product. Its failure created untold hardship, both in the crop failure itself and in the political and economic upheavals that followed. How is this seen as a possible parallel to the events of the late twentieth century and our dependence on computer technology? What lessons can we draw from these two examples?

2. Many simple systems other than the dripping faucet discussed in the example exhibit catastrophic shifts brought on by small incremental changes over time. Think of five simple systems that exhibit this type of behavior. Determine what the small incremental changes are that bring on the stress and the nature of the eventual reaction. How does this lead to a new state of equilibrium?

3. The concept of unintended consequences is universal. Give an example of this process in each of the following areas of day-to-day life: (a) the economy, (b) politics, (c) a family vacation, (d) a championship football game, (e) the decision to put a man on the moon, (f) your decision to go to college, and (g) the impeachment of President Clinton.

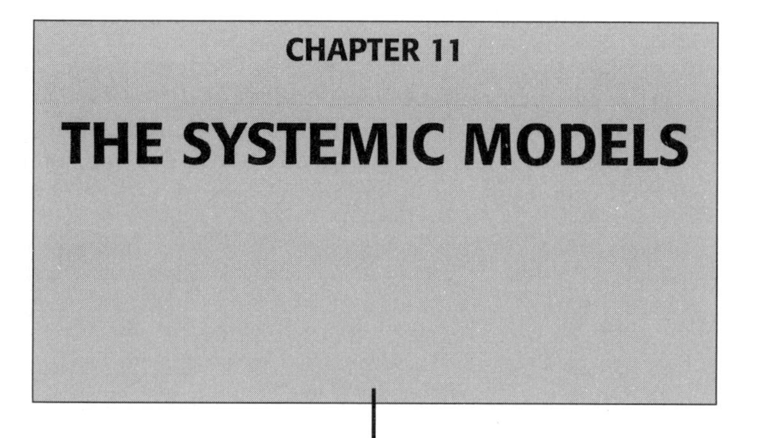

CHAPTER 11

THE SYSTEMIC MODELS

The following discussion centers on a list of questions designed to be a systemic outline for the study of technology and social systems. It has been created using the basic methodology discussed earlier and is presented as an outline of study for any technological change or group of changes. The questions asked and the categories chosen are designed to survey possible areas of social structure that may be affected by changes in technology as they occur. It is hoped that the outline approach will enable you to focus on the most likely future effects of any given technology.

However, a caveat is in order. As with any discussion of social structures, particularly speculative discussions of future conditions, the one thing of which we may be certain is that no matter what we discover of the probable effects of technological change on the culture through the manipulation of this set of categories, what actually happens will be somewhat different from what we predict. There is no method available to investigators, no matter how skilled, that allows them to be accurate in prediction. Yet by predicting, manipulating social factors, speculating, and then comparing speculations with reality, investigators improve the quality of their predictions. Only through experiencing the "art" of futuristic prediction can you as an investigator be able to improve your accuracy.

Please note that the factors to be considered are subsets of the cultural whole. By applying this method to historical subject matter, that is, by investigating how extant technologies have related to the culture, we gain insight into what should be expected in the future. What does it matter if we are not absolutely correct? Our purpose is not to find a specific definitive answer that does not exist. It is to improve our ability to deal with the future. If we can improve our accuracy, we have gone a long way toward understanding our society, our technology, and the place of both in the larger system that is our world. Even physics, one of the most definitive and precise sciences, is quick to admit that it deals with only probabilities in the physical world. Who are we to attempt more with something as nebulous and multifaceted as the sum total of experience of the entire human race?

In this chapter, we present three systemic models for the use of students and researchers studying the connections that exist between society and technology. The first is a general categorical model; it notes specific areas of interaction between society and technology. The second, known as the PERSIA concept, is based on a scheme for studying historical events. It too divides the culture into a list of areas of influence, but the list is somewhat shorter and more to the point. The third model is anthropological and relies on the same basic divisions for studying technology as anthropology does for studying cultures. It contains some elements missing from the other two models and is designed to take into account such fields of study as myth and semantics, as well as the more immediately identifiable points of interaction among cultural elements. Any of these models will yield an organized format for investigation. The choice is up to you, the investigator, based on your individual interests, questions, and motivations for study.

GENERAL CATEGORICAL MODEL

Technology's Effect on Commerce

The model first deals with the economic and commercial consequences of the technology in question, whether it is an existing historical technology or a burgeoning one that is just beginning to affect the society. This is not chosen without considerable thought. The commercial aspects of technology have already been extensively covered in Chapters 4 and 5. By this time, it should be apparent to the student that the impact on this aspect of human culture is a major one.

Most technological changes begin in the economic realm. Technology is a key factor in supporting and developing an economy, in securing and maintaining jobs for the population, and most certainly in determining the level of economic welfare experienced by members of the society. What is the effect of the new technology on business and commerce? Does it represent new goods and services? Are we dealing with new products resulting from technological change? If so, how will the new products affect the economic structure?

One source of new technology is the search for increased economic efficiency, a sincere desire to reduce the cost that the society has to pay for the availability of goods. Whether the purpose of the technology is to improve its effective use of available natural resources or to increase or alter the supply of available resources, it will affect what people buy, what they choose to do with their time, what jobs are lost due to changes in the overall economic mix, and possibly the price of other goods that compete with the technologically changed function for raw materials, labor, and the limited capital resources that are available.

As a secondary effect, how will the new technology create changes in unrelated or distantly related markets? The computer was a fantastic new technology. Indeed, it is still so. The changes that have taken place in the business world reach far beyond the immediate impact anticipated. On the surface, it was not difficult to realize that the computer would affect the market for mechanical devices designed to do tasks that a computer does, such as adding machines, typewriters, or even automatic mechanical control mechanisms. Yet this consideration does not begin to deal with the other changes that have taken place in the business world as a result of computer technology. An entirely new industry had been born in the form of the microcomputer, a stepchild of the larger computer industry available only to big businesses with big dollars to buy big computing power. New skills and new opportunities for employment have come about as a result. The expansion of the computer market and proliferation of microcomputers into the mainstream of American life have increased the ability of the homemaker to run a household effectively; to shop and cook more efficiently; to learn about the true nature of family finances; and to create more leisure time for spending money, for education, for hobbies, for part-time productive jobs, and for a host of other things. The nation has, in effect, stepped out of the industrial arena into the information arena. Here is a single technology that has so changed our ability to gather, store, manipulate, and disseminate information that the entire economic structure has been transformed. And all from a single technological change, albeit with a highly sophisticated and extensively proliferated collection of applications. The computer and its impact on the economic structure is no less startling and dramatic than that of the steam engine, electric power, or, for that matter, fire and the wheel. Through the computer, the efficiency and expansion of the economy have been so greatly accelerated that we are hard pressed to keep up with the changes. It is indeed revolutionary in nature and explosive in the speed at which it is altering our economic lives. Answer two questions: Ten years ago, how many people did you personally know who had access to a computer or dealt with one? How many do you know now? If you answer honestly, you will be astounded at the manifold changes that have taken place in just a single decade. And the next decade promises to outstrip the last by far.

Technology's Effect on Social Systems

The extent to which a technology affects social systems has been briefly discussed throughout this book. Specifically, the emphasis of this question deals with basic patterns among social groups and the changing patterns of needs and need fulfillment resulting from technological changes.

When the Industrial Revolution came about, and particularly when the industrialization of America took place in the nineteenth century and first half of the twentieth century, a number of social factors changed that would affect not only those directly involved in the process but also those who chose not to be involved in the industrial boom.

Workers in an industrial setting are able to command higher wages than farm workers. This is a fact of economic life. It is the result of the efficiency of labor in an industrial setting compared with the efficiency and productivity of farm workers, on which, at the time of the first industrializing moves in America, the country's economy was based. Economic systems recompense workers in accordance with their productivity rather than how hard or how long they work. It is their production level that determines how valuable they are. For industrial workers, whose level of productivity working in a mill or production plant is as much as ten times what their farmer counterparts could achieve, this meant ten times the wages for the same amount of work by simply shifting from farming to industrial work. Thus the mass migration from the country to the city occurred, with its accompanying rapid rise in urban population.

The industries were located in urban areas near supplies of raw materials, centers of transportation and communication, and markets. Chicago rose as the center of the meatpacking industry. New York grew through its transportation, communication, and financial centers, as did Houston and Atlanta at a later time. Pittsburgh, in the center of the coal and iron ore belt, was a center for foundries as early as the 1850s. Hundreds of other examples support the same argument. Industrial concentration and bigness, being the way to achieve efficiency, meant concentrating industry in small areas, which led to the packing of population in and around those locations where high-paying, high-productivity jobs existed.

With these heavy concentrations of large numbers of people came all of the attendant problems of urban life that were heretofore no more than mere annoyances to the general population. Families lived closer to one another. The unavailability of living space created multistory and multifamily living. Families found themselves in close proximity to neighbors, unable to depend on themselves for food and simple tools, dependent instead on supplies bought in local neighborhood markets. With crowding came an increase in crime, an increase in disease, and an increase in stress on the family unit.

City life, with its compacted physical structure and rapid pace, replaced the easygoing, steady pace of rural communities. One no longer knew everyone in the neighborhood or in the community. People came and went more frequently. Mobility increased for some and decreased for others. Time telescoped as efficiency in business spread as a concept to efficiency in lifestyle. Dispersion, particularly among generations, tended to decrease the level of interaction among members of the extended family and to increasingly isolate the primary family unit.

New social institutions have arisen as a result of the industrialization process. Unions arose as groups of workers fought for their collective rights. New groups within the work environment have risen to satisfy or thwart many of the needs of individual workers who are no longer able to obtain need fulfillment through the extended family. Identification with cliques, social groups from the work environment, or the company itself forms social structures for the benefit of the urbanized worker, a condition neither necessary nor possible in the older, agrarian culture of preindustrialized America.

These processes are still taking place in the late twentieth century as our technology alters our perceptions and our patterns of living. Television and the telephone have replaced more personalized forms of communication, again isolating us from one another and reducing the opportunity for interaction and the need to form social structures through traditional channels. The "high-tech high-touch" concept of John Naisbitt is no myth. It exists and promises to have a strong influence on our social behavior in the foreseeable future.

The key is to determine how a technology will affect the *opportunity* and the *probable form* of social systems as it alters our day-to-day lives. As an evolving species, humans should expect change, and the changes in social structure that result from technological innovation will determine to a large degree the quality and kind of life available to us in the future.

Technology's Effect on the Environment

Technology's effect on the environment has received much attention in recent years, mainly due to (a) the reduction in lead time perceived between the instigation of a new technology and the serious effects on the ecological balance of our world that may possibly occur as a result of it and (b) the greatly increased control over the environment that modern technology represents.

Environmental considerations are among the most obvious areas of importance to us in dealing with technology, because they are uppermost in our minds. Every technology affects the environment to some extent, just as it affects every physical entity to some degree. Many of the technological advances brought forth in recent years have been viewed as detrimental to ecological balance. Acid rain threatens wide ranges of forests and farmland. Pollution from nuclear tests is suspected of causing cancer in victims unfortunate enough to be exposed to it. Chemical content in rivers and streams destroys communities and reduces the productivity of already overworked soil. Diversion of water from natural sources feeds towns and cities, only to create shortages elsewhere. Leisure use of natural habitats destroys landscapes and threatens the homes of wildlife. Pipelines are purported to damage the ecologically delicate balance of permafrost environments and to disrupt the migration patterns of elk. Oil spills pollute our oceans, the main source of oxygen for the planet, leaving a trail of tar and oil solids from one continent to another. The list goes on and on . . .

Yet it would be inequitable to consider only the negative impacts of technology on the environment, though they are serious matters of concern. Technology also creates positive effects. It is through our understanding of nature and the manipulation of its laws that we can construct dams and spillways to bring life to the deserts of the Near East. It is through our understanding of nature that we can prevent disaster by destroying diseases detrimental to wildlife or save endangered species facing starvation and extinction, not from the hands of humankind but from the pressure of natural droughts. Science and technology can reclaim natural wilderness areas as well as destroy them, protect the integrity of ecological systems as well as disrupt them, and prevent catastrophic occurrences as well as create them. As with any other human system, it is the use to which technology is put that determines its desirability, not the nature of the technology itself.

Technology's Effect on Individual Psychology

Our attitudes, opinions, approaches to problem solving, and psychological balance are all affected by changes in technology. The world in which we live, including technological change, includes all the inputs we use in developing our personalities. Experiences teach us what to believe about the constitution of our world. Observations shape attitudes about social interaction and what is and is not considered appropriate. Threats from external sources create the need for a host of adjustment mechanisms, which collectively add up to a considerable part of our behavior patterns.

The vignette in Chapter 2, which described a hypothetical meeting of Luddites, illustrated the pressures of industrialization in early nineteenth-century England as the cause of a mass movement. Violent action resulted from the inability of the members of the movement to cope with the changes that were taking place around them. This is an example of how technological change can affect the otherwise stable thinking patterns of a human being.

In modern times, the degree of specialization and separation of workers from their tools inherent in the "big business" approach to industry results in feelings of alienation among workers. Personal satisfaction of needs drops as workers become less and less attached to the finished product with which they are working. There is no feeling of accomplishment, no personal interaction, and therefore little identification with either work or company. Patterns of behavior that would be considered unthinkable in a more personal form of endeavor become commonplace, including a reduction in pride, a reduction in honesty, and an increasing dissatisfaction with work as a whole.

There are also the feelings of frustration, fear, anxiety over the unknown, and dislike toward others that result from the interaction of the individual with the technologically changing society. With the proliferation of television sets in America, the methods by which the population internalizes information changes. The presentation of the Vietnam conflict in detail and, at times, within hours of actual occurrences, is a case in point. For the first time it was possible to sit in one's living room and be a part of the carnage, the fear, the death. Such a strong input was beyond the capacity of many in the society to cope with, and the result was a peace movement that eventually brought about an end to U.S. involvement in that war. Countercultural elements, whether representing a peace movement seeking an end to carnage or merely malcontents not willing to interact with a world that is changing too fast for them to internalize, are the result of changes in psychology or the result of an inability in people to change attitudes in the face of changing times.

This phenomenon has not been lost on military and government officials. In the Desert Storm operation of 1991, there was even more "instantaneous" coverage of events, ranging from the first air raids on Baghdad to the SCUD missile attacks on Israel. Yet these live reports did not elicit massive demonstrations or other negative reactions from viewers. The higher degree of military control over reporting during the operation and the brevity of the conflict contributed to this milder reaction from the U.S. public, but the true cause was the fact that enhanced coverage by media of violent events has become rather commonplace in the past quarter century. The nation has become accustomed to the idea, and judging from trends in feature film and television violence, the population has also developed a taste for it. Exposure desensitizes.

How will a given technology change the psychology of a nation or a given group of people within a culture? How will a society react to a new methodology? Will it ignore it? Will certain people attack it as the Luddites did on the basis that it is "stealing" jobs? What about society's attitudes concerning technological advances in the past? Will a person who grew up on science fiction movies be predisposed to fear robots? Do people view new technology with suspicion because of the Frankenstein and Dr. Jekyll stereotypes? How do they handle the creation of artificial life? What has the death threat of nuclear war done to their attitudes and ideals? Will future technology have a similar effect or will it act to ease tension, alleviate our anxieties, and usher in a new era of confidence? These are the kinds of questions that need to be asked concerning technology and individual psychology.

Technology's Effect on the Rate of Change

Again the prime example of this issue in the modern world is the computer. Never have we been able to manipulate and have available to us such huge stores of information. With this availability comes a heightened possibility for progress and development of still other technology. With a sharing of knowledge comes a sharing of wealth that can only create more wealth through its use.

But this is certainly not the only example we could cite. The oared galley, the railroad, the automobile, and the airplane are all examples of technological changes that brought about a flowering of progress in individual societies. In the case of oared galleys, the ancient vessels so equipped were instrumental in initiating international trade by opening up foreign lands in the ancient world as they made their way across oceans and seas, disseminating knowledge and spreading local technologies to other lands. Until the railroad opened up the West, the movement of population outward from the coastal areas was extremely slow in the United States. Railroads meant faster, safer travel. They were a means of expanding markets by shipping greater amounts of goods from one area to another. Markets were no longer local. Perishables could be sold over long distances. The movement of people and raw materials decreased dependence on limited, highly favorable habitats. The automobile opened up the entire country to the population, connecting towns and villages with cities. Whole new industries sprang up, creating still other industries in an orgy of growth that did not slow down until midway through the twentieth century. The airplane meant fast travel for passengers. This was accompanied by increases in the efficient use of time, particularly in business, and in the growth of holiday travel, which caused still other industries to flourish. In each of these cases, the speed with which progress was made was magnified. By facilitating the spread of technology, whole societal structures were created.

Technology's Effect on Institutions

Technology reshapes a society. In that capacity, it must necessarily have an impact on the social institutions of the society. Social institutions are like other social constructs—they are in existence because they fulfill some purpose for the population. If the introduction of a technology into a society changes the needs of the population or alters the availability of the institution to perform the function, then the institution itself is affected, either by altering its form or by disappearing altogether. Some examples will clarify this point.

Religion is one of humanity's major institutions. Yet it has often clashed with changes in the structure of society and the shifting paradigms that cultures embrace. During the Renaissance, Western Europeans reevaluated their worldview, and this expansion effectively destroyed their traditional understanding of the world. The Church as a political, religious, and administrative institution found itself in the thick of battle over many of the new technological and scientific discoveries being made. Leonardo Da Vinci wrote his discoveries in code, using a mirror-image script, for fear of being killed as a heretic or sorcerer. Copernicus brought forth the idea of a heliocentric system in his works, but arranged for them to not be published until after his death, mainly because he feared the consequences of announcing such radical ideas, ideas that flew in the face of Church doctrine. Galileo turned his telescope on the heavens and discovered the Jovian moons and other celestial wonders. He also espoused the Copernican heliocentric view of the structure of the solar system. When his work was published, it resulted in censure by the Inquisition, and he was put under house arrest for the remainder of his life. Threatened with torture and excommunication, he formally recanted his theories, although he never ceased to believe them.

The Catholic Church was certainly not the only religious institution to oppose new views. In Switzerland, Giordano Bruno was burnt at the stake by Protestants in 1600 for supporting the Copernican view of the universe. This philosopher and founder of the Carthusian Order discovered that the idea of a heliocentric universe was equally repugnant to his conservative brethren, despite the fact that the Protestants are viewed as part and parcel with the new freedom of thought that arose in the Renaissance. Martin

Luther awoke to the inconsistencies within the Church hierarchy and created a new approach to Christianity, which became the Protestant Reformation. His philosophy was in part the result of his increased knowledge of the world through reading books—technologically oriented objects, printed on printing presses, technological devices made by artisans using technological techniques of the day. Printing and the resulting infusion of material, mostly translated from ancient Greek and Latin through various Arabic texts, presented a great variety of opinion and information to Renaissance Europe, and with much disruption, weeping and wailing, and gnashing of teeth, the face of the world changed forever.

Indeed, the practical, worldly nature of the Protestant ethic—the moral code that so inspired the founders of this country and helped create the mercantile empires of Europe—had as one impetus the technologically fruitful age—the Industrial Revolution. Religion tends to change as the consciousness of people changes. And technology can do much to change people's consciousness. Religion also benefits from technology. The spreading of the Word of God was greatly enhanced by the invention of the printing press, as was the ability to become involved in comparative religious studies as the availability of books on other religions increased. And in the modern age, religion uses the communications systems of today to spread its message through television, radio, and computerized information systems. Scholars share information and search out obscure answers more quickly. Religious leaders and pilgrims travel with greater ease to the shrines of their faith. Whole new vistas of understanding open up in the face of a scientific community whose leading edge more and more approximates the mystic concepts of religions. Since science asks how and religion asks why, they cannot be one, yet religion has the capacity to avail itself of modern technology like any other social institution.

And this is certainly not the only example that we can find. What about education? With the changes in technology leading to central heating, electric light, and the mass distribution of energy, the whole concept of education changed. As a result, the shift from the rural social structure to the urban one took place, necessitating a shift from the traditional, one-room schoolhouse of the last century to the mass education institutions of today. Indeed, education is ever in the throes of technological change. One hundred years ago, a person might spend five to seven years in the same schoolhouse, learning from the same teacher. Less than fifty years ago, education meant larger, more centralized schools, with more facilities shared by a larger number of students and, it is hoped, extended opportunities for gaining knowledge. Today, there is a shift away from the centralized approach, with college-by-television available to many students so that they need not even leave their homes in order to receive the education that they desire. The institution changes again.

And what of government? The political structure that a society chooses is also a social institution, whether it is the town meeting, diet, parliament, or

Congress. These are all political institutions, designed to carry out a single set of functions, that is, to govern (creating rules, supplying public goods and services, and maintaining order). Yet how does a structure change? Feudalism was killed by a population that settled as a result of the invention and implementation of the plow as a means of increasing productivity. Of course, feudalism took several hundred years to die once the deathblow was struck; nevertheless, it was the lack of mobility on the part of the population that led to its decline. The rise of the middle class through trade and the expansion of knowledge created the right of the people to have a voice in their government. Although self-rule was not a new idea—indeed, the Greeks had a pure democracy long before the Romans ever thought of becoming an empire (I wonder how the Greek slaves felt about Greek democracy!)—the true rise of the middle class to political prominence came with the economic clout developed through its dominance of the technology of the age. Do we have democracy in the United States? Technically, the United States is a republic with a representative government rather than a direct government. People are elected to make laws rather than it being done by the population at large. Yet with changes in technology, such as television, computers, and rapid long-distance communications, the idea of every citizen voting on an issue before Congress is not as farfetched as it once was. What would be the effect on the political structure if everyone had a button on the television set to instantly vote on an issue?

And what of privacy? Privacy is a privilege that we take for granted in this country, yet it is strongly threatened by advances in technology. The ability of political and economic institutions to discover private information about individual citizens is awesome. There are satellites capable of focusing on a single individual on the ground. They can do so with such precision that the dial of a watch can be read from orbit. There is the capability, with the proper authorization, to screen telephone calls for certain key words and then record those conversations for later study, or to check on what people buy, what they owe, to whom it is owed, and whether they live beyond their means. These are capabilities considered by many to be threatening to the institution of privacy, a social structure long valued for its social value. Even the right to private property is altered by such simple technological devices as the photocopier and the tape recorder. How does one protect copyrights when it is so easy to gain illegal access to the fruits of one's labors?

Technology's Effect on Individual Freedom

Are we more free or less free by virtue of our technology? The answer is not a simple yes or no proposition. As with most aspects of culture, many shades of gray exist between the black and white extremes. Technological innovation both expands and curtails individual freedom. Depending on the use to which a technology is put, it can be either freeing or enslaving, or both at the same time. We are doomed to be dependent on technology as long as that technology is allowed to shape our world. The technology of building modern homes creates comfort, safety, beauty, security, and a host of other positive benefits that free us from the fears of our distant ancestors. Yet technology can enslave us to a lifestyle that depends on this element being in our lives. How would it be if the houses were no longer there? How would we survive a cold winter? How would we protect ourselves from the elements? How would we maintain our privacy? We are dependent on housing in the forms that are available to maintain our lifestyle.

A more serious example is our dependence on the automobile as a means of basic transportation. This dependency is one that was brought home with shocking clarity in the last quarter of the twentieth century as a result of shortages and rising fuel prices. When the fuel crisis of the 1970s began, many Americans were unaware of their dependency on the automobile. It was simply a fact of life, a technological device that was taken for granted in modern American society. But within weeks, motorists found themselves stranded in lines to buy gas, paying black market prices for fuel, wondering how they were going to cope with the crisis and how they could alter their lifestyle. Seventy years of dependency on a reliable means of transportation had locked them into its use. In the long run there were solutions to that dependency. One could buy a foreign car that did not burn so much gas. One could buy a motorcycle or begin riding the local mass transit systems (a far more efficient form of transportation anyway) or take up bicycling or jogging. But in the short run, when the crisis first arose, all that could be done was to bite the bullet, dig a little deeper into the old wallet, and pray that the pump did not run dry before your turn.

As an experiment, consider the following. There is a simple device that is available to all of us. It is a technological miracle that has given us a tremendous amount of freedom in our lives, yet it has created a huge dependency as well. It is called the light bulb. To drive home just how freeing and how enslaving this simple device is, take note for the next few hours just how often you use one. Note each time you switch on a light in your home or office. Consider how often that light is on and that you have no control over how long it lasts. Consider street lights, the little light in your car that automatically goes on when you open the door, or the one in your refrigerator that always goes off when the door closes. How often and in how many ways does the lowly light bulb free you, yet make you dependent? Think about it.

One additional concern needs to be addressed here, and it is one that deals with our individuality, freedom, and psychology. Current technological advancement has presented us with an unusual situation where the nature of the technology itself forces us to reduce our level of *independence of thought*, not only changing our way of thinking but forcing us to regiment our manner of living. The main thrust of current technological change for the individual centers on communications and data manipulation, that is, in the computer, the Internet, and personal communications devices, which are becoming the standard *modus operandi* for all of us. These are wonderful devices, increasing our level of interaction, expanding our personal control over our lives, and allowing us to react to increasing change with increasing speed. However, this technology carries with it a specific way of operating. If we are to use it, we are forced to operate within the boundaries of the technology's protocols. If you want to use a piece of software, you have to follow specific instructions. If you want to use specific forms of communication, there are rules and regulations for how to go about it. Further, the protocols are gaining an ever-increasing level of standardization. This is necessary to establish an efficient framework for all those using these devices and software, yet it also reduces our freedom to act independently.

Because of the rate at which technology is changing, we are forced to upgrade our skills continually and learn new computer tools in order to keep up. Yet to do so, we increasingly have to depend on others to tell us how. We read the instructions, we follow those instructions, and we learn how to manipulate the device or program. This makes total sense if we want to develop our skill set and remain on the leading edge; however, it leaves little room for individual thought. The proper way to use new technology is very specific and proscribed. We depend on those who know how it works to tell us how to proceed *and how to think about* the device or program. Our psychology is changed to think in the same terms as those who provide the technology, and the more we do this, the more we start to depend on others to tell us how we should behave. The techno-experts and the technology itself become the authority on life. There is no room for autonomous decision making, and there is a reduction of individual thought.

This may all seem very fanciful—even fantastic—but consider the effect of the process over time. Small children learn to use computers by playing games and running useful programs. As they grow up, they play increasingly more complicated games and gain expertise in manipulating the computers on which they work. They do so not so much by trial and error as by following the instructions that come with the software or from following the instructions received from online help sites. It can't be done by trial and error as people used to do with typewriters, cars, etc. You have to know how to get into a program and know what the proper commands are, information that you can only get from others. Furthermore, just because you have a general understanding of those commands, there is no guarantee that when you begin to use another program, the same exact protocols will be followed. Even if they are, there are always new and different possibilities for new software or new devices, and the techno-experts are the only ones who can tell you what they are.

If this process continues as a child grows up, and it does, there is an increasing dependency on the expert to define what is and is not possible, what does and does not work. The only logical way around this is to become one of the techno-experts and begin telling others how to think and how to operate. The autonomous nature of the individual slowly gives way to the dependence on the expert, and the tendency toward original thought can actually disappear. This actually happens. Have you ever received an essay question on a test or as an assignment and asked the professor, "What are you looking for in this paper?" You have then deferred to another person to do your thinking. Have you ever called on techie friends to help you install some device you are not familiar with rather than reasoning it out for yourself? I certainly have, and have been grateful for the expert advice. The trouble is that if we do this exclusively, we reduce our capacity to operate independently in the world. If you will go back to the introduction to this book and read the passage dealing with Freeman Dyson's childhood experience, you will see what I mean. Technology is indeed a double-edged sword, capable of freeing and enslaving. This is as true now as it has ever been.

Technology's Effect on Our Perception of Reality

An individual's perception of reality is created by the observations that are made by that person. If someone were forced to grow up in a room that was totally dark at all times and was never allowed to see any light, that person's perception of reality would be seriously distorted in that he or she would lack any content involving seeing. If another person were never allowed to experience kindness, that person would grow up believing that kindness in humanity is a myth. What we observe to be true, we believe to be true. In fact, we build our world in the image of those beliefs.

But what if the world we experience is altered by technology? What if we experience something new that does not fit into our contextual framework? This is the effect that new technology can have on a person's sense of reality, and it is no more than what was earlier discussed in terms of altering the paradigm with which we work. I can think of no more dramatic or elegant example of this phenomenon than the following, which is presented as the only example to be given.

Before the second half of the twentieth century, no known person had ever viewed the planet Earth from a height of more than a few tens of miles. We were well aware of the planet's makeup, of what constituted its surface, and where the various planetary features could be found, but no one had ever seen the planet from afar. Then we entered the age of space exploration, which brought with it, among other things, the first pictures of our own neighborhood. From the first moment that first camera took the first picture of that huge blue marble in a velvet void, the life of every person on the planet was changed. It was possible for the first time to see that we are all aboard the same ship, traveling at some 250 miles per second through a sea of cosmic flotsam, and that what happens to one of us happens to all of us, particularly as concerning the planet itself. This is one ship that it is difficult to get off. It is this realization among an increasing number of people that holds the greatest hope for stability in the world of tomorrow.

Technology's Effect on Our Mutual Dependence

Buckminster Fuller once recommended ending the threat of nuclear holocaust by connecting the electrical grids of the United States, Canada, and the USSR to create a single, huge electrical system. He theorized that no one is crazy enough to blow up the other half of his or her own electrical grid.

Regardless of the merits of the idea, it illustrates an important aspect of technology and what life is like in a technological world. Just as there is an increase in our dependence on technology, there is also the possibility of becoming more dependent on one another because of technological involvement. The United States is highly dependent on Middle Eastern oil producers for the supplies of crude oil needed to run our economy. Without oil, we would be hard pressed to maintain supplies of fuels, plastics, and a host of chemicals, to mention only a few items. In a similar manner, much of the world depends on the United States for food. Because of technological innovations in agriculture, less than 5 percent of our population is capable of feeding not only our own population but also millions and millions of others. Nevertheless, if we are to have coffee, we must import it. If we are to have rare earths and exotic metals, much of the supply must come from elsewhere. We live in what is, as Marshall McLuhan has said, a global village. This is a *world* economy that we are involved with, and it is that involvement that makes us dependent on one another for what we need to survive.

Technology can both create and alleviate that dependency. In the absence of certain goods (such as, for instance, our dependence on the supply of natural rubber from Southeast Asia during World War II), technological innovation can create new substitutes. Likewise, a dependence on a certain technology may result in a dependence on a specific commodity, such as our insatiable thirst for oil stemming from combustion engine technology.

The key is to consider the consequences of a technology in terms of its tendency to increase or decrease mutual dependence of sociopolitico-economic groups and, from that, ascertain the probable outcome of a technology's introduction.

Technology's Greatest Effect

What sector of the population will be most seriously affected by a given technology and when? If the technology is one that can be generalized over the entire society, then it will probably have an impact on everyone, but if it has localized application, then what form of effect will exist and for whom?

Historically, any example that deals with a technology being localized for some reason would suffice to demonstrate the necessity of studying this issue. In the ancient world, about the time of the rise of the Mesopotamian city-states, such as Ur and Lagash, in the Fertile Crescent, it was the technological innovation of agriculture that created cities where trade took place. City-states arose around the farmlands of the Tigris and Euphrates river valleys, and because of the availability of water transportation (another example of technology), they were able to grow prosperous and powerful. Thus, the citizens of these areas were most greatly affected by the introduction of agricultural technology, though tribal peoples from along the rivers were involved in the trade that resulted from the sedentary lifestyle the agricultural people practiced.

A more contemporary example is the age of "king cotton" in the southern United States. With the cotton gin, cotton manufacture became cheap and profitable. By technologically solving the problem of how to comb the seeds from the cotton fiber rapidly and cheaply, the desirability of cotton as a crop rose dramatically, and there was a ready market for those goods in England, where the textile industry depended on cotton and wool for its livelihood. The people most directly affected by the technology were those in a position to take advantage of it, that is, the Southern states, where the climate was perfect for growing cotton and where the employment of slave labor was productive enough to be profitable. These conditions did not exist in the North, and, as a result, people living there were relatively untouched by either the cotton industry or the slave labor method of operation, since Northern industry depended on different inputs to create goods and services. Eventually, the schism in social systems represented by the localized nature of the cotton industry technology led to the Civil War and to the end of an era. After years of political and economic dominance by an agricultural South, the nation was dominated by a more efficient and more productive North, where, it should be noted, there was a technological advantage in manufacturing industries because of the abundance of raw materials and transportation.

Whom a technology affects is as important as how. This can lead to moral questions as well. What of the wonder drugs that could be manufactured or developed but are not because the number of patients requiring them is too small to warrant the costs? What of the ability to save lives through expensive operations such as heart transplants or mechanical hearts? Who receives them and who does not? Who pays for them? Are there too few people affected to warrant continuing the technology? These are some of the questions that arise in considering this aspect of technological development.

THE PERSIA CONCEPT*

The PERSIA concept is similar in outline to the general categorical model in that it divides the society into specific elements and then analyzes technology in light of its interaction with these elements. Where it diverges is in the historic nature of the criteria and in its economy of structure. The PERSIA concept uses six rather than ten criteria—*P*olitics, *E*conomics, *R*eligion, *S*ocial interaction, *I*ntellectual content, and *A*esthetic, or artistic, endeavors.

*I am indebted to Professor Sharon Rodriguez of De Vry Institute of Atlanta, Georgia, for her assistance with my explanation of the PERSIA concept, which was developed during her association with the late Dr. John Alexander of Johns Hopkins University.

Using the techniques of historical study, we can use the PERSIA concept to analyze the future as we have the past. By viewing technology as an interactive part of each of the six aspects of culture, we can create a future history just as we have a past one. Here the emphasis is on the repetitiveness of historical events through time, reminding us that, as the saying goes, those who do not learn from the past are doomed to repeat it.

Common elements from experience serve as obvious possibilities for new technologies when viewed in this manner. The rise and fall of economic and political power, the development of cultures through stages of technological advancement, and the importance of specific technological innovations on the development of countries, empires, and regions of the world all provide historical perspectives about technological development. Communications, for instance, is as much a key to modern social order and political control as it was when the Romans constructed their roads for the movement of troops and information. Opening new markets and introducing new ideas can still change a society and determine the direction in which it moves, just as during the Renaissance, the age of discovery, and the European exploitation of the Far East.

The Six Aspects of PERSIA

Political

The influence of technology on the political process in industrial and postindustrial societies is well represented historically. Railroads and the development of the industrial power in the North in the mid-nineteenth–century United States can be viewed as key precursors to the American Civil War. Although slavery was the primary social issue, the actual conflict centered upon the political struggle between the industrial North and the agricultural South. Indeed, the very issue of slavery is one of human power versus machine power. If machinery had existed to do the agricultural work, slavery might never have occurred in the United States. Conversely, if the North had had a slave-based economy, the abolition movement would have had a much more difficult time finding supporters.

In modern American politics, there is little doubt as to the importance of communications in mounting and winning a political campaign. Without adequate television coverage, a candidate has little chance, which means that creating image and opportunity for publicity is as important as issues and political platform. Quite often, it is more important. Political control and political change caused by technology is evident throughout history, from the rise of city-states with the development of agriculture, through the feudalism of the technologically limited Middle Ages, to the rise of modern political states based on oil, gold, or productive capacity in electronics. Analysis of how this comes about yields clues as to what to expect as new technology develops. Political processes are affected by technology as is everything else. What would the political makeup of this country be without the census, compiling information about the population with reasonable accuracy and speed, affecting choice of issues, political boundaries, and the importance of political control in different areas of the country? How could those who advocate political positions operate effectively without methods of both obtaining information and disseminating it to the public? How many of our decisions as a nation are dependent on our understanding of issues as presented by the media? Ray Bradbury in *Fahrenheit 451,* George Orwell in *1984,* and others have warned of the power connected with the capacity to control information. When a new technology arises, how will it affect this political structure? How will it serve to shift political and economic power within the body politic? What will its effect be on our thinking, our goals, and our outlook on our national destiny?

Economic

Here we are dealing with an overall historical outlook for the economy of the society rather than with the individual consequences of a technology. We have emphasized the economic importance of technology several times in this volume, but here we compare historic trends connected with that development and the effects of new technology. The creation of canal systems and railroads in the United States led to a coast-to-coast population expansion. Indeed, if it had not been for railroads, we would not have developed local standard time or our time zone system, we would have not been able to maintain control of the west coast in the nineteenth century (a lesson learned from Britain's attempts to control their colonies across the Atlantic in the eighteenth), and we would not have had the rapid economic growth of the past century and more.

By studying the economic history of the world since the Industrial Revolution, it is possible to determine trends that can be expected to continue as new methods of production, distribution, transportation, and communication continue to develop. What changes are in store in our market system, for instance, with the development of computer communication nets? More and more people are shifting to computer- and television-based purchasing. Catalogs are everywhere, promising consumers the opportunity to do their shopping without leaving home. Transportation systems have made localized shopping islands the rule rather than the exception, and the decentralization process continues to show an increasing tendency among the population to not only live in suburban areas but also depend on them for work as well. Without the necessity of centralized markets, how will our economic structure change? How will our tastes and needs differ? These are all economic questions.

Religious

In this aspect of society, technology serves to alter not only the form of religion but also the content. We already have noted that the Protestant

Reformation was as much a matter of printed communication spreading ideas as it was one of religious conviction. In addition, note that often resistance to wholesale changes in society stemming from technology is expressed in religious ideas. ("If God had wanted man to fly, he would have given him wings!") Beyond this, technology can alter the very concepts with which we work. The interaction of societies on new levels or the meeting of civilizations not formerly in contact can cause serious dislocation of religious patterns. Defense of tradition can result in everything from war to the destruction of a society. The Aztecs believed that Cortez was a returning deity, Quetzalcoatl. This belief allowed the Spanish, with cannons and steel and ships, but with less than a thousand men, to gain control over an empire. It was the development of television and automobiles that led to a resurgence in religious fervor in the 1960s through broadcasts and drive-in churches. It is the fear and feeling of helplessness that some people feel in the face of overwhelming technological change that supports religious fundamentalism, in this country, in the Middle East, and elsewhere.

As we have new experiences and our worldviews change, our religion changes to accommodate us. Even the arguments over the ordaining of women into Catholic priesthood and the general role of women in the religions of the modern world are a reflection of the freeing of women from the drudgery of their pre-twentieth–century lives, a freedom brought about by technology. Finally, there is an emergence of practitioners of the old religions based on nature and natural processes as a result of the threats of humans and their technology to the ecosystem. This heightened awareness of humanity's connection with the whole structure of Earth, and new respect for its fragile, nurturing power, is epitomized by the increasing numbers of people who express reverence for Gaia, the name given to our planet perceived as a single, complex, sentient, living entity. Technology allows us to suddenly face our moral dilemmas with euthanasia and ethical suicide and abortion as religious belief comes into conflict with our capacity to terminate or extend or create life on this planet. Historically, religion is a driving force in societies, and societies that do not have a religious-moral basis for decision making flounder in the face of challenge by those that do. It is important as well as interesting to analyze what this meant for us at the end of the twentieth century.

Social

Here there is little that needs be said. Much of the content of this book deals with the effects of technology on the social structures of a culture. As technology changes the content and context of our lives, it changes the social relationships that exist among the members of societies. Family structure, loyalties, and the makeup of groups and their behavior are all deeply intertwined with the technology they use. The very need for society as a whole varies with the degree and type of technology available, controlling to some degree everything from the number of offspring a family has to the degree of dependency the members of the society feel for each other.

Intellectual

Technology is intellectual. It is the result of ideas and the application of those ideas to solve problems and to create. Science and engineering concepts form a large part of the content of the intellectual capacity of a people. Beyond that, there are the paradigms that a society operates under, which create its potentials and limitations and define what it terms important and worth knowing. Literacy, the degree to which education is general in the population, and the capacity and ability of the population to involve themselves and their curiosity in the exploration of the unknown all depend on technology to create the time, resources, methodology, and wealth to pursue intellectual development. This is seen in the contrast between the Dark Ages and the Renaissance, and between the poverty of agricultural nations and wealth of industrialized societies. Intellectual development is also apparent in the lines of research followed and the choice of infrastructure a society adopts. Today, as before, technology is aiding, rewarding, and encouraging intellectual development. The advent of the Internet and its accompanying freedom for computer literate individuals to study anything from nuclear physics to how to build terroristic weaponry is an example of the intellectual freedom technology affords. Not since the Renaissance has there been so much available information for the population to study and absorb. And as this increases the number of possible approaches to a given problem or issue, the complexity of the system is magnified and the unpredictability of the future increases. Historically, however, it would appear that we can expect a second renaissance, a new age of discovery and intellectual growth unparalleled in human history, and one that will be filled with unrest and social turmoil as the traditionalists fight to maintain their world. Again, to obtain clues as to what may come, it is important to study the process as it happened before, such as with the Helenization of the ancient world, with the Renaissance itself, and with the scientific revolution of the nineteenth and early twentieth centuries.

Aesthetic

The arts are the creative expression of humanity raised to its highest level. Art requires freedom and methodology, and never before has so much opportunity existed for humanity to be artistic. New ideas are in need of expression. New fears and emotions need to be explored. New innovations in media, from computer art and virtual reality to views of the universe we could only dream about before, are all now fair game for the creative artist. The need for this art is acute. It is in the nature of the species that an expression of feelings and of ideas accompany growth, to put events in perspective and serve as a statement of the collective psyche. And those who are on the edge of that expression are the pathfinders to the future. Just as technology is as much art as science, the very art of living requires creative expression and groups of individuals willing to present those new ideas and evaluate the changes that occur in our lives. All music was changed forever with the discovery of electronics. The visual arts have expanded more in the past one hundred years than in all history before that time. The future is an image until it becomes a reality, and it is the writers and artists, the movie makers, and the fantasists that present the alternative possibilities to us now, so that we can explore them, learn of them before they arise, and discover their possibilities. In the end, life itself is an art, not some deterministic formulated construct, and it is through our technology that modern societies learn to do it well or learn to destroy themselves.

THE ANTHROPOLOGICAL MODEL

Anthropologists view culture as a constant battle of forces seeking to reproduce the culture on the one hand and to transform it on the other. In technological terms, this expresses the tendency toward homeostasis, resisting the transformation of the society through the development and exploitation of new technologies while attempting to maintain the status quo. Culture is, as far as we know, unique to human beings. It is a learned process that is passed down from generation to generation, and as each succeeding generation embraces the knowledge of the last, it changes it to fit current conditions. Thus the cultural system reacts and adjusts.

The systemic view of culture views all cultural systems as complex, from the simplest extended families of the Aleuts to the megacomplexity of the world culture. Culture is viewed as a multilayered structure, beginning at the local level and proceeding in ever-expanding structures through the regional, the national, and the international, each seen as a systemic element affecting and being affected by the activities and changes that occur in each of the others. To study the phenomena that occur within cultures, anthropologists look at culture from the perspective of its constituent parts to discover how the content and context of each is woven into the whole cultural fabric. For our purposes, the list of cultural elements studied includes the following:

- Government—the methods by which the culture creates order, decides issues of power, and establishes authority
- Economics—the subsistence strategies of a culture by which the community provides itself with the necessities of life
- Kinship patterns—the interpersonal relationships through which the members of the culture establish lineage, family membership, authority, obligation, and succession, while ensuring viable gene pools and sufficient diversity to allow for individuality and variety

- Linguistics—the study of language in its cultural and social context and how it changes through time and from one location to another
- Concepts of time—the way the culture views time, how it is measured, how it is utilized in ritual, political, and daily life, and how this structure of time limits and defines the culture's behavior
- Concepts of space—the ways in which the members of the culture view and use space and how this relates to the culture's overall behavior and beliefs
- Religion—the process by which the culture attempts to make sense of its world, particularly in the absence of direct understanding and evidence
- Myth—the stories and legends of the culture as they relate to the history of the culture and the moral principles by which it operates
- Ritual—the symbolic methods employed to signify events, teach morality, and mark the passage of life events, particularly rites of passage and religious rites
- Play—the manner in which the society utilizes recreation as a method of instilling culture, as practice for adult life, and for relieving the pressures of daily life

We have already discussed several of these aspects of culture, in some cases extensively, and there is little need to reiterate here. The other aspects benefit from elucidation.

Kinship Patterns

Kinship patterns may seem rather cut and dried to people in our society, yet there are many ways to define kinship. Some cultures pass property and lineage through the female, or distaff, side of the relationship. Others, particularly traditional Polynesian societies, may direct property from generation to generation through an uncle–nephew relationship. In Western culture the striving of women for equality with men has led to such kinship changes as the use of hyphenated last names, families without resident fathers, and a reversal in economic power among the members of the nuclear family. The nuclear family itself is under attack from modern life and is now seen as only one of many family patterns. In each case, the individual pattern of kinship is associated with what works best in a given cultural situation and is therefore appropriate to that culture. Seeing cultural behavior from the point of view of the culture studied rather than judging it from the point of view of the investigator's own society is called *cultural relativism.* It is an absolute necessity if any sense is to be made from what is observed. Among the Pharaohs of Egypt, for example, and among the Inca rulers of Peru, incest was not only common but also, for the most part, subscribed. Yet incest is one of the strongest cultural taboos that exist. The reason was a practical one. In both cases, the individual rulers were superb organizers and administrators, and the genetics of their lineage tended to breed offspring of similar genius. By practicing incest, the traits were maintained, in spite of the rise of negative recessive traits, such as epilepsy. The only reason the process worked for them at all is that they were fortunate enough to have a genetic makeup relatively free of serious recessive traits, although negative traits crept in. For the common people of these societies, the admonition against incest remained absolute.

In some Polynesian cultures, property is passed down to the son of a man's sister. This passing on of property to a nephew makes sense in light of the sexual behavior of these cultures, in which copulation among a number of partners was common. It was seen as a method of ensuring that the property and name passed to someone with the genetic material of the giver, since one's wife may have a child by some other man, but the offspring of a sister is guaranteed to carry the same gene pattern as the brother, regardless of who fathers the child.

Changes in technological societies create similarly strange-seeming patterns when compared with traditional ones. It is now possible to have a family where the father lives on one side of the country and the wife on the other while the children can live with either parent or with both, though obviously not simultaneously. Multiple marriages, through divorce, are now common, with over half of all marriages ending in divorce and most divorcees remarrying within three years. It is possible, therefore, to have multiple fathers or multiple mothers culturally, although not biologically. This process of multiple marriages is often referred to as *serial monogamy.* Without transportation and communication technology, this multifamily approach would be difficult to maintain, and were it not for the extended opportunities and rapid rate of change inherent in modern industrial and postindustrial societies, it would probably not be culturally acceptable or possible.

Linguistics

Linguistics, the study of language, is fertile ground for the technologist. The language of a culture is a rich source of information, and as the culture changes, so does the language spoken. Language is full of information about both the technology of a culture and how people use and perceive it. For example, Indo-European is one of the major world language families. Words with an Indo-European root are words that have existed in the languages in this family for the longest time. Yet no one has ever found direct evidence of an "Indo-European" culture or group of cultures itself. Everything we know about Indo-European comes from the remnants of the language. Simply put, if there is an Indo-European word for a technology, the original speakers knew of the technology. If not, they did not. We know, for instance, that Indo-European cultures had domesticated animals and carts, as there are words for these technological elements with Indo-European roots. But whether they used mortar or brick in building is doubtful, as no root words for these items are known to exist in the Indo-European base.

Today, the English language is constantly changing, and doing so at an increasing pace. Slang expressions, the meanings of certain words, and the creation of new terms often result from technological change. The expression "the real McCoy" stems from a single invention of the nineteenth century that allowed for the automatic oiling of steam engines and related machinery. When cheap imitations occurred, the manufacturers encouraged the public to look for a genuine McCoy lubricator instead of the other ones. Today we often use computerese in a general context, talking about data links in general social situations, telling someone to beep us when we do not actually carry beepers, or saying there's a "bug" in some process unrelated to computer equipment. These usages are clues to the importance and generalization of not only present but also future technology. We therefore study changes in language through time to gain insight into the effect of technology on our thought processes and lifestyle.

Time and Space

Concepts of time and space are also determined to a large degree by technology. We have mentioned that without railroads, the concept of local standard time and time zones would have been much later in coming. Beyond this, time must be understood to consist of two elements, one being objective time, time as measured by the clock, and the other subjective time, determined by what the individual experiences. Five minutes is five minutes, but it is either a long time or a short time, depending on what the individual who experiences it is doing. When you're waiting for a bus, five minutes can seem like an eternity. If you're reading an enjoyable book, it can pass in the blink of an eye. The very concept of what constitutes a long time or, for that matter, a long distance is culturally determined by the technology involved. With automobiles and superhighways, supersonic jetliners and bullet trains, fifty miles is a short distance. One hundred years ago it was easily the farthest most people traveled from home in their entire lifetime.

Our sense of space also shifts for other reasons. In a research project the author undertook in 1994, the concepts of commuting drivers regarding time and space were studied in detail with interesting results. Among other findings, it appears that drivers on highways regard their world as consisting of their immediate vicinity and the other automobiles surrounding them. They gauge relative speed and safety not in terms of miles per hour but in terms of the envelope of surrounding vehicles with which they are traveling. It is a static system to them, shifting only as they approach or distance themselves from vehicles around them. The fact that the entire group of cars may be traveling at sixty or seventy (or more) miles per hour is only peripherally part of their reality. The surrounding countryside and passing buildings are seen as a surrealistic backdrop. In fact, reality itself shifts when people drive a vehicle. Drivers often perceive everything beyond the windshield as surrealistic, like a giant wraparound television screen, where the real world is inside and outside is just a passing show. This is evident to anyone who observes the behavior of fellow drivers, particularly at streetlights. People do things sitting in a car that they would never consider doing if they were walking on the street or sitting on a park bench!

Myth

The myths of a culture are designed to act as guidelines for the population in deciding what is and is not valuable or moral. Myths provide information, explain the unexplainable, help justify the behavior of the society, and provide a spiritual and theoretical structure upon which to draw. Myths are inevitably linked to the nature of the culture in which they appear. We have numerous technological myths in American culture, including Casey Jones, the engineer, and John Henry, the steel-driving railroad worker killed in a contest with a steam drill. As with these examples, myths are often about historical people whose exploits are enlarged into legendary proportions through folktales, as with George Washington's cutting down of the cherry tree or Benjamin Franklin's discovery of electricity by flying a kite (a process I would not recommend, as several now-deceased individuals discovered while trying to duplicate his experiment). This process is no less true of today's heroes and heroines than those of the past. People such as John Glenn and Alan Shepherd have become as much the stuff of legends as of reality, as have the exploits of Mother Theresa and Albert Einstein. These are real people who have performed real feats for which we revere them, yet many stories about them are more apocryphal than real.

The importance of myth to our purposes is that the subject matter and the behavior a society reveres reflects its values and the kind of information it deems suitable for passing down to future generations. Quite often either the content or the context of these myths contain seeds of technology and help explain how to react in that society's world.

Ritual

Ritual is an integral part of culture that is affected by technology. All cultures have rituals, to establish custom, to explain morality, and, particularly, to mark the passage of life milestones with what are called *rites of passage*. These milestones are often inculcated in religious ceremony as a further attempt to separate them from everyday events and to note the solemnity of the occasion. Thus we have passages into adulthood marked by first communion, confirmation, or Bar Mitzvah. Some African cultures hold group circumcisions, and others stage ritual combat and ritual torture to symbolically kill the child so that the adult may appear.

In the United States, we have the senior prom, a driver's license at age sixteen, and the importance of reaching age twenty-one. Equally important, as we become more dependent on technology and therefore take longer to mature in terms of being a fully functioning adult member of society, there is graduation from college as a milestone filled with pomp and ritual. There is the military ritual graduation from boot camp, when the recruit is transformed into a full-fledged soldier, and the ritual titles connected with the rise of individuals in organizations through various degrees of vice presidencies or secretariats in government.

What all of these rituals signify is a change in *status*. As technology changes, the nature of the rituals to signify that status and what the status is also change. Reaching the age of mobility (when one gets a driver's license) is meaningless until there are automobiles. Receiving your own computer or perhaps your own account on an Internet connector is equally unimportant without this technology of the communications age.

Play

Play is an extremely important element in a person's life. It is no accident that it is often referred to as *recreation*, since its primary purpose among adults is to re-create their lives—to take time out to put things in perspective and deal with the fragmentation caused by high pressure, high living. For children, the purpose of play is to teach cooperation and coordination and to prepare them for their adult roles. As such, a study of both games and toys yields incredibly valuable information about the technology, ideals, and occupations of members of a society. For many years, archaeologists who studied the Peruvian Incas thought that they had no knowledge of the wheel, since it was not used as a technology and no one had ever found evidence of one in the culture. Then an artifact was unearthed that had wheels—but it was a child's toy. At the time, it caused quite a stir. As it turns out, the Incas knew of the wheel, but because of the mountainous terrain, it was useless as a transportation technology and was simply never used!

By viewing a society's games, we may discover what is important and expected of children as they mature. Games of chance, such as card games, teach children how to deal with uncertainty and risk and how to develop successful strategies in the face of incomplete information. Group sports teach teamwork and cooperation and how to develop group strategies for dividing a difficult task into manageable pieces and coordinated efforts to achieve success. Video games teach strategy and skill with the tools of the next generation, and channel aggression. Thinking games, such as guessing games, quiz games, and puzzles, sharpen the mental capacity of individuals. Even the simple entertainment shows of television provide information and present hypothetical scenarios to prepare people to face possibilities that may or may not arise. A study of games and toys will therefore yield a better understanding of how the technology of a culture fits into its overall structure.

CONCLUSION

We have presented three different schemas for the study of the interaction of technology and culture. Each has its own strengths and weaknesses and has points in common with the other two. Each is a systems approach, breaking down the whole into a set of constituent parts and then analyzing those parts and how they interact to create the whole. You must choose which model to use for a specific study, as each offers its own unique view of the world, its own unique perspective. Whatever approach you choose, the key is to be aware that all elements of a culture are interrelated and that we are constantly being created and re-created through our evolving technology. The problems to be solved are legion, and the issues to be addressed are manifold. It is the purpose of this methodology to make that process simpler and more manageable.

Following this chapter we present three examples of society in the face of technological growth. The first two are papers by students whose purpose was to investigate the effects particular technologies have had and continue to have on our lives. They are similar to the papers you have already encountered in this book. The third presentation consists of excerpts from the diary of a German immigrant to the United States in 1848. They are taken from the original diary, now in the possession of the author, and are offered not only for contrast to our modern world but also to illustrate how true the saying is that the more things change, the more they stay the same.

KEY TERMS

Aesthetics
Anthropological Model
Cultural Relativism
Culture
Environment
Freedom
General Categorical Model
Government
Kinship Patterns
Linguistics
Mechanization
Mutual Benefit
Myth
Nuclear Family
PERSIA Concept
Play
Privacy
Recreation
Rites of Passage
Social Institutions
Social Systems
Time and Space

REVIEW QUESTIONS

1. Define the three systemic models presented in this chapter.
2. What is the PERSIA concept and what does the acronym stand for?
3. What are the elements of the anthropological model?

4. How does the anthropological model differ from the general categorical model?
5. Which of the three models presented do you prefer and why?
6. Give an example of the effects of technology on each of the following concepts: (a) ritual, (b) both time and space, (c) economic systems, (d) religion, and (e) mutual dependence.

ESSAY QUESTIONS

1. Discuss the relative merits of the three systemic models offered, noting the strengths and weaknesses of each.
2. Discuss the inherent shortcomings of modeling complex, real-world systems as the inherent strengths of the modeling process. Include how the process is and is not useful to our understanding of our technological world.
3. Using one of the three systematic models, pick a single element and show how technology has molded and changed that part of human culture. Use current and historical examples.
4. Using one of the three systyemic models, pick a leading edge technology—preferably one broad in scope—and discuss the effects that the technology will have on our culture. Include both positive and negative consequences of the development and use of the technology, as well as potential catastrophic effects and perceived chaotic issues.

THOUGHT AND PROCESS

1. Choose some simple technological device and speculate on how it has affected your life and the lives of people in your society. Pick something that seems insignificant, such as a pencil eraser or the common zipper. Itemize as many uses and applications for the technology as you can think of. Check your list against the list of general categories given in the chapter and see how many you can find.

2. Choose an institution (such as education, religion, politics, the family, the extended family, or your own group of friends), and think about how technology has created that institution in its present form. Is this form different from those of previous times? Are you personally aware of any changes that have taken place in your experience of the institution? Is the institution better the way it was or the way it is, or is it just different in the face of new conditions?

3. Ask a small child, a young adult, and an older adult about how they feel regarding (a) the computer, (b) space exploration, and (c) modern medical techniques. How do their responses differ? Who shows the highest degree of acceptance? Who shows the least? After reading this far in the book, how have your own feelings about modern technology changed, if at all?

4. The following problem is "food for thought." One of the underlying scientific principles with which engineers, physicists, chemists, and others work every day is the circle of 365 degrees. Trigonometric functions, esoteric topology, vector analysis, astronomical measurement, and industrial quality control are all areas of investigation that require some sense of understanding on the part of the investigator of how this circular system of measurement works. It is identical in action to the sweeping hands of a clock. Inasmuch as reading clocks is a symbolic–conceptual process by which we gain expertise in visualizing and manipulating the informational content of a circular coordinate system (that is, we learn to understand the circular coordinate system by an intimate familiarity with clocks as a symbol of that system), how do you think the growing predominance of digital clocks will affect future generations in their attempt to learn about angles and trigonometric functions?

5. Imagine the availability of a vehicle that operates on the principle of levitation, one of about the same size and cost as the ordinary car. It can operate at up to ten feet above the surface and is in all other ways comparable in performance and cost to the normal automobile. In other words, the only severe change in design in the vehicle is the antigravity mode of suspension and propulsion. Speculate on how it will change life in our society, using the list of PERSIA concept elements given in this chapter.

World War II is the single most influential event of the twentieth century. It is absolutely pivotal in producing and explaining the shift from the old industrial age to the age of information. It is the ultimate organizing and structuring event of the late Industrial Revolution, when all of the political, economic, technological, and cultural forces of the modern world came to fruition and competing approaches to operating in a world of mechanical devices were able to challenge each other for dominance. What emerged from this great conflict was a new world, unlike anything that had come before, and one that has spent the next fifty years sorting out the details, shifting from the Industrial Age to the Information Age in an attempt to redefine who and what we are. The amount of scientific and technological progress during the years between 1935 and 1945 is staggering. The wholesale reorganization of political units worldwide is without its equal in world history. The redefinition of what it is to be human has never been more complete or more profound than during this short period of time in the middle of the twentieth century. For these reasons, we now look at the technology of the era and how it has influenced our times. As before, note the systemic nature of the approach used by those creating the document and their particular paradigm in studying it. What other paradigms, you may ask, could have been used and how would that shift have redefined the conclusions reached by the researchers?

World War II: Technologies and Social Implications

ERIC ROBINSON AND RUSSELL PENNINGTON*

World War II officially ended on the Battleship *Missouri* on September 2, 1945. It would be unrealistic, however, to assume that the effects of the war halted at the same time. In fact, we all live under the shadow of World War II, the greatest source of change in modern history. Virtually all technological developments have their roots in the war years. But even beyond this, humanity's philosophy of life has been drastically altered. All in all, the effects of World War II are much more dramatic and widespread than the postwar population can imagine.

Before we examine the effects of World War II on the modern world, let us first scrutinize the war itself and the ways in which this war was different from any other. The differences we are going to examine are not by any means inclusive; a discourse on every difference between World War II and any other war would run several thousand pages. What follows *does* represent the major developments through the war years. The major elements analyzed will be

1. **Increased mechanization**
2. Increased mobility
3. Increased role of science
4. Mobilization of the civilian sector
5. Improved communications
6. Total nature of war
7. Use of atomic weapons

INCREASED MECHANIZATION

World War II, unlike previous wars, was fought by human beings and machines, rather than simply human beings. This means that an army consisted as much of the factories at home producing the machines as of the soldiers in its ranks. The ability to produce effective weapons became the ultimate goal of the production lines. This led to a staggering growth

*Eric Robinson and Russell Pennington, "World War II: Technologies and Social Implications" (unpublished paper presented at De Vry Institute of Technology, Atlanta, Georgia, April 17, 1984).

in the use of sophisticated weaponry. Tanks and planes were employed as never before; the attack on Pearl Harbor was a far cry from the almost benign reconnaissance planes of World War I. The planes had far greater range and mobility, making bombing a favorite type of attack with a high chance for success. Tanks were no longer simply armored vehicles with thin metallic shells. Now they were as heavy as 36 tons, with 62-millimeter armor, and they could reach speeds of 25 miles per hour. Some tanks were armed with a massive 75-millimeter gun.[1] The numbers were staggering as well. In 1940, the United States produced 300 tanks, 6,100 planes, and ships with a total displacement of 52,600 tons. Production reached maximum levels of 29,500 tanks (1943), 96,300 planes (1944), and 3,176,800 tons of shipping (1944).[2] Some mechanization had no precedent—for the first time the world saw jets, rockets, amphibious units, and the atomic bomb. The increased mechanization was almost inconceivable to the people of the time and still boggles the imagination.

INCREASED MOBILITY

The increase in mechanization led directly to an increase in the mobility of both the military forces and the population in general. But the sharpest increase in mobility at the time was in the area of the ability of the military forces to move large quantities of soldiers, machines, and supplies. During World War II soldiers could be transported rapidly by rail, motor car, plane, or ship, and canned food could be brought to them. In earlier wars, only 1 percent of a given population could be mobilized at any one time. In World War II, there were mobilizations totaling 10 percent of some populations with as much as 2.5 percent at the front.[3] This meant that a much greater proportion of the people was directly involved in the actual fighting. The advances in technology created better means of communication, which enabled the armies to be virtually anywhere on the face of the earth and still be in close contact with their superiors at home, thereby giving the units more actual mobility. This increase in mobilization had many effects on the military and on the rest of the world, now and then.

INCREASED ROLE OF SCIENCE

The advances in weapons and mobility would not have been possible without a corresponding increase in the development of technology. These developments would have been impossible without science. As the war raged on, so did the race for the fastest, strongest weapons. Each army depended on its production lines and its scientists and engineers to keep it on the leading edge of technology. An example of need dictating development is the growth of the United States tank force. The earliest models of the war were virtually useless against the German Panzers—the guns could not penetrate the armor nor could the carriages outrun the heavier, faster German machines. Later models, such as the Sherman tank, surpassed the Panzers in every way. The improvements

[1]Richard Collier, *The War in the Desert* (New York: Time-Life, 1977), p. 113.

[2]Edward Jablonski, *A Pictorial History of the World War II Years* (Garden City, NY: Doubleday, 1977), p. 294.

[3]John G. Burke and Marshall C. Eakin, eds., *Technology and Change* (San Francisco: Boyd and Fraser, 1979), p. 308.

occurred in a period of only four years. It can be said that World War II spawned a "blitzkrieg" of technological developments and scientific discoveries.[4]

MOBILIZATION OF THE CIVILIAN SECTOR

Since a greater proportion of the population of any country could be mobilized at any given time, a greater proportion of the population at home was required to be mobilized into military support services. For any army to be successful in its campaigns, the civilian sector must be dedicated to its support. This dedication is required during the war and in times of peace as well. The organization of the civilian sector by the military has become necessary as a preparation for war. This has resulted in the armed forces no longer being a self-contained unit apart from the general population. However, the military does not overwhelm the civilian sector. For this reason it became necessary for the military to make use of propaganda to sustain the greatly needed morale of the population. This use of propaganda extends government control into virtually all aspects of the society, including the economy and public opinion. This in turn has led to the increase of the autocratic totalitarian state and the subsequent elimination of free speech and free economy in some countries. History has proved that a free-market system is less adequate than a military-based economy for the support of the war effort.[5] It is ironic that a war fought in the name of freedom has led to the reduction of freedom in many areas of the globe.

IMPROVED COMMUNICATIONS

As mentioned previously, the increased nationalism due to World War II increased the importance of propaganda. Propaganda is only effective when distributed to the masses—the greatest words of wisdom are wasted if no one is listening. It was therefore important to each government to ensure that every citizen have some way of "listening." This led, in part, to an explosion in the communications industry. Since armies were spread out further, better communications were also required to coordinate the movements of the armies. Improvements were made in radiography, radar, cryptography, and mass communications technology.

TOTAL NATURE OF WAR

The modern military technique called for the mobilization of the civilian sector. Since it is directly involved in the support of the military, it is obvious that the civilian sector would now become military targets and therefore be liable to attack from the enemy. The part of the population involved in the contribution to the war effort is not safe from attack. It should be apparent that the enemy has little or no way of distinguishing the difference between the military-supporting population and the nonmilitary population. This lack of distinction has led people to realize the total nature of war. The civilian population of World War II saw starvation, bombardment, confiscation of property, and terrorism, which have been considered since that time to be applicable against any enemy population as

[4]Collier, *War,* p. 113.

[5]Burke and Eakin, *Technology,* pp. 308–10.

a whole. The advent of a possible attack against a nation's general population has caused the methods of war to become more ruthless and indiscriminate. Today the methods of war call for the use of terrorism, which is a direct result of the total nature of war.[6]

USE OF ATOMIC WEAPONS

The total nature of war is a startling and frightening concept. But the ante was upped just before the war ended. Humanity discovered an entirely new way to be inhumane to itself—the atomic bomb. The atomic bomb represents a whole new way to attack the enemy—mass destruction of property and people utilizing a single bomb. The atomic bomb had an additional impact in that it made defense essentially obsolete. The only defense became retaliation. This had the further effect of making the old war machine essentially fickle. A disquieting result of all this is that the smallest country can intimidate the biggest country if it has the "big one!" Yet this bomb was supposed to be a peace-keeping force. As Hans Bethe said, "If two opponents armed with hand grenades face each other in a six-by-nine-foot cellar room, how great is the temptation to throw first?"[7]

Unquestionably, World War II still affects us today. To simplify analysis, we will attempt to discuss aspects of society and technology in discrete parts. Naturally, all of the elements are actually interactive, and nothing affects one aspect while leaving the others untouched. The specific areas we will deal with are

1. **Economics**
2. Ecological–geographical
3. Sociopolitical
4. Discoveries
5. Ethics and morality
6. Psychological

ECONOMICS

Economics can be viewed in two ways: the immediate effects during the war and the long-range effects after the war. The war pulled our nation out of the deepest depression in its history and put the postwar economy in a boom period that lasted for the next twenty years. The main change in the economy during the war came in the form of the billions of dollars in government contracts given to the corporations of the United States. This, in turn, gave our nation the possibility for the staggering production that it demonstrated so well to the world. Another change during the war was the increased employment of women and the increased use of assembly-line techniques. The economic necessities of war benefited not only women, but black Americans as well. By 1943 the number of skilled black workers had doubled, and the number of semiskilled workers rose even more steeply. Nearly two-thirds of the one million blacks who took war jobs

[6]Burke and Eakin, *Technology,* p. 310.

[7]Karl Jaspers, *The Future of Mankind* (Chicago: University of Chicago Press, 1961), p. 59.

were women. "More women, especially married ones, worked for wages. The rate of the female work force participation rose 24 percent, peaking in 1944 with 19,370,000 working women. Women took jobs customarily allocated to men. They worked at steel mills, shipyards, airplane factories, and railroads."[8] The heavy industry jobs paid much more than the service jobs usually allocated to women. The war aided all aspects of business, but the large corporations received the largest benefit. The administrators of the federal war production efforts were usually the heads of large corporations. These administrators usually gave their own companies the government contracts whenever possible. Thirty-three corporations received one-half of $75 billion in contracts between June 1940 and September 1944. The war centered the power of the corporate administrations, and as a result corporate profits rose from $6.4 billion in 1940 to $10.8 billion in 1944. The war's boom was so intense that even the most severe strike wave in the history of the United States couldn't stop it. The strikes of 1945 to 1946 involved 4.5 million workers in 6,000 separate strikes. During this period corporations won the right to pass on wage concessions to the consumer.

In Europe, the economy was on the verge of total collapse. Despite U.S. aid to Western Europe and Japan totaling $14 billion beginning in 1945, by 1947 Europe was economically on its knees. The loans were given mainly to Great Britain and France. Great Britain received $3.7 billion. France received $1.4 billion. These amounts were merely a drop in the bucket; neither France nor Great Britain were even close to their prewar production levels. To make matters worse, the most severe winter in many years hit Europe in 1946 to 1947. Production fell 50 percent in Great Britain alone. The United States realized that something had to be done, so the Marshall Plan was born. This plan called for long-range economic aid to be given to the war-ravaged countries of Europe. The plan called for the continent of Europe to be treated as a whole, not as individual countries. The Soviet Union objected to the Marshall Plan for that reason. This disagreement contributed to the cold war between the United States and the USSR for the next twenty years. If any one thing was affected by the war, it was the economic condition of the world, now and then. Even today and for many years to come, we find our economic situation stemming from decisions initiated as a result of World War II, such as the Marshall Plan.[9]

ECOLOGICAL–GEOGRAPHICAL

World War II changed the face of the earth. Also, anyone who bought a map prior to the war tried valiantly to get a refund, until it was discovered that it was a collector's item. Germany was divided into four sectors immediately following the war. This division has been a source of tension ever since it was made. In general, the shape of many European countries changed. Beyond the way the land was represented on maps, there were changes in the land itself. Europe, Japan, and Russia were all wartorn. When they rebuilt, they neglected to replace farmland. Throughout the world, there was a marked decline in

[8]Melvyn Dubofsky and Athan Theoharis, *Imperial Democracy: The United States Since 1945* (Englewood Cliffs, NJ: Prentice Hall, 1983), pp. 5–6.

[9]Dubofsky and Theoharis, *Imperial Democracy,* p. 22.

the amount of farmland, and a still greater decline in the number of farmers. Both of these conditions stemmed from the increased role of factories and automation, and the resultant decline in human labor. Finally, humanity created for itself another environmental concern—that of radioactivity. By the end of the war, people around the world began to realize that as destructive as World War II had been, the next war could be the last.

SOCIOPOLITICAL

The startling worldwide realization that a single country could conceivably destroy the world led to a greater interdependence of nations. That is, the Monroe Doctrine was declared null and void. Isolationism would no longer work in a constantly shrinking world. Following World War II, the United Nations, NATO, and the Warsaw Pact were chartered. Reluctance to join was based on the fact that these would be merely entangling alliances, not unlike those that were the root cause of World War I. Yet the peoples of the world, as a whole, needed some symbol of unification following the traumatic war; the UN represented a compromise between the two factions. As one would well expect, the tragic war led to an aversion to war; this unwillingness for foreign involvement at least partially accounted for the "fall of China." This led almost directly to the infamous "Red Scare." After the brutal leadership of the Axis powers in World War II, it was easy for the American public to group all aggression together as a "plot." Certainly, the space race has its roots well entrenched in World War II. The Red Scare and technology gained from the war provided both the motivation and the means for space exploration. Finally, the horrors of war, and especially the Holocaust, led to a new emphasis on the fundamental rights of all humans. A part of the United Nations' charter discusses the issue of human rights, and that all humans have certain inalienable rights. This is where the civil rights and women's movements have their roots as well.

DISCOVERIES

World War II put the world in a mad scramble for technology. The echoes of this scramble are still being heard today. During the war, there were many new discoveries. They included jet engines, rockets, synthetic fuel, plasma, penicillin, advanced pesticides, and nuclear power.

The Germans were losing the war, but they did not quit. They worked furiously at coming up with some new device that would save them from the advancing Allies. One of the things that they developed was the jet engine. Obviously, the jet has changed our world. The jet has made global travel convenient and economical. It has made war machines that can fly three to four times the speed of sound; without the jet our world would be totally different. This is not to say that without World War II the jet would not have been developed; the war merely sped up the technological research required for that development.

The rocket was developed almost exactly in the same fashion as the jet. Hitler's scientists designed the rocket as a vengeance weapon. The V-1 "buzz bomb" and the V-2 rockets were the weapons that would periodically obliterate one or possibly two city blocks in London. The rockets were not mass produced, so the Germans could not really exploit the power of this technology. The rocket technology proved invaluable in the years following the war. President John F. Kennedy employed Hitler's chief rocket scientist in the beginning of the space race. The roots of our space program and the roots of our nuclear missiles are in the early German rocket research conducted by Wernher von Braun and his colleagues.

The war saw many shortages, but the most drastic shortage as far as the military was concerned was a shortage of fuel for internal combustion engines. In fact, the Germans lost the Battle of the Bulge as a result of a severe fuel shortage. Scientists on both sides were working on the development of a synthetic fuel substitute. Toward the end of the war, one was discovered, but the problem was that the cost proved to be too high for practicality. This synthetic fuel never saw use, yet in the future, when the world will have used its supplies of crude oil, the synthetic fuel research of this period can be used. Today research is in progress to improve on the fuel formulas of this era.

Any war is going to have casualties. World War II was no different. The doctors were trying to find better ways to deal with the wounded. There were several medical developments during the war. The top two discoveries were the use of blood plasma and penicillin. Both were great medical innovations, possibly the greatest that we have seen in this century. The medical breakthroughs could have come at any rate, but undoubtedly the war sped up the research.

Part of the research toward the war effort was in the area of generally useful chemicals for many applications, military and otherwise. There were many chemicals developed, but the most useful from this area were in the field of pesticides. Some of the pesticides developed were later proven to be environmentally unsafe, yet the research proved to be useful in many areas, such as in the development of plastics and rubber.

The greatest development of the war, or the most infamous, depending on one's perspective, was atomic power. The immediate effects are well known—the annihilation of thousands of Japanese people. The long-range effects are not yet fully understood as the controversy over atomic and nuclear energy rages on. The only thing that can or should be said is that the issue affects every single human being today and in the future.

None of the developments mentioned above was due singularly to the war, but all of them were altered by the war. During the war, the development of all of them was sped up, and the research on them was intensified.

ETHICS AND MORALITY

World War II raised some serious questions in the minds of many people. In this portion of the paper we deal with the ethical and moral impact of the war. The implications in this area are literally infinite in number, as each person has his or her own questions or considerations concerning the impact of the war. Some of the questions that we feel are applicable to this paper deal with the technological implications in the realm of ethics and morality.

Where are science and technology leading us? This is a very important consideration. Do we know where science is leading us? After the war, many people came to the realization that the technology around them was virtually in control of their lives. Humankind and its machines were inseparable. Never before in all of the history of the human race had there been a time like that following World War II. Suddenly people discovered that the whole globe could be destroyed in a matter of days. This was, to say the least, a difficult concept to grasp. The "explosion" of technological research during the war demonstrated to the world's peoples that technology was growing at a rate much faster than their ability to comprehend it. The conclusion that many people reached was that they were not in control of their own existence. This was one way of looking at the concept, but more than likely the problem was that, for the first time, humankind was totally in control of its destiny.

The human race was startled to realize that technology, its creation, was leading the world, not the other way around. Of course, this is only one perspective; the other would say that we are the masters of technology. The latter is much easier to deal with in the rational human mind. We made it, therefore we can control it. Since the first concept is more difficult, we will deal with it here. Do we want to go where technology is leading us? Many people wonder where the world is headed and if the scientists and engineers are suited to determine the world's destiny. There has always been a resistance to change, but the changes in the period of and immediately following the war were of a magnitude never before experienced. There are no simple answers. There are no answers at all in the usual sense! These issues are simply things that the person of our time must consider.

If humanity was to decide that it needed or wanted to control technology, could it? For the Allied powers to win World War II, they pushed for massive scientific research. They did not question its worth to society in general. The only thing that they cared about was, Would a particular research end the war a day sooner? The lasting effects of the technology were not considered; it was simply a necessity. This "open season" on research led the world to a point of no return. The more technology human beings have, the more they crave. This has become a vicious circle. Should or shouldn't we try to halt technology is not the question; the question is can we? We could or would not voluntarily go back to a more primitive way of life. The right or wrong of the issue is irrelevant. The fact remains that human beings are destined to continue in their technological development.

One of the more ordinary questions resulting from the war was that of human rights. This idea is not new. People were forced to look at the way they thought about other people and how other people viewed them. People began to wonder if their race could be the Jew of the next Nazi Germany. Thus the idea of human rights was renewed. Today we believe that people—all people—have certain rights that are not uncommon to any race or nation. The interest in the concept of human rights sparked the beginnings of the modern civil rights movement in the United States and has altered the entire world's thought patterns.

PSYCHOLOGICAL

It comes as no surprise, then, that as human beings have changed their thought patterns about the world in which they live, they have suffered psychological repercussions. The realization that the world could be destroyed any time has led to the idea of instant gratification. The idea that since the world can go at any time, the thought, Why shouldn't I do just what I please? has led to the moral decline of the population. Typically, all people follow a natural progression. As children, we seek instant gratification. As we gain maturity, we begin to see that we cannot always have what we want; sometimes we have to wait for it. But the shadow of the "big one" has made us less willing to wait. As learning patience is a major part of maturity, the society's development has been interrupted by the harsh realities of the world around us. Many people cannot handle the lack of control that they feel around them. They seek escape in drugs, alcohol, or any other thing to suppress their fear of what they cannot control. Others react differently. Feeling that they have no control, they assume the attitude that they do not care, that they are not concerned with what happens to them or with the world. Or they believe that their input is so valueless that there is no point in giving it. Voting records indicate that since World War II, there has been a marked increase in the "undecided" or "what difference does it make?" response to voter surveys. Yet another reaction, and probably a more healthy one, is the realization that what happens on one end of the globe can affect people on the other end. This makes the population hungry for information as to what is happening. They have a fear of the unknown. They demand more information, faster and more accurate. They want to see it from the source. This has led to the explosion of what is commonly referred to as the "Information Age." Naturally, there are other aspects influencing the processing and distribution of information, but its roots are at least partially in the postwar reaction to the war with Germany.

It would be understating the issue to merely contend that World War II was a dominant force in every person's life forty years ago. We believe a more accurate statement is that World War II could be thought of as a motivating factor driving the general population to a period of increased productivity and innovation. It could well be contended that the impetus was a little overzealous, but this point must be considered by the readers. It is our purpose merely to serve as messenger, bringing up the issues for each person to consider on his or her own. For, in the end, each person will form his or her own evaluation of the implications of World War II, and the way it has affected that individual's life.

BIBLIOGRAPHY

Bauer, Yehuda, *The Holocaust in Historical Perspective.* Seattle, WA: University of Washington Press, 1978.

Burke, John G., and Eakin, Marshal C., eds., *Technology and Change.* San Francisco: Boyd and Fraser, 1979.

Collier, Richard, *The War in the Desert.* New York: Time-Life, 1977.

Dubofsky, Melvyn, and Theoharis, Athan, *Imperial Democracy: The United States Since 1945.* Englewood Cliffs, NJ: Prentice Hall, 1983.

Hoehling, Allan, *Home Front U.S.A.* New York: Thomas Y. Crowell, 1966.

Jablonski, Edward, *A Pictorial History of the World War II Years.* Garden City, NY: Doubleday, 1977.

Jaspers, Karl, *The Future of Mankind.* Chicago: University of Chicago Press, 1961.

Solar energy is considered a new and innovative technology, yet it has been around since the first time someone used the sun to dry fruits or dry clothing. It is in effect our ultimate source of energy, since all other forms of energy, with the exception of nuclear, come from this one single source. The difference is that today we strive to tap this energy more directly, through photocells and passive solar heat collection. Think back to Chapter 5, An Idea Whose Time Has Come, and consider the reasons that we may now be paying more attention to the idea of solar energy using modern technology. Is it just that for the first time methods of tapping this source are available or is it a matter of necessity, the importance of solar energy growing as other means of supporting our voracious appetite for energy consumption wane in availability and effectiveness? Perhaps it is simply a matter of environmental wisdom to embrace this very old "new" technology. After all, humanity survived quite nicely for over ten thousand years of technological activity with only fire and the sun as energy sources. Are our present efforts to harness the power of the sun to create electricity really new technology or just a new way to use a very old source of power?

Solar Energy and Its Social Consequences

KEVIN BAGWELL AND WAYNE ERGLE*

ABSTRACT

Solar energy is and has always been an abundant source of alternative energy. Furthermore, the sun will continue to shine whether we choose to take advantage of the energy or not. Several astounding innovations have already taken place in the solar energy industry. These include the development of a solar-powered automobile, a solar-powered generating plant, solar heating for the house, and solar energy in space.

The social and economic issues of solar energy are also quite important. The use of solar energy would result in a cleaner and safer earth. Unfortunately, cost is still a major setback in the field of solar energy. In order to maintain the rate of advancement in the solar industry, the government and society must choose to provide the researchers with the funds needed to develop new technology that will cut the costs of solar energy. Solar energy will be a primary alternative source of power in the future. The natural resources of the earth are being consumed at an astounding rate. It is up to the present society to

*Kevin Bagwell and Wayne Ergle, "Solar Energy and Its Social Consequences" (unpublished paper presented at De Vry Institute of Technology, Atlanta, Georgia, May 2, 1984).

give the green light to further the research in the solar energy field before all of the natural resources are depleted.

INTRODUCTION

This paper contains a discussion on solar energy and its social consequences on contemporary society. As is the case with all alternative sources of energy, scientists have just begun to realize the true potential of using the sun as an efficient source of power.

In order to properly discuss these social consequences, several factors are introduced throughout this paper. Some of these factors include:

1. The technical innovations developed in the solar energy industry within recent years—These innovations will directly affect the way the society comes to depend on solar energy both today and in the future.
2. The social issues of solar energy based upon contemporary society—These issues will include the government's views on solar energy, the public's reaction to those views on solar energy, and the advantages and disadvantages of a solar-reliant society.
3. The economic impact of using solar energy for an alternative source of energy—This section will include a discussion on the effects of solar energy on the physical environment, the rate of development of a society based on solar energy, and how solar energy will provide for the advancement of economic structures.

DEFINITION OF SOLAR ENERGY

Solar energy has always been a readily abundant source of alternative energy. The concept of solar energy applies to every life-sustaining organism known to humankind. However, what is solar energy? Is it the visible light spectrum seen by the human eye? Or is it the invisible radiation resulting from the solar activity of the sun? By deriving a general definition of both solar and energy, the term *solar energy* can be thought of as the "natural energy vigorously exerted by the action of the sun's light and/or heat."[1]

THESIS STATEMENT

Life has depended on solar energy since the creation of the earth. The primitive human beings possibly discovered the use of solar energy by laying on a sunlit rock for warmth. Or perhaps they found that their clothes would dry much faster when exposed to the sunlight. The ancient civilizations increased the use of solar energy by using the sun to bake the clay bricks used for building materials. However, to write a paper discussing all of the aspects of solar energy since the beginning of time would be a monumental task. Therefore, this paper will be restricted to the issues of solar energy in contemporary society. This will allow the discussion to focus on how solar energy technology has affected

[1]David B. Guralnik, ed., *Webster's New Collegiate Dictionary* (Cleveland, OH: Collins, 1975).

the present generation. Through the technical innovations of solar energy within recent years, contemporary society has become more conscious of the reality of solar energy as a primary source of alternative energy.

TECHNICAL INNOVATIONS

Some of the more useful innovations of the solar energy industry are geared toward contemporary society. These innovations include solar heating, solar-powered airplanes and automobiles, solar-powered generating plants, and solar technology for space equipment. The following paragraphs discuss these innovations in detail and describe the feasibility of each innovation from a societal viewpoint.

Solar heating actually consists of two types, active and passive. An *active* solar heating system actually uses solar collectors to heat a substance (usually water or air) that is circulated through the house via heat registers. When using an active system, a storage area is needed to store the unused heat. Usually the storage area can retain the heat for three to five days of cloudy weather. The active system operates by using solar collectors (commonly located on the roof or adjacent ground) to focus the sunlight on a system of pipes contained inside the collector. The pipes contain the heating substance, which is then transferred into the storage area until heat is needed. When the need for heat arises, the substance is pumped out of the area and circulated through the house in a manner similar to a normal HVAC system. The substance is then pumped back through the solar collector and transferred to the storage area once again, until heat is needed.

One of the more obvious setbacks to the active system is that if cloudy conditions exceed more than five days, a backup source of energy (such as propane or natural gas) is necessary to provide the desired heat. Another setback is the initial cost of an active system. A totally dependent solar house can be extremely expensive to build. Fortunately, federal and local governments of many states offer grants or tax credits to offset the initial investment in the system.

The *passive* solar heating system retains the same basic concepts of the active system without the need for expensive equipment. Basically, the passive system uses large windows located on the east and west exteriors of the house to allow the sun to warm walls and floors constructed of masonry or slate. This process provides heat for both daytime and nighttime hours. At night, insulated shutters are closed over the windows allowing the radiant heat contained within the walls to be released, thereby heating the rooms. This concept is very practical in cold weather areas that experience a great deal of sunshine.

Most people think of a solar house as a house consisting solely of apparatus used by the sun. However, in most cases, solar heating should be combined with other equipment to provide additional efficiency and savings. Heat pumps, wood burning stoves, and well-insulated walls and floors all help to provide a truly energy-efficient home.

Israel is one of many countries to experiment with the development of a solar-powered automobile. The engineering department of Tel Aviv University has developed a solar-powered car named the *Citicar.* The Citicar was designed by a group of students under the direction of Professor Arye Braunstein.[2] This vehicle is capable of traveling at speeds of up

[2]"Israel's Solar-Powered Car," *Mother Earth News,* vol. 65 (September–October 1980), p. 120.

to forty miles per hour and has a maximum range of fifty miles for each charge. The car operates on a two-step DC system (24/48 volt) utilizing two solar panels mounted on the roof and hood of the car. These panels consist of 342 solar cells and provide a peak power of 400 watts and a charge to the battery of 48 volts. The entire weight of the car including batteries and solar panels is approximately 1,320 pounds. Improvements are planned to provide an additional ten miles per hour to the top speed and to double the effective range of the car per charge. Since Israel is located in one of the sunniest regions of the world, the Citicar should prove to be beneficial to the society. Unfortunately, no estimated cost for this vehicle was discovered during the research of this topic.

In 1981, the Solar Challenger (a solar-powered airplane) successfully completed a five-hour and twenty-three-minute flight over the English Channel. This was the first time a solar-powered plane had completed a flight using only the sun as its source of energy. The Solar Challenger was designed and constructed by Paul MacCready of Pasadena, California.[3] The Challenger generated its power from 16,128 solar cells located on its wings and horizontal stabilizer. These cells powered two electric motors connected to the propellers of the plane. Using tough, lightweight plastics (Kevlar, Mylar, and Lucite), the Challenger weighs only 217 pounds unoccupied. Unfortunately, the cost of the flight exceeded $725,000; flying by solar power commercially could be a long time off into the future.

Solar One, located near Barstow, California, is one of the first ten-megawatt solar-powered generating plants in the world. The plant consists of "over 1,800 heliostats, sprawled in a fan-shaped array covering 70 acres, focused on the central receiver, a 45-foot-tall metal cylinder atop a 300-foot tower."[4] The plant was designed and built by the McDonald Douglas Corporation and is operated by the California Edison Company under a government contract. The cost of the plant totaled over $141 million. Solar One doesn't turn solar energy into electricity. Instead, the plant uses solar thermal technology, which uses heliostats (concentrated mirrors) to focus the sun onto the central receiver containing a vapor. This vapor is heated to 700°F, which is fed through turbine generators. These generators then produce the actual electricity that is sent to the power grids. This system is now undergoing a five-year test program. However, with the favorable results that have already been recorded, California Edison is planning to construct a larger 100-megawatt plant to be in operation by 1988. Like any new innovation, there have been minor setbacks. Some of these setbacks include nonfunctional heliostats, unexpected thermal gradients, and buckling of the panels on the central receiver. A storage area (consisting of a bed of rocks or high-temperature oil) can retain heat for a period of four hours after the system goes down to continue the production of power. An automatic control system is being designed so that the entire plant can be operated with a crew of four employees. With time, engineers should solve the problems to date, making the idea of a solar-powered generating plant a costworthy investment.

Using solar energy for space vehicles is one of the most efficient concepts available. Most of today's satellites contain solar cells or panels that provide the energy needed to

[3]"Icarus Would Have Loved It," *Time Magazine,* vol. 118 (July 20, 1982), p. 45.

[4]"Solar One: Sun, to Heat to Electricity," *Popular Science,* vol. 221 (October 1982), p. 114.

operate the onboard equipment. Both the original Skylab and the present space shuttle contain solar panels that generate the power used to operate the craft. With the successful repair of the satellite on a recent shuttle mission, it is now apparent that solar energy has become even more feasible for the satellites of the future. Now for a fraction of the original cost, companies will be able to recover their nonoperational satellites and implement repairs on these satellites in space, saving the companies millions of dollars.

Scientists are now working on an idea known as the solar power satellite (SPS). This concept is a satellite that would contain either a photovoltaic array or a thermal power plant. The satellite would transmit the electrical power generated to antenna receivers located on the earth. These receivers would then transfer the power to power grids, which would distribute the power to consumers. There are a number of positive arguments for such a system. The system would eliminate the hydrocarbon combustible products, radioactivity, and second-layer thermal pollution that is now occurring on the earth. Since the satellite would be in space, it would have an unobstructed sun and would also be immune to factors on the earth such as earthquakes, severe weather, and so forth. However, for every positive argument there appear to be two negative arguments. The earthbound antenna receivers will have to be massive in size. The system will theoretically introduce water vapor and microwaves into the upper atmosphere. The largest argument against the system is the astronomical cost. A prototype satellite is estimated to cost $1.3 trillion under the most favorable conditions. Perhaps in the near future scientists will find a way to reduce the cost of the system. The general attitude of the government at this time is that the system is just too costly for such a gamble.

SOCIAL ISSUES

There are numerous issues concerning the society that arise when discussing solar energy. Perhaps the primary issue is, What are the social consequences of accepting solar energy as a source of power? By placing the cost of solar energy aside for a moment, we can observe several beneficial aspects of solar energy technology. First of all, solar energy is available to every geographic region of the world. It does not require any special resource found only in a particular region. Naturally, some regions of the world are sunnier than others. However, with the implementation of the solar power satellite system, the actual amount of sunshine that a country receives becomes irrelevant. There is a great possibility that solar energy will allow a country to become self-sufficient in energy resources. If this becomes true, the resource-poor countries will not have to rely on the resource-rich countries to provide them with power at an exorbitant price. This concept can also be related to the individual level. It is not theoretically impossible that an individual will own and operate his or her own power plant based on personal needs. This concept has already become evident in solar heating. This would also allow the individual to become more self-sufficient as far as energy needs are concerned.

The U.S. government's views toward solar energy were at one time favorable. In the United States, there are still grants and tax credits available in some states that encourage the implementation of solar equipment. A solar bank was proposed by the Jimmy Carter administration providing subsidies and the lowering of interest rates for solar energy construction.

On the proclaimed "Sun Day" in 1978, Carter called for the "dawning of the second solar age."[5] However, in the Reagan administration, officials began to cut the Energy Department's budget for solar energy research to half its former level. In 1982, they unsuccessfully tried to eliminate the tax credit available for investing in the solar energy industry. Fortunately, even with the resistance by the Reagan administration, homeowners and investors still feel that the solar energy industry is worthy of their investments.

ECONOMIC IMPACT

Solar energy is one of the cleanest and most abundant sources of energy available to humankind. The sun will continue to shine regardless of whether we choose to take advantage of the energy or not. One of the few drawbacks of using solar energy is that, theoretically, the amount of water vapor and microwaves will increase within the upper atmosphere. However, the advantages of solar energy far outnumber the disadvantages. First of all, solar energy is clean. There are no byproducts or hazardous wastes encountered when using the sun. Perhaps best of all, the physical environment of the land is not harmed in any way. The most fundamental change in the ecological base would be that a cleaner and safer countryside would result. Without the combustion engines that pollute the atmosphere, much of the air pollution would cease to exist. Since the demand for natural resources such as petroleum and coal would decrease, the physical environment would not continue to suffer from strip mining or oil spills that occur today.

It is extremely likely that solar energy would provide countries with a new set of goods and services. Designing, developing, and maintaining solar equipment would come into demand. Products such as solar energy equipment for the house would become popular. However, other industries would suffer. Any industry in the field of producing internal combustion products would find the demand for its services severely decreased. Many other industries would have to adapt to the new energy source. Automobile manufacturers, power companies, and the oil industry would have to change their products to supply the new needs of the society. The oil industry has already begun to prepare for the future. Many of the oil companies have invested in various areas of the solar technology industry.

Unfortunately, these economic impacts are strictly theoretical in the present society. It seems probable that countries will continue to deplete and deface the natural environment until all of the natural sources of energy are exhausted. Only then, it appears, will countries seriously consider the alternative of solar energy. Let us hope that the natural environment can withstand the abuse until this decision is finally made. Until new technology finds a way to dramatically decrease the cost of using solar energy, a totally solar-reliant society is far from reality.

PREDICTIONS FOR THE FUTURE

The future of the solar energy industry can only be based on personal opinion due to the unknown future advances of technology. If the government decides that it is a worthy investment to support solar energy technology, then the expansion of the field is unlimited. On the other hand, if the government continues to cut back on the funds provided

[5]"A Possibility, Not a Novelty," *Time Magazine,* vol. 114 (July 2, 1979), p. 27.

for solar energy research, the field will probably experience limited advancement until all other sources of energy are near exhaustion. Therefore, it is up to the government and the society as a whole to determine if solar energy research is technically feasible, cost efficient, and worthy of future research. In a strictly hypothetical situation, based on the supposition that advanced technology and cost efficiency have been discovered, the following ideas are not entirely impossible.

It is possible that the solar power satellite system could provide the earth with enough solar power to make petroleum-powered generating plants obsolete. However, these plants could be converted to accept the thermal power of the SPS system, since the concepts are basically the same, allowing the plants to continue operation. I believe that the majority of housing will become solar reliant. In Shenandoah, Georgia, entire subdivisions are being constructed using both active and passive solar energy systems. The recreation center in Shenandoah uses the sun to completely operate its HVAC system. This also includes the freezing of the ice for the skating rink. It is possible that entire communities will own and operate their own solar power stations. To carry the concept further, individual homeowners who live in isolated parts of the region could implement an antenna system (similar to a communications system) that would allow the homeowner to receive electrical power directly from the satellite.

There appears to be little doubt that many of the items used by consumers will use solar energy for power. There are already numerous items on the market, including pocket calculators and small novelty items, that do so. It does not seem impractical to expand this idea to larger household items such as the lawnmower or any other motorized machinery. Solar-powered automobiles will become more evident in the future. These vehicles will provide the society with a safer, more efficient, and cleaner form of transportation.

CONCLUSION

Solar energy is an alternative source of energy that seriously needs to be considered. There will come a time in the not-so-distant future when the physical resources of the earth will either be too expensive to recover or completely exhausted. Solar power allows for a cleaner environment and reduces the chances of catastrophe that might occur with other sources of energy such as nuclear energy. However, research in the solar energy field will be limited unless the U.S. government decides to give its support to the idea that solar energy is a solution to the problem. With the technological advances that occur every day, there appears to be little doubt that research in the solar energy field will dramatically lower the costs of solar equipment. This paper has discussed some of the technical innovations resulting from the extensive research in the solar energy industry. The automobile, solar heating, and space technology are all actually in use today. With these results already established, one can only imagine what the future holds in the industry.

The social consequences of solar energy are primarily beneficial to the society. The use of solar energy would increase and strengthen an individual's independence. Implementation of solar products would decrease the costs of the power and provide a cleaner and more efficient way to operate the home. Quite possibly the best result of solar energy will be a cleaner and healthier Earth.

Therefore, the society must make a relatively straightforward decision: Is it worth the extra research expense in the field of solar energy to increase the chances of a cleaner environment? Or is it more beneficial to the society to continue the exhausting consumption of natural resources, contend with the increasing amount of pollution, and hope that the resources will not be exhausted until after we have left the earth and let the future generations contend with the problems we have bequeathed to them?

BIBLIOGRAPHY

Guralnik, David B., ed., *Webster's New Collegiate Dictionary.* Cleveland, OH: Collins, 1975.

Hayes, Denis, "Environmental Benefits of a Solar World," *Vital Speeches of the Day,* vol. 46 (March 1, 1980), pp. 306–10.

"Icarus Would Have Loved It," *Time Magazine,* vol. 118 (July 20, 1982), p. 45.

"Israel's Solar-Powered Car," *Mother Earth News,* vol. 65 (September–October 1980), p. 120.

Lee, Al, "Passive/Active Solar," *Popular Science,* vol. 219 (December 1981), p. 123.

"No Profit in Politics," *Environment,* vol. 24 (March 1982), p. 25.

"Oil Industry Buys a Place in the Sun," *Science Digest,* vol. 221 (August 26, 1983), p. 839.

"A Possibility, Not a Novelty," *Time Magazine,* vol. 114 (July 2, 1979), p. 27.

"Solar One: Sun, to Heat to Electricity," *Popular Science,* vol. 221 (October 1982), p. 114.

"Solar Plane Soars," *Popular Mechanics,* vol. 155 (April 1981), p. 127.

"Solar Power Satellite: A Plea for Rationality," *Science,* vol. 203 (February 23, 1979), p. 709.

"Tax Breaks Help Fuel Solar Power," *Business Week* (April 26, 1982), p. 36.

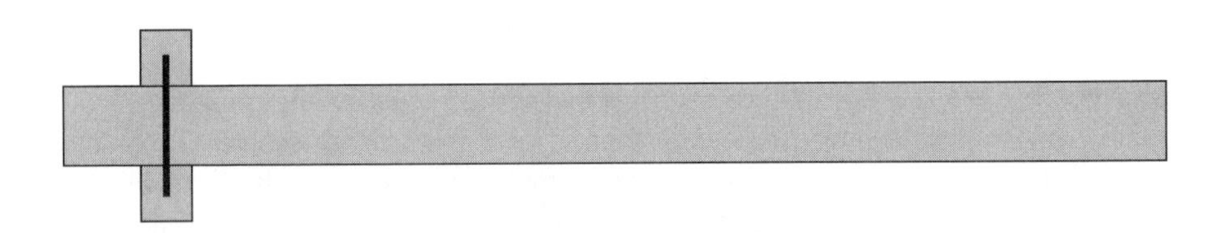

Dear Parents, Acquaintances, and Relatives

EXTRACTS FROM THE DIARY OF AUGUST KETTERER*

The following extracts are from the diary of August Ketterer, a German immigrant who came to the United States in 1848. He traveled from Freiburg in Germany to the United States by rail and then ship, promising to write home and inform those left behind of his experiences and discoveries in the New World. The diary contains a wealth of detail and example of life in the mid-nineteenth–century United States, particularly as regards the

*The diary extracts were translated by Donald D. Hook and are reprinted from the original unpublished document, privately commissioned in 1948.

technology, as August was a miller by trade and therefore intensely interested in the mechanical devices that he saw. He was also interested in architecture and other facets of American industrial life. His mechanical interests seem to be inherited—his ancestor was the inventor of the cuckoo clock.

Of particular note to the reader in the present context are the similarities and differences among the problems that we face today and those of a century and a half ago. Although the different eras present different challenges, it is obvious that people today still face the same basic issues.

UPON ARRIVING AT NEW YORK AFTER THE VOYAGE FROM EUROPE

I am not capable of describing the beauty and splendor of this harbor and the region roundabout in nice weather. This panorama is just too beautiful. On the one side, Brooklyn, Williamsburg, New Jersey; on the other side New York with Hoboken, the favorite place of New Yorkers. There are gardens and parks, then the Castle-Garden with other fortifications in the vicinity of New York. In the background, the green, forest-covered mountains with decorative castle-like houses and farms of the rich Americans and in front of us the forests of masts of the ships in the three cities with countless small boats, and the peal of bells. All this makes a remarkable impression on a person and immediately he feels God's omnipotence. . . . Several stopped on the dock by the ship and the others looked around for places to eat. Still many others went quite soberly to the railroad stations and steamboats and offices and bureaus, in order immediately to travel into the interior of the country. That I also had in mind. However, I ate lunch and when I returned to the docks, it was too late to travel further and I was forced to stay here all day. After I had bought a suitcase, I took the others to the hotel. Thursday, a friend resolved to go with me and his brother to Ohio along the Erie and bargained as I did in Hare to Sandusky.

New York is, so far as the main streets are concerned, as beautiful as Paris. Only in the suburbs and outlying parts stands a muddy little peasant town. The German quarter distinguishes itself greatly, resembling for all intents and purposes as late middle ages by its industry. It's very beautiful here—the horse-drawn railroad is particularly fascinating. For five cents one can ride through the whole city, even for seven miles around the city. However, from one house to the next, it costs the same as the entire city tour. Along the streets are, as in other American cities, beautiful trees planted between the sidewalk and street itself, that blossom out beautifully in spring and summer. I visited a theater here for the second time in my life. In America, theaters are very small and of little interest to me. . . .

Buffalo is one of the most beautifully situated cities on the Erie and very healthy. From here it's only about eighteen miles to Niagara Falls, which I was not able to visit. While we were in Buffalo, two sailing ships burned up in the harbor and from midnight to the next morning we could see the divers with diving bells going along on the bottom of the sea to save what they could. If this city is very pretty on the sea side, so are its suburbs. Like the other old American cities, it had little one story houses which indeed, since they are painted white or white washed, look very pretty from the distance with the green trees and little churches, very much resembling large villages built on to each other. In these little towns it is very unusual to see a brick building at all. The style of architecture also differs. Stone houses are not smeared with chalk as they are in Germany, but exhibit the raw, red brick. The art of the masons of this country consists of leaving it natural. To illustrate the speed with which houses are built there, one can go past a deserted, uncultivated place and come back eight weeks later to find a beautiful house of brick where there was nothing before. Brick and wood houses are intermixed and the plans of the houses are quite simple due to the regular construction of the city. One seldom sees a very large private house. I also experienced something here that is seldom seen in large Eastern American cities, that being the sight of cows, pigs, and geese running around everywhere loose. Some people here have fat swine or a good milk-cow running around in the absence of good fodder, much less a stall. As the American proverb states, the animals here must help themselves and eat what they find. The cows here will often nourish themselves with horse manure, and though there are undoubtedly street cleaning laws here, they are not observed. There is, therefore, filth in over flowing quantities and this is true everywhere in the cities. Yet this is not true of the sidewalks, which are kept clean by each separate house owner.

We took a steamer to Cleveland. These steamers are three or four times the size of Rhein boats which look like simple skiffs compared to these. They are all old and built of wood. I do not recall seeing a single iron one either here or in any other city. On the second day of the trip, we ran aground and were able to free ourselves from the sandbar only after some four hours of effort. We then proceeded to Cleveland, which had originally been two cities but is now united into one. One of these original cities, Ohio City, is still called by that name. After two days, we travelled to Pittsburgh, though we had been offered employment by the railroad in Cleveland. We were determined to travel on.

Travelling by foot and covering some twenty-four miles on the first day, we spent the night in an inn, then boarded the Railroad express to Rochester, some twenty-eight miles from Pittsburgh. Everything was still forest along by the railroad, at least in the northern part of Ohio. Farther south, one comes through prettier regions. Moreover, we took quite a long way, since it was much better going on the highway and more thickly populated. These lead over hill and dale and are not at all difficult. I would not really recommend anyone to go by rail. Not only pedestrians but also riders and wagons travel on the rail line. The single warning that is offered is the ringing of a bell when dangerous areas are approached, or the engineer sees something. No railway house is to be seen, no one to watch the tracks, and even the railroad station is the most ridiculously built thing in the world. It is almost lying on its side and then the train goes roaring by with enormous speed. If it were not for the cow-catcher, it is doubtless that more serious tragedies would occur.

When we arrived in Pittsburgh, we found that the city is divided into main parts, those being Allegheny with Manchester and Pittsburgh with Reisville, and Bagardstown and Birmingham with South Pittsburgh and Brownstown. Together, these all form one of the largest industrial cities in America due largely to excellent deposits of coal. pittsburgh [*sic*] itself is situated in the middle of the mouth of the two rivers and is really the center of commerce and trade, and seat of the government of the United States. The county court and jail are also here. The streets of Pittsburgh are straight and beautiful. The most remarkable buildings are the courthouse on a hill, down which the simple but beautiful and tasteful Cathedral with its pontifical buildings is located, and the post office at the foot of the same hill, along with the different banks and the Iron City or trade-society with a high school. These different buildings have cast-iron front walls. In Reisville, there are the two principle hospitals and many brick-kilns and beautiful churches. Moreover, there is a cannon foundry, many other kinds of foundries, nail factories, and iron works here. There is also a copper foundry and hammer works. Equally remarkable are the enormous railway depots, the wire bridges over the Monongahela, the four other bridges over the Allegheny of which over one the railroad passes and over the other the canal passes across the river so that while steamships go below, locomotives and canal boats go above. On the whole, there are no less than six enormous railroad depots in Pittsburgh. Also in the neighborhood, in the little city of Lawrenceville is a kind of fortress by the same name in which the United States soldiers here are stationed.

Allegheny is indeed more thickly populated but not as pretty as Pittsburgh and more the seat of workers and factories. There are four wool and cotton mills, carriage, shovel, and axe factories together with many sawmills. Also, there are nail, nail-keg, flour barrel, and sack factories. Traditionally, we find here the Allegheny water works, the Catholic orphanage under the supervision of the sisters of Mercy, the two Catholic graveyards, the poorhouse, and three large steam mills together with an enormous whiskey brewery in which one hundred feet or nearly twelve thousand gallons of whiskey can be made daily with four stones in order to rough grind the grain.

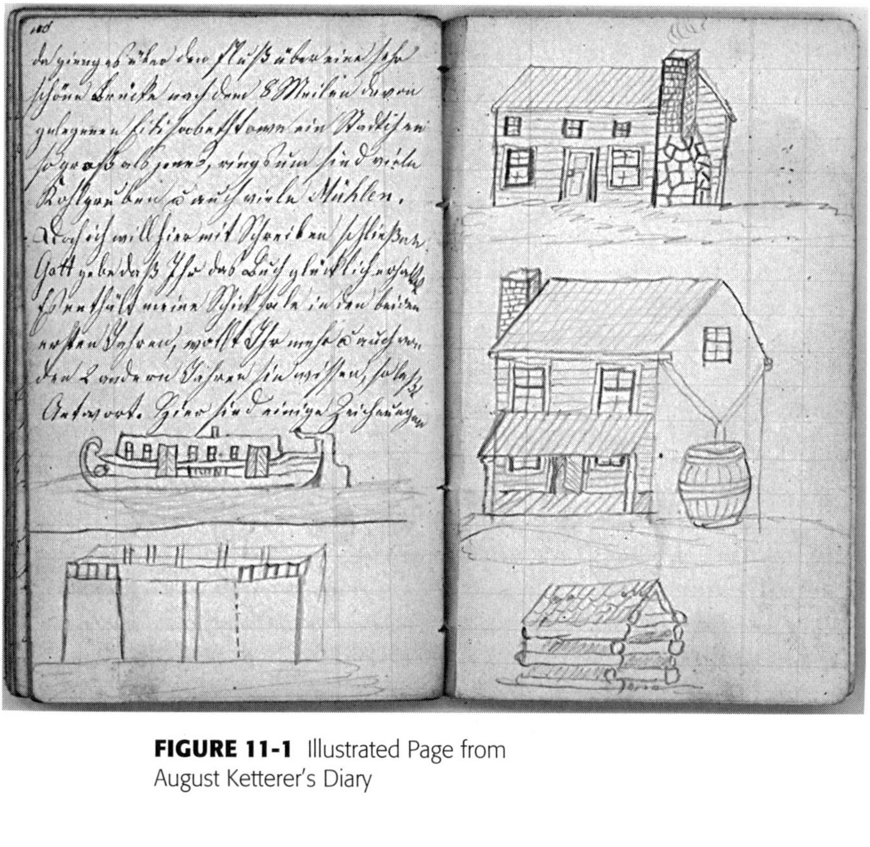

FIGURE 11-1 Illustrated Page from August Ketterer's Diary

Birmingham is situated south of Pittsburgh along the Monongahela on the slope and at the foot of a mountain that runs along by the river. Because of the mountains roundabout, there are seldom floods here. In Birmingham there are glass-houses and iron works mostly. There is a single main street on which are beautiful shops. It is probably four miles long, for Temprenzville and Ceatsburg with Saw Mill flow almost into the city. It is unfortunately difficult to actually see the city. Located here are most of the coal mines which extend to the source of the river. Several flow almost into the city itself. Nowhere is more iron and glass worked and it is no accident that Pittsburgh is called the "Smokey City," since one seldom gets to see the clear sky. . . .

Eight days before New Years I traveled through Connelsville to Pittsburgh once again. It was nearly as cold or colder than the year before. The snow was again three to four feet deep. I went over the Lahrer mountain, a high and well known mountain in Pennsylvania. It was about eight miles to Pittsburgh. As I travelled toward my destination, I asked for work at every mill I passed, which is not customary, but about six miles from Pittsburgh, I was engaged by a large mill at which I presently work. It is located at a place known as Six Mile Fair, or Hope Church. The region around the mill is hilly like all of Pittsburgh, but because of the many coal mines, very thickly inhabited. From here to Pittsburgh, one comes through one village after another each being no more than half a mile distant from the next. These coal mines are operated by rich companies and owned partly by the companies, partly by rich individuals. One of these owners is the father of the miller for whom I now work. This same man owns eight to ten farms of forty to three hundred acres and a brewery in which fifty bushels of rye, which we furnish rough ground, are daily used up. Out of one bushel of rye they make three gallons of whiskey, each gallon selling for fifty cents while the fruit per bushel costs only about forty-five or fifty cents. Moreover they keep three hundred pigs which they feed the scraps and leftovers from the brewing process. They even have thirty oxen and five horses, ten cows and two mules, one of which costs more than two horses since they can work so much harder and longer. The coal mines and farms bring in several thousand dollars a year, but the old man and his youngest son operate only the mill and whiskey brewery. The family is one of the richest around Pittsburgh. The name is Risher and therefore German. They do not, however, speak the language at all. The older son is a merchant and also operates a mine of his own. He even has several farms and on one of these he has a kind of donkey-mule-horse "factory" in which he has bought over one hundred old mares and donkeys to fill it up. . . . Risher's Mill, where I work, can be operated with water or steam or with both at the same time, since in the summer there's sometimes such a little bit of water that not even the sheep have enough and that's certainly not enough to make steam. This is true of all water mills. When I had been in this mill just fourteen days, all the water was gone except two big barrels and we had to dig a trench some two thousand feet long and one and one half feet deep in order to get more water. Some streams which usually swell in the spring and cause floods were so dried up that one can't even get water from deep ditches. With the steam engines (we had two) the mill went at double speed in threefold connecting gears. These gears were pretty much constructed like ours with conical wheels, a main wheel, and several spur wheels. The main rod to which was attached the large wheel went through the top of the mill. The bags and everything hang down from it. We had three cylinder boilers and the mill was over three and under four stories high.

The first cylinder is a double where the bruised grain goes first. From here it then goes over into the second cylinder and finally into a series of compartments. Each compartment has a tube leading from it into other tubes, each of which makes the grain whiter than the one before. Usually one makes three kinds of grain. The worst grain is called "extra family flour" and is sold from a box. Another type runs into a special container and is fairly pretty to look at. To grind it finer one lets it run on. If the grain is too coarse, it is run through again and is called "extra super-fine." Finally, the first two kinds are used by many bakers for rough work and the other kind is used for fine baking. The second cylinder was on the second floor while the first was on the third. The bruised grain went first into a conveyor, from there driven along by a system of belts. This carried it to the fourth level into a little room called the "cooler" and here a simple machine put it into the first cylinder by means of a tube of pipe. In the case of the second it fell off the belts into the cylinders. This set of pulleys was just twenty feet long. The meal then had to be sacked, which is necessary to the farmers. The third is the rye-sack and like the second. The waterwheel is twenty feet high and eight feet broad. I cannot give an account of the other specifications. The steam machines make twenty-two and the stones one hundred fifty revolutions per minute. Everything possible here is made of iron. Nothing is wet. The fruit from the wagon was weighed and emptied into containers. From there they were carried up and left in a box. From this one could place them in the machine to the ground. The windmill grinds the grain on the rough side of the stones. Below is another windmill that discards the powder and dust and from this the fruit runs through a belt into a burnishing machine, from this into a box with two divisions. There are two bells here and when it is empty, one can notice the fact by the sound of the machine. In the morning at five, I got up, made a fire and in an hour and thirty minutes had steam enough to begin. Then I would grind until sundown. Since there were just two of us and I was merely the second miller, I had to do other jobs that did not please me when we were not grinding. I did not get enough food at these rich people's house and so went away after I had saved about seventy dollars. . . .

The mill in Somerset is nearly entirely of wood and where extra iron doesn't have to be, everything is wood. The axles for the wheels lone are of iron. Most of the mills here are constructed like many houses in the Black Forest; one can go up one side of the house and down the other to the ground. Above on the second floor things are unloaded; on the first they are loaded. It is situated on the slope of a mountain and, below, the rocky cliff forms a wall on one side. It is walled in below, and above it is blocked up with logs and overshot. The waterhouse and mill are under one roof, or one can also say the water comes into the mill. In the attached diagram [Figure 11–2] (a) is a big cog-wheel; this leads to an erect beam (b) on the axle-tree (c) on the same one is the cog-wheel (d) which is attached to the beam (e) made of iron. On the big cog-wheel (a) is, in addition the horizontal wheel (f) with gears. The winch goes up by means of this which by means of the wheel (h) the machine operates. Above the wheel (h) is the wheel (i) with gears. This is attached to the wheel (j) on the axle-tree (k). Onto this is the wheel (1) with gears which is attached to the wheel (m) so the connection gear or cylinder works. Since old mills are constructed and furnished with wood, they creak and crack so that the miller or a man who is not used to it or has not seen one before could think that it was going to fall to pieces. Later I saw the same thing with steam and water and

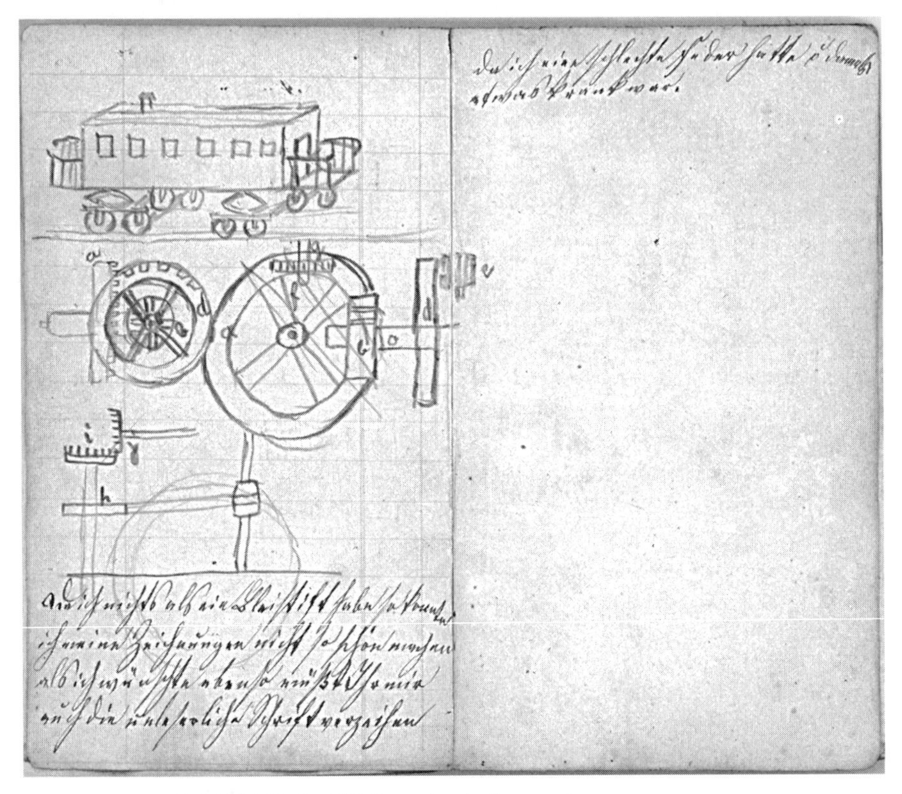

FIGURE 11-2 Mill Cog-wheel Diagram from August Ketterer's Diary

water alone. At the same time, they had kettles that were worse than old tubs and steam machines from which the water and fat runs off together into streams. One can hardly hear himself think in one of these.

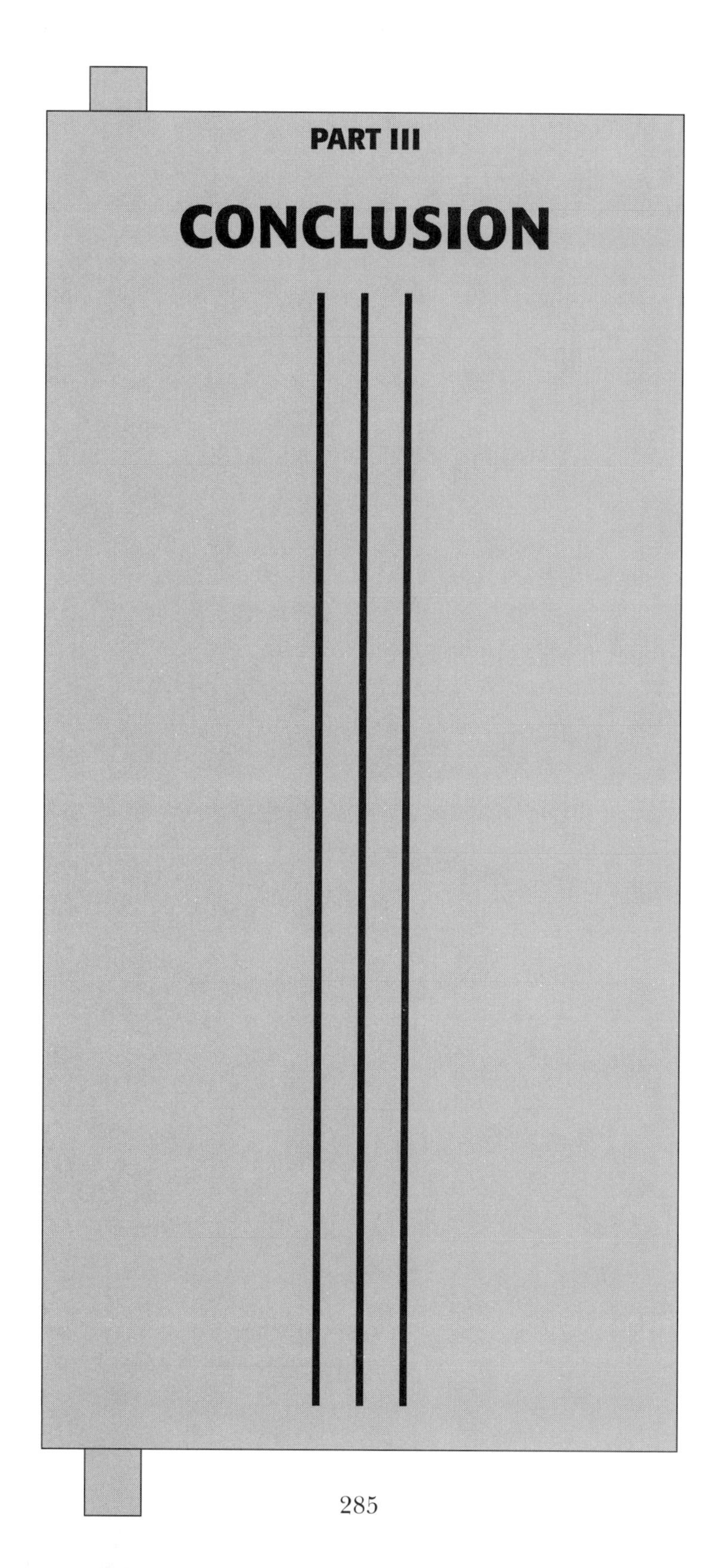

PART III

CONCLUSION

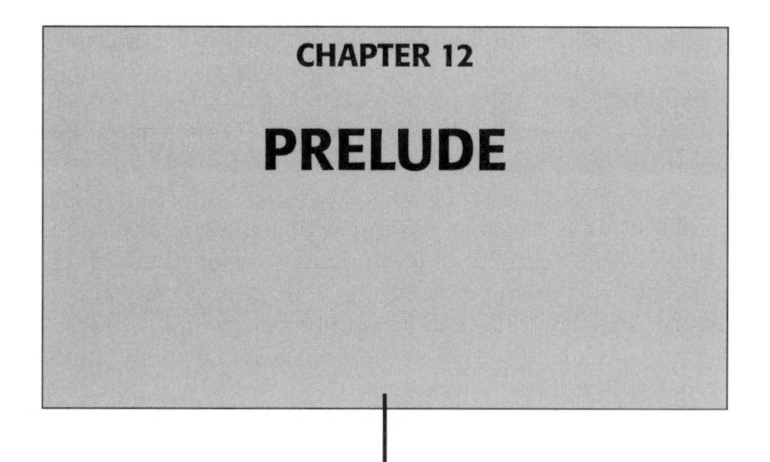

CHAPTER 12

PRELUDE

Most books begin with a prelude rather than end with one. Yet due to the nature of this work, the prelude comes at the end as a summary of what has come before and as an introduction leading to what will come in the future. Since where we have been is no more than a preface to what will be, this book ends at the beginning of the next step, and the next step is the reader's.

Unlike descriptive texts, the purpose of this book has been to acquaint the reader only with enough primary information to start the search for knowledge. Like technology itself, that search is always new, always just beginning, and always punctuated with the imagination and hopes of the investigator. Technology is not a "thing" per se. It never has been. It is a process, a natural consequence of who and what we are. It is a way of surviving, growing, and thriving in the face of the ever-changing environmental conditions that challenge our ingenuity to the limit. If it were not so, then we would surely have succumbed to the pitfalls of this planet long ago.

In the first chapters of the book, we dealt with the nature of technology, searching for a definitive understanding of what it was we were studying and analyzing the "nature of the beast," in an attempt to give ourselves direction. We considered the artificial nature of technology, the fact that it consists of constructs that are unnatural manipulations of the physical laws of the universe in which we operate and that, in spite of this unnatural content of device and artifact, the *context* in which the artifacts were held was completely natural.

We create technology by natural design. We manipulate the laws of nature in accordance with our understanding of those laws and bend them (they cannot really be broken) to fit our needs. Technology is a mirror, an extension of humanity, created in its image to increase the species' capacity to function. The wheel is nothing more than the foot taken to its logical conclusion, then pushed further by compounding the technology of the wheel with that of the piston, a mechanical arm. Likewise, steam or gasoline power is an extension of solar energy in a transformed state to provide the desired results—cheap, efficient power and mobility for the human race. The camera and the lens are no more than specialized eyes, yet very special inorganic eyes that can see farther than the eyes of the sharpest eagle and detect movement more adroitly than the wisest cat. And ears are extended through such devices as microphones, amplifiers, and sonar equipment. And the bird's secret is harnessed in the airplane, the jetliner, and an Atlas missile in flight. And the dolphin is emulated through submarines, the turtle through tanks and mobile homes, and the ant and the termite through cavernous structures lying both above and below ground through which we pass in constant streams to carry out the collective and cooperative duties of a modern society in full operation.

And yet in each of these cases, the emulation is only superficial, humanity adapting the lessons of nature to be learned from other living creatures and from ourselves to push a principle artificially through technology as far as it will take us. We are builders. We are dreamers. We are the ultimate simulator in nature, borrowing from everywhere to combine what nature does through selection and then purposely structuring the pieces into a cohesive whole for our own use.

Thus we build our technology. And in so doing, we speed up the process of evolution many thousandfold, doing more in fifty years with the concept of flight than nature was able to do through natural selection in seventy million. With such an increase in speed comes hazards, and with the hazards the need for a check to balance the growth with caution. And homeostasis provides us with that check. By resisting change, and by limiting our acceptance of the new and tempering the fever for growth with the fear of the unknown, humanity has been able to select those technological innovations that create value without too much inherent cost. The balance is maintained, at least for the present, through the conflict between the desire to extend ourselves and the fear of rushing into unknown realms. And so it has been for thousands of years.

Along with this dual concept of innovation and homeostasis, we viewed the nature of the creativity that feeds the process, the condition of insatiable thirst for knowledge that seems to permeate every aspect of human life. We considered curiosity. We considered the changing paradigms. We considered the revolutionary nature of an animal ruled not by genetic coding and uncontrolled conditioning but by a freewill choice to restructure the normal into the new. Like any good scientific investigators, we studied the "how" of the process, and then, in a contextual shift, we considered motivation, the "why" of technology, as well.

Economic impact is a single segment of a total social pattern that includes technology and is both created by and creates the physical mechanisms by which we are able to maintain our lives. The why and the how of technology are initially economic in nature, expressing our desire to increase the overall welfare of the society through the production and distribution of goods. And as we saw, those goods are the result of production using *scarce* resources, resources that are not easily replaced and that increase in the effort required to obtain them. Technology allows us not only to expand our productive capacity, filling the society with an ever-widening variety of consumption items, but also to continue producing in the face of dwindling supplies. Efficiency—the capacity to produce with a minimum of expense, effort, or time—is the why of technology, as is consumption. Economically, everything that we have in the way of goods and services is directly attributable to some aspect of technology.

In the second part of the book, we discussed different methods of study, including a look at the usefulness of cause and effect in constructive thinking and the use of models in developing an understanding of the nature of the processes taking place in the complex modern world. The systems view was presented as one logical way of organizing information and structuring our understanding of technological processes, emphasizing the interrelationships that exist among the various facets of society as it reacts to technology and the consequences of our own ingenuity. We introduced the idea of chaos as a natural mechanism limiting our understanding and our ability to predict events.

If we are to understand the place of technology in our world, we must have some means, no matter how unreliable, of predicting, of describing the expected consequences of our actions in an attempt to deal with the obvious pitfalls of technological progress before they catch us unaware. Yet we must remember that no matter how effective we are at prescient descriptions of future consequences, we still must handle the unseen and unconsidered results as they arise. That we are not perfectly predictive in our attempts at prescience is an indication of our humanness. So is our felt need to be as predictive as possible, to weigh the odds in our favor by considering as many possible future worlds as we can and by preparing for those contingencies that we can imagine and deem likely enough and urgent enough to require our attention now, before they become a major problem in our lives.

This brings us to the here and now, the so-called prelude. Why the prelude? Why not the epilogue or the afterword? Why end a study with a beginning? Because.

Or more precisely, "be-cause." To this point we have defined the nature of a technological society and discussed methods for study. But the actual study of the subject is yet to begin. That is no accident. It was never the intention of this book to tackle the immense content of technology and society, to pour over the essence of weighty tomes, searching for some significant body of truth to pass on to the reader, as if by the magic of the summary of ten thousand years of history, any book could capture the meaning and nature of social change. To do so would require a detailed understanding of mechanics, physics, biology, botany, medicine, engineering, anthropology, history, religion, economics, philosophy, archaeology, and a hundred other disciplines. And in the end, the only presentation that could be made is the opinion of the author, a contextual understanding of technology's impact on society, restricted to the paradigms embraced by some group of investigators. So I repeat: "Be-cause."

Be the cause of your own experience of the subject. Involve yourself in the content of technological change and of humanity's use of technology. Find your own answers to your own questions, and *be* the *cause* of the future. Not only is this chapter a prelude, but the entire book is a prelude. It is the beginning of the reader's own experience of what it is like to live in and understand technological societies. I hope that through personal investigation, some sense of the immensity of the subject and some understanding of the beauty and fascination inherent in the discipline will begin to implant themselves in your mind. Tomorrow is always before us, and we must be always preparing for its arrival. To know what is now true is to be left behind. As John Naisbitt has indicated, the proper frame for success today is the future. What will you be doing five years from now? Where will you be working? Where will you live? What will you think, and what will you wear? These are very real questions, and for the vast majority of the population, they are questions without adequate answers. What will your reaction be to the technology of the future? How will you behave? How will you be restricted by the technology that controls your life? How will you control your life through that technology? Will you even exist five years from now? How do you know?

Obviously, you do not. But an understanding of the nature of the future is the best bet that any of us has to help ensure that future and our place in it. Thinking that emphasizes where we are going is necessary. Just as a jet pilot must think five or ten minutes into the future in order to keep up with her or his craft, so must we think five, ten, or even twenty years into the future to keep up with our world. And the progress of the world is one thing that will not wait for any person, should she or he fall behind.

By looking at the past and finding the patterns that repeat themselves, we can adequately operate in the present. By understanding the conditions under which we are now living, we can prepare ourselves for the near future. And by studying the technology of today and the developments of tomorrow, we can cause the future to happen. We can form it and guide it, leading our lives into channels of success and away from the unseen dangers of dead-end social behavior that could leave us stranded as the culture passes us by.

And what are the possibilities of that future? We stand on the edge of a great adventure, one that will take us within ourselves, within our planet, and outward toward the stars. We have the capacity to explore the reaches of near space and the ability to reach and learn from the deepest recesses of the ocean floor. Underway right now is research that can give us the ability to produce needed chemicals through biological machinery created through genetic splicing and biological manipulation. We are on the verge of curing a dozen killer diseases, including significant gains in the battle against cancer. We have the ability to feed the people of the world, yet in the face of that ability, millions of people suffer from starvation and drought. We can reattach a severed finger or arm with some hope of the patient regaining partial or total use of the limb. We can replace hearts and kidneys, eyes and vital tissue. We can make deaf children hear and watch them experience their own laughter for the first time. We have the capacity to reach out and mine the asteroids, to look across the universe to the beginning of time, to span an entire continent in a few hours, or to calculate the probabilities of success in a horse race while sitting at a terminal in our own home.

We have other capacities as well. We can kill our fellow human beings more efficiently than ever before. We can strike with colorless, odorless gas or with viruses against which there is no defense. We can incinerate entire populations or create a winter that would outlast the last human being. We can turn the planet over to the insects and the mosses, and we can do it all in less than a single twenty-four-hour day. And we are improving our capacity to do all this as well.

How we choose to spend the rest of our lives depends to a large degree on how we perceive that future. A Chinese sage, famous for his study of war and the art of warfare, said some 2,500 years ago that we should know our enemy and know ourselves. To do so is to enter into battle without fear. Not knowing one or the other is disastrous in the extreme. The future is our battlefield, and our technology is what we must know. To know it and understand its potential is to give ourselves the power to extend the development of the human race to heights not yet dreamed of. We are the future. What we decide to do in terms of technology is what will create that future for us all. It is a responsibility and a legacy that we will all leave to those who follow. To ignore the consequences of future actions until they are upon us is to guarantee disaster, for ourselves and for all of our fellow travelers on this journey. Neglect is tantamount to imposing an unknown sentence on us all.

The important issue is whether or not to "be-cause." It is a choice of watching society progress, being involved in society's progression, or causing that progress to whatever degree we choose. For this reason, and with hope for the future of us all, the prelude ends. To you, the reader, is left the opportunity to fill in the first word of the first line of the first paragraph of your own . . .

CHAPTER ONE

TOMORROW AND BEYOND

Appendix: Topics for Discussion and Analysis

The following is a partial list of suggested topics for study in the field of social issues in technology:

Acid rain
Age discrimination—the result of improved health care?
The age of steam and its cultural consequences
Agriculture and the rise of city-states
Air transportation and its place in modern societies
The appropriateness of a technology
Architecture and technology
Artificial intelligence and chaos theory
Artificial intelligence and the nature of creativity
Bioengineering
The causes and cures of air pollution
The causes and solutions to water pollution
Chaos theory and changes in technologically advanced societies
Chemical warfare—an alternative to the bomb?
Chinese technological advances
The computer age—a new beginning?
The computer and systems thinking
The computer revolution and changing cognitive processes
Computers and the changing face of the American culture
Computers in education
Countercultures as homeostatic reactions to change

Creativity and freedom
Dam building techniques and the spread of culture
Death of traditional universities—alterations to education
Determinate randomness in the Malthusian curve
The development and cultural impact of metallurgy
The development of the electrical automobile
The discovery and cultural impetus of microwave technology
Discrimination and technological change
Economic consequences of twentieth-century medical technology
The effect of the rise of Japan's world markets on American life
Energy conservation in a technological society
Euthanasia—result of improved health care?
The future of cryogenics
Gene splicing and designer genes
Geothermal power
Global urbanization and demise of rural living
Global warming and the role of technology as cause and cure
History of glass in human culture
Housing, technology, and culture
Hydrophonics
Hydroponics
Hypothetical technology in a utopian society
The importance of freedom of expression in creating technology
Internet—the printing press of the twenty-first century?
Internet—the source of freedom or intellectual slavery?
The laser—a solution looking for a problem
Leonardo da Vinci's life and work
The library of the future—form, content, and accessibility
Medical technology and quality of life
The microscope and the microcosm in human thinking
Modern agricultural technology
Nanotechnology and its future implications
The new electronic library
Nikola Tesla—genius or charlatan?
Nuclear energy
Nuclear waste—its disposal and its threat

The oil embargo and floods in Bangladesh
Oil wars of the twentieth century
Passive and active solar energy systems
Photography and changing perceptions of reality
Photovoltaics
Plastic and the changing face of society
The plow and Medieval society
Pollution and the Malthusian proposition
Population growth and technology
The postindustrial society (information-based economies)
The potential consequences of biotechnology
The printing press and its logical consequences
The process of urbanization
Radio and alteration of the culture
Railroads and the building of North American society
Refrigeration
Renaissance science and the Church
Robotics
The role of lighter-than-air craft in the twenty-first century
The role of randomness and diversity in technological innovation
Roman legions and engineering
Science fiction—protection against future shock?
Shipbuilding and culture since the age of steam
Shipbuilding prior to the age of steam
Smart houses
Social and cultural implications of space colonization
The social impact of Charles Steinmetz
The social implications of the Internet
Social implications of orbital manufacturing
Social implications of space exploration
Solar power
Street people and technology
Sustainable technology
Swidden agriculture—the first creator of societies?
The technologies of cooking
Technology and changing ethical patterns

Technology and concepts of space and time
Technology and its effects on the art of industrial design
Technology and music—fundamental changes and consequences
Technology and terrorism
Technology and the changing nature of language
Technology and the nature of play
Technology and world epidemics—cause and cure
Technology and world theologies
Technology as a cause of gentrification
Technology as a function of sexuality
Technology as a reflection of available resources
Technology as an adaptation to environmental change
Technology in World War I
Technology in World War II
The telescope and changing cultural paradigms
Television and alteration in the culture
The tertiary effects of fuel choices
Thomas A. Edison—his inventions, life, and approach to research
Time, technology, and society
Toys as a method of studying the technology of cultures
Waterways and technology—the transportation impetus
Wind power and its applications

Glossary

Adaptability A characteristic an organism is said to have if it can change its structure to meet the needs of a changing environment, or alter its method of operation to fit new circumstances in that environment. The more adaptability an organism has, the greater its capacity to survive in changing environments and the wider the range of environments that it can inhabit.

Adjustment In psychology, the act of modifying, changing, or adapting individual behavior so that it conforms to generally acceptable societal norms.

Analog model A model that behaves in some manner similar to the reality that it is designed to represent.

Anxiety In psychology, a vague feeling of apprehension and hope for the future resulting in tension and stemming from some external stimuli or internal, unresolved conflict.

A priori knowledge Knowledge thought to exist in the mind exclusive of knowledge. Knowledge from a valid law to a particular incident.

Artifact Any object made by a human being that is particularly designed to be used in some way that increases efficiency and reduces hardship for the user.

Balance In a system, the concept that the primary goal of all systems is to achieve and maintain a state of balance, which the system will define in terms of its goal. Achievement of the individual goal of a specific system is the achievement of balance for that system.

Beatnik Any member of a counterculture existing primarily in the 1950s, characterized by subculturally distinct speech and clothing, and professing a dislike of accepted social behavior patterns.

Biological growth curve The sinusoidal curve form representing the growth and decline of living organisms through time.

Bohemian Any member of a counterculture existing primarily in the 1920s and early 1930s, exhibiting behavior considered at the time to be irresponsible. The phenomenon is thought by some to have been the result of an inability on the part of the members of the subcultural group to cope with sociological and technological changes beginning with World War I.

Capital investment The investing of savings from individuals in the means of production in order to increase the capital base of the economy and thus create economic growth and an expansion of available goods for final consumption. It is a method of foregoing present consumption to ensure higher levels of consumption later.

Capitalism An economic system distinguished by self-interest, consumer sovereignty, capitalization, competition, private property, and utilization of the market system to solve the problems of production and distribution within the economy.

Catastrophe theory An occasion when a small, gradual change in the environment of an object brings a sudden and abrupt response, or when increasing tension from minor elements within a system reach a point of critical mass, triggering a sudden shift to a different state of equilibrium.

Catastrophic probability A shift in the stability of a system such that due to the level of complexity within that system, the chance of a catastrophic event occurring under increasing degrees of tension shifts from possible to probable.

Catastrophism An evolutionary theory that saw the world's development as a succession of worlds, each separated by brief transition phases when catastrophic events occurred inexplicably, causing major changes.

Cause and effect A relationship between actions and events in which the events are the result of the actions.

Chaos theory A general body of theoretical information dealing with nondeterminate, nonlinear systems dynamics. It postulates the existence of random behavior in all systems, no matter how ordered.

Competition The concept in capitalism that both consumers and producers must vie with each other for the right to scarce resources and finished goods and that through this process, a free market system is able to properly produce and distribute goods and services through the society.

Conditioning In psychology, the process of acquiring or learning certain modes of behavior through reward and punishment, repetition, or other learning methodology.

Consumer sovereignty The concept that the consumer has the final authority in deciding what will be produced, who will produce it, how it will be produced, and for whom it will be produced in a free market economy.

Content The physical reality of something. Content is the real-world constituent makeup of a given object or situation and is measurable and discernible in physical terms.

Context The view held by an individual concerning the meaning, condition, or characteristics of the content of some situation or physical object. It is the individual "idea" that a person attaches to a physical reality by virtue or his or her own opinions, beliefs, experiences, and so forth.

Copernicus Renaissance astronomer who developed the theory of a heliocentric system, in contrast to the geocentric theory supported by Church doctrine.

Counterculture Any subculture whose behavior patterns, mores, and customs are counter to those of the general culture within which they exist. It is usually a reaction to some change in the societal structure with which the countercultural members are unable or unwilling to embrace.

Creativity The condition of causing to come into being as an original and non–naturally occurring phenomenon. To produce a unique and novel device, idea, or work.

Custom-made model A model that is created for a specific need, designed to perform specific functions as a model of the reality that it is designed to represent.

Descriptive model A model designed to describe characteristics or conditions of some reality that it is designed to represent.

Deterministic model A model designed to offer a definitive answer to questions concerning some reality that it is designed to represent. This type of model is generally, though not exclusively, mathematical in form.

Deterministic randomness Another name for chaos, the concept of a deterministic system that is directly ascertainable in terms of progression but not in terms of regression. That is, cause and effect may be determined from formulas, whereas analysis of original causal states is impossible due to the multiple paths by which a given end state can be achieved.

Diversity The concept in ecology, chaos theory, and dynamic systemics that supports the importance and presence of many modes of behavior to the general well-being of growth.

Early majority That part of the majority of the buying public that purchases a new product that has survived the test of the innovative minority early in its introductory stage. The early majority purchases during the phase of the product life cycle represented by sales rates that increase at an increasing rate.

Economics The study of how people choose to use scarce resources for the production of goods and services and how they choose to distribute those goods and services to the general population for their consumption.

Economic trade-off The concept that states that because of the scarcity of resources, the choice of any given economic activity reduces the opportunity to carry on alternative economic activities by an amount equal to the resources flowing into the chosen activity. That is, for each choice to produce and consume a given product, we trade the opportunity to do other things with the same resources, labor, and capital inputs.

Economic welfare The overall economic well-being of a sociopolitico-economic system. It is a nonquantitative factor, though it can be described in quantitative terms, dealing with the quality of life experienced by individuals in the system.

Economies of scale The principle that as production rises, the cost of production per unit declines due to increasing efficiency with which inputs are used. That is, as production rises, costs rise at a decreasing rate.

Efficiency The quality of producing or working with a minimum of effort, time, and energy.

Entrepreneur An individual who is aware of an opportunity to make a profit by buying low and selling high in the market. These are the risk takers of the society who instigate economic behavior and production of goods and services, seeking a profit.

Evolution The continuous adaptation of species to changing environments through selection, mutation, and hybridization.

External evolution The adaptation of a species to changing environments through means external to genetic change, as in humanity's use of technology to create adaptive change external to the human body.

Feedback Information returned to an instigator of change that reflects the results of that change, allowing the originator to correct for undesired results.

Frustration In psychology, the condition of thwarted drives, in which attempts to move toward the satisfaction of some drive are denied or attempts to escape from some negative drive are denied. The result of frustration is often anxiety and the need to adjust.

Galileo Galilei Renaissance astronomer and discoverer of the Jovian moons. First known astronomer to use a telescope for astronomical observations. Advocate of the Copernican heliocentric theory of the solar system.

Hierarchy of needs Maslowian proposition that humans are motivated to satisfy needs, with the strongest motivations for the most basic needs. Maslow arranged human needs into categories; beginning with the most basic and, therefore, most strongly desired, those categories are physiological, safety, social, self-esteem, and self-actualization.

Hippie Any member of a counterculture existing mainly in the mid 1960s to mid 1970s whose members exhibited characteristics of pacifism and a denial of generally accepted societal goals. Some believe this countercultural movement to be the result of collective revulsion of public policy surrounding the Vietnam War, particularly as affected by the extensive television coverage of events in Southeast Asia during this period.

Holistic thinking A nonlinear approach to thinking in which events and environments are viewed as a whole rather than as discrete parts. Philosophically Eastern, this type of thinking deals with the concept of everything being a part of the whole rather than discrete individual physical realities unto themselves.

Homeostasis Resistance to change. The tendency of humans to resist change due to fear, a desire to reduce work, or an inability to understand the change in question.

Iconic model A model designed to resemble a physical reality, though not to behave in an analogous manner. These models look like the physical reality that they are designed to represent.

Industrial concentration The degree to which productive capacity is concentrated in some small number of firms within a given industry. It is usually measured by the concentration of the top four or eight firms involved, with a monopoly of one company, with all productive capacity representing maximum concentration and pure competition representing minimum concentration.

Innate ability Inborn traits, such as aptitudes or physical characteristics, giving rise to specific dominant capabilities in the individual. These abilities are genetic in source.

Innovation The introduction of something new or the re-creation of some established technology into different or novel forms.

Innovators Those who innovate. In marketing, those members of the society willing to try new products because of their newness. The least homeostatic of the consuming population.

Input That which is introduced into a system to be processed for purposes of producing some output. Input can be in the form of raw materials, labor, capital, information, or any other phenomenon that carries some form of content.

Instinct An inborn pattern of activity or tendency to some specific form of behavior common to all members of a given species.

Late majority That sector of the consuming public willing to purchase a product only after it has become an accepted item. This portion of the population purchases goods near the maturity point of a product's life cycle, when sales are increasing at a decreasing rate. This is the most homeostatic sector of the consuming public.

Law of diminishing returns After a given point, additional units of input will result in progressively smaller quantities of additional output, with total output first increasing at a decreasing rate and finally decreasing absolutely. That is, past a certain maximum level of input efficiency, each additional unit of resource input to the productive process will be less efficient than the unit prior to it.

Law of downward-sloping demand There is an inverse relationship between the price of a product and the amount of a product that will be demanded in the marketplace, all other things being equal.

Law of supply In general, there is a direct relationship between the price of a good and the amount of a good that is offered in the market by producers.

Leonardo da Vinci Renaissance Italian figure, famous as a painter, sculptor, architect, and technologist.

Linear thinking A mode of thinking that views reality as a set of discrete physical objects and events, whereby the rule of cause and effect is followed. This is step-by-step thinking incorporated into Western logic.

Lorenz, Edward American meteorologist and early experimenter in chaotic behavior. Noted for the butterfly effect, which states that it is impossible to predict the weather with accuracy.

Luddite Any of a group of workers in England between 1811 and 1816 who roamed the countryside destroying machinery and technological innovations in the belief that this machinery was responsible for destroying jobs. The movement receives its name from one Ned Lud, an eighteenth-century worker who first exhibited the destructive tendency due to frustration caused by an inability to deal with automatic equipment in the textile mill in which he was employed.

Malthus, Thomas The first political economist. An eighteenth-century economist who investigated the relationship between economics and population. It is after Thomas Malthus and David Ricardo, a contemporary, that economics receives its title of the "dismal science."

Malthusian curve The sinusoidal curve form representing Malthus's contention that due to the manner in which food supplies and population grow, the human race was doomed to cease growing and possibly decline as a result of starvation, disease, and war.

Malthusian proposition The theory presented by Thomas Malthus that indicated that since population grows geometrically and food supplies grow arithmetically, eventually population would outstrip the food supplies and starvation would ensue.

Market equilibrium A condition in a free market that is said to exist if buyers and sellers agree on the quantity of goods to be bought and sold and on the price at which they will be bought and sold, and agree in addition to go on buying and selling that quantity at that price.

Maslow, Abraham Twentieth-century psychologist responsible for developing the hierarchy of needs to explain human motivation and behavior.

Maslow's hierarchy of needs *See* Hierarchy of needs

Mathematical model A symbolic manipulative representation of reality designed to describe relationships among certain factors of the reality that it is designed to represent. The method uses numerical representations to describe the reality in question.

Memory trace The connective "highway" created among brain cells as a given memory is created. A memory trace is reinforced whenever the memory is activated and weakened when not reinforced through time.

Model A copy of a physical structure or a concept that is designed to demonstrate certain characteristics of that physical structure or concept in accordance with the purposes of the modeler.

Motivation In psychology, that which prompts or drives a person to exhibit a given pattern of behavior.

Natural selection A process occurring in nature by which only those organisms with traits particularly favorable to survival are able to survive in the competitive environment. They are therefore "selected" for survivability through the process.

Negative externalities Costs incurred by third parties external to the economic decisions made by buyers and sellers of some product or service and incurred because of the market interactions of these groups.

Oligopoly An economic market structure characterized by high industrial concentration, difficulty in entering and leaving the market, and the sale of a uniform product or product group. This form of market appears to be a natural result of high production efficiency inherent in technological societies experiencing economies of scale.

Optimizing model A decision model designed to present the best solution possible to the modeler, that is, how to "optimize" results in line with certain predetermined criteria. It is often, though not necessarily, mathematical in nature.

Output The result of the production process. In systems theory, that which results from the processing of inputs. It is the goal that the system has as its purpose for existing.

Paradigm A belief system that limits action by creating outside parameters within which one is allowed to operate. These are descriptions of reality based on observation that are used to explain that reality, excluding alternative explanations until replaced with some broader or different understanding of the phenomena in question.

Physiological needs In the Maslow hierarchy, those basic needs that enable a person to maintain life, including warmth, food, water, shelter, and so forth. They are the lowest level needs in the hierarchy.

Primary effect The immediate and obvious effect of some action or event, particularly in relation to simulation building and systems analysis.

Principle of unintended consequences The principle that holds that every action or purposive change imposed on a system will have consequences that were not intended and not predicted.

Private property The characteristic of capitalism that holds inviolable the right of individuals to hold the means of production as personal property and the right to do with that property as they wish.

Probabilistic model A nondeterministic mathematical model designed to predict the most likely outcome of some set of activities. It is designed to create a "best guess" of the probable behavior of the reality that the model is designed to approximate.

Process In systems analysis, that factor of the basic systems format that refers to actions taken by the system to convert inputs to acceptable and sought outputs. It is the method by which the goal of the system is achieved.

Production possibility curve A graphic representation of economic trade-off, indicating the opportunity costs of producing a given product in terms of the amount of another product or product group that must be foregone in order to do so.

Product life cycle The normal progress of a product in the market through four stages of life—introduction, growth, maturity, and decline. The cycle approximates a sinusoidal curve.

Psychophysiological restructuring The process of relearning in which old patterns of habitual behavior are replaced with new ones. The restructuring occurs of necessity if homeostasis is to be overcome due to the established brain cell connections, analogous to memory trace patterns, causing a resistance to change.

Punctuated evolution A modern theory of evolution akin to catastrophism that views evolution as a combination of slow growth through processes of mutation and periodic violent shifts due to unexpected global environmental changes in conditions.

Qualitative model A model designed to encode inexact concepts numerically through the use of statistics to produce probabilistic descriptions of the reality that the model is designed to represent.

Quantitative model A model designed to produce discrete values for characteristics of the reality that the model represents through the process of manipulating formulas.

Ready-made models General models designed to produce answers to questions that are repetitive in application and often encountered.

Reciprocity The concept in systems theory that what is put into a system is equal to what comes out of the system, or that input is the exact determinant of output.

Safety needs In Maslow's hierarchy of needs, the second level of needs in which the individual is concerned with protection of self and avoidance of danger.

Satisfice To satisfy a need to the point that other needs offer more compelling motivation for fulfillment, at which point the original need is abandoned in favor of the more important ones. In Maslow's hierarchy, the act of relieving the pressure of a need and moving on to the next one, even though the other needs are not totally satisfied.

Scientific method A systematic method of investigating real-world phenomena designed to ensure consistency of approach and maintenance of the integrity of evidence and findings. It is the rational method for explaining real-world observations.

Secondary effects In systems analysis and simulation theory, those effects stemming from an action that are not immediately obvious or are caused by the primary effects becoming a causal factor.

Self-actualization In Maslow's hierarchy of needs, the highest level in the need hierarchy—the need to express who and what one is. It is the ultimate level of need fulfillment, at which a person is considered to be balanced and whole.

Self-esteem In the Maslowian hierarchy of needs, the need to feel good about one's self and what one does—the need for self-respect and a feeling of self-worth. It is directly below self-actualization in the hierarchy and is considered to be one of the developmental needs that humans possess.

Self-interest The capitalistic principle that states that each person in the economy acts in such a way as to maximize his or her own interests and that it is because of this that the free market system produces the most goods at the lowest possible price and of the highest possible quality, making those goods available to the largest number of people.

Shortage In a market structure, when demand exceeds supply at the prevailing price of the market. In a purely competitive market system, this results in a rise in price toward equilibrium.

Simulation A model that copies the behavior of some aspect or aspects of reality. Its purpose is to describe and predict behavior in real-world situations within the confines of the parameters of the simulation model itself.

Social needs In the Maslowian hierarchy, the central level of needs in the five-tier structure incorporating the belonging needs of individuals, including group membership, the need to succor, and the need to nurture. They are both physical needs and developmental needs, thus representing the boundary between the more basic and the higher needs in the structure.

Specialization In biology, differentiation in order to adapt to a changing environment or some special set of circumstances.

Subsystem A system that is part of a larger system and therefore represents a discrete element in that larger system. It represents a subclassification of activities in an overall systems design.

Surplus A case of supply exceeding objective demand in a market system. When surplus exists in a purely competitive free market system, the usual result is a drop in price back toward equilibrium.

Survival of the fittest The concept that those life forms best fitted (suited) to prevailing environmental conditions tend to survive, whereas those related organisms that are less fitted to survival tend to become extinct. It is an expression of the natural selection process of evolutionary theory.

Survival trait Any trait that an organism possesses that tends to improve that organism's chances for survival. A trait that adds to survivability.

System A collection of physical and nonphysical interactive parts all having connection through a mutual purpose of achieving some goal.

Systems analysis The method of analyzing a system by breaking it up into its constituent parts and relationships in an effort to understand its structure and its ways of functioning, and to predict its behavior under given sets of conditions.

Systems approach A method of studying a real-world phenomenon that centers on constructing a model of the reality in a systems mode, that is, describing a real-world phenomenon as a system. The advantage is to create an orderly structure and predictive ability through an understanding of the nature and goals of the phenomenon. The shortcoming is that it tends to create a paradigm that disallows possible explanations outside the orderly systemic form utilized to study the phenomenon in question.

Technological specialization The process of narrowing one's scope, either as an individual or as a society, toward specific technological specialities, creating a high degree of division of labor; a separation of the worker from the process of work; and the accompanying psychosociological problems such as boredom, psychosomatic illness, dissatisfaction with job, apathy, reduced productivity, and a general decrease in psychological health.

Technologize The act of creating technology. Although the resulting artifact is artificial by definition and not a natural structure, the act of creating is natural to humans and is a survival mechanism innate to the organism.

Technology That whole collection of ways in which the members of a society provide themselves with the material tools and goods of their society. The collection of artifacts and concepts used to create an advanced sociopolitico-economic structure.

Verbal models Models that are designed to convert thoughts and concepts into language, to establish relationships and restrictions of real-world systems, and then to organize them.

Vinci, Leonardo da *See* Leonardo da Vinci

Y2K The disruption of normal societal function in the year 2000, due to the inability of some computer programs, imbedded chips, and computer controlled systems to effectively cope with a date shift to the two digit year designation of "00."

Bibliography

GENERAL

Grun, Bernard, *The Timetables of History.* New York: Simon & Schuster, 1979.

Trager, James, ed., *The People's Chronology.* New York: Holt, Rinehart & Winston, 1979.

CHAPTER 1: TECHNOLOGY: A NATURAL PROCESS

Ardrey, Robert, *African Genesis: A Personal Investigation into Animal Origins and Nature of Man.* New York: Dell, 1961.

Birdsall, Derek, and Cipolla, Carlo M., *The Technology of Man: A Visual History.* London: Wildwood House Limited, 1979.

Cooke, Jean, Kramer, Ann, and Rowland-Entwistle, Theodore, *History's Time Line: A 40,000 Year Chronology of Civilization.* London: Grisewood & Dempsey, 1981.

Dobzhansky, Theodosius, "Man and Natural Selection," in *Classics of Western Thought,* vol. 4, ed. Donald S. Gochberg. New York: Harcourt Brace Jovanovich, 1980.

Fincher, Jack, *Human Intelligence.* New York: Putnam's, 1976.

Jastrow, Robert, *The Enchanted Loom: Mind in the Universe.* New York: Simon & Schuster, 1981.

Matson, Floyd, *The Broken Image: Man, Science, and Society.* Garden City, NJ: Doubleday, 1966.

CHAPTER 2: RESISTANCE TO CHANGE

Alland, Alexander, *The Human Imperative.* New York: Columbia University Press, 1972.

Childe, V. Gordon, *Man Makes Himself.* New York: The New American Library, 1961.

Freud, Sigmund, *The Future of an Illusion,* trans. W. D. Robson-Scott. Garden City, NJ: Doubleday, 1964.

Hays, Samuel P., *The Response to Industrialism: 1885–1914.* Chicago: University of Chicago Press, 1957.

Schaffer, Lawrence Frederic, and Shoben, Edward Joseph, Jr., *The Psychology of Adjustment: A Dynamic and Experimental Approach to Personality and Mental Hygiene*. Boston: Houghton Mifflin, 1956.

CHAPTER 3: CREATIVITY AND INNOVATION: THE CRITICAL LINK

Bradley, Alan, *Your Memory: A User's Guide*. New York: Macmillan, 1982.

Bronowski, Jacob, *The Origins of Knowledge and Imagination*. New Haven, CT: Yale University Press, 1978.

Hassan, Ihab, *The Right Promethean Fire: Imagination, Science, and Cultural Change*. Chicago: University of Illinois Press, 1980.

Kuhn, Thomas S., *The Structure of Scientific Revolutions* (2nd ed.). Chicago: University of Chicago Press, 1970.

Malthus, Thomas, *Principles of Political Economy*. Fairfield, NJ: Kelley Augustus, 1936.

Mote, Frederick, *Intellectual Foundations of China*. New York: Knopf, 1971.

Sahal, Devendra, *Patterns of Technological Innovation*. Reading, MA: Addison-Wesley, 1981.

Thomson, George, *The Inspiration of Science*. Garden City, NJ: Doubleday, 1968.

Zukav, Gary, *The Dancing Wu Li Masters*. New York: Morrow, 1979.

CHAPTER 4: ECONOMICS AND CULTURAL IMPETUS

Allingham, Michael, *General Equilibrium*. New York: John Wiley, 1975.

Miller, Roger LeRoy, *Economics Today* (4th ed.). New York: Harper & Row, 1982.

Rasmussen, David W., and Haworth, Charles T., *Elements of Economics*. Chicago: Science Research Associates, Inc., 1984.

Smith, Adam, *An Inquiry into the Nature and Causes of the Wealth of Nations*, ed. Edwin Canaan. New York: Random House, 1937.

CHAPTER 5: AN IDEA WHOSE TIME HAS COME

Bowden, Elbert V., *Economic Evolution*. Cincinnati, OH: South-Western Publishing, 1981.

Malthus, Thomas, *Principles of Political Economy*. Fairfield, NJ: Kelley Augustus, 1936.

Stein, Philip, *Graphical Analysis: Understanding Graphs and Curves in Technology*. New York: Hayden, 1964.

CHAPTER 6: THE CAUSE AND EFFECT OF TECHNOLOGY AND SOCIETY

Bedemeier, Harry C., and Stephenson, Richard M., *The Analysis of Social Systems*. New York: Holt, Rinehart & Winston, 1962.

Roberts, Nancy, Anderson, David, Deal, Ralph, Garet, Michael, and Shaffer, William, *Introduction to Computer Simulation: The System Dynamic Approach*. Reading, MA: Addison-Wesley, 1983.

Rubenstein, Moshe F., and Pfeiffer, Kenneth R., *Concepts in Problem Solving*. Englewood Cliffs, NJ: Prentice Hall, 1980.

Springer, Clifford H., Herlihy, Robert E., and Beggs, Robert I., *Advanced Methods and Models*, Vol. 2, Math for Management Series. Homewood, IL: Richard D. Irwin, 1965.

CHAPTER 7: MODELING, SIMULATIONS, AND GAMING

Bertalanfy, Ludwig von, *General Systems Theory: Foundations, Development, Applications.* New York: George Braziller, 1968.

Davis, Morton D., *Game Theory: A Non-Technical Introduction.* New York, Dover Press, 1997.

Lynch, Robert E., and Swanzey, Thomas B., eds., *Examples of Science: An Anthology for College Composition.* Englewood Cliffs, NJ: Prentice Hall, 1981.

CHAPTER 8: SYSTEMS BEHAVIOR: THE UNIVERSAL LAWS

Fuller, R. Buckminster, and Applewhite, E. J., *Synergetics: Explorations in the Geometry of Thinking.* New York: Macmillan, 1975.

Robertson, Donald W., *The Mind's Eye of Buckminster Fuller.* New York: St. Martin's, 1974.

CHAPTER 9: DIVERSITY, RANDOMNESS, AND SYSTEMIC INTEGRITY

Barnsley, Michael, *Fractals Everywhere.* San Diego: Academic Press, 1988.

Briggs, John, and Peat, F. David, *Turbulent Mirror: An Illustrated Guide to Chaos Theory and the Science of Wholeness.* New York: Harper & Row, 1989.

Casti, John L., *Complexification: Explaining a Paradoxical World Through the Science of Surprise.* New York: Harper Collins, 1994.

Cohen, Jack, and Stewart, Ian, *The Collapse of Chaos.* New York: Penguin, 1994.

DeVaney, Robert L., *Chaos, Fractals, and Dynamics: Computer Experiments in Mathematics.* Menlo Park, CA: Addison-Wesley, 1990.

Gleick, James, *Chaos: Making a New Science.* New York: Penguin, 1987.

Waldrop, M. Mitchell, *Complexity: The Emerging Science at the Edge of Order and Chaos.* New York: Simon & Schuster, 1992.

CHAPTER 10: CATASTROPHE THEORY: THE PLAGUE OF TECHNICAL COMPLEXITY

Castrigiano, Domenico P. L., and Hayes, Sandra A., *Catastrophe Theory.* New York: Perseus Press, 1993.

Proctor, Tim, and Stewart, Ian, *Catastrophe Theory and Its Applications.* New York: Dover Press, 1998.

Woodcock, Alexander, and Davis, Monte, *Catastrophe Theory.* New York: E. P. Dutton, 1978.

CHAPTER 11: THE SYSTEMIC MODELS

Bel Geddes, Norman, *Horizons.* New York: Dover, 1977.

Galbraith, John Kenneth, *The New Industrial State.* Boston: Houghton Mifflin, 1967.

Graham, Loren R., *Between Science and Values.* New York: Columbia University Press, 1981.

Mumford, Lewis, *The City in History.* New York: Harcourt Brace Jovanovich, 1961.

Naisbitt, John, *Megatrends: Ten Directions Transforming Our Lives.* New York: Warner, 1982.

Richter, Maurice N., Jr., *Technology and Social Complexity.* Albany, NY: State University of New York Press, 1982.

Toffler, Alvin, *The Third Wave.* New York: Morrow, 1980.

CHAPTER 12: PRELUDE

Bronowski, Jacob, *A Sense of the Future,* ed. Piero E. Ariotte. Cambridge, MA: MIT Press, 1980.

Burke, James, *Connections.* Boston: Little, Brown, 1978.

Dyson, Freeman, *Disturbing the Universe.* New York: Harper & Row, 1979.

Hall, Edward T., *The Hidden Dimension.* Garden City, NJ: Doubleday, 1969.

Morris, Desmond, *The Human Zoo.* New York: McGraw-Hill, 1969.

Naisbitt, John, *Megatrends: Ten Directions Transforming Our Lives.* New York: Warner, 1982.

O'Neill, Gerard K., *2081: A Hopeful View of the Human Future.* New York: Simon & Schuster, 1981.

Toffler, Alvin, *Future Shock.* New York: Random House, 1970.

Index